PIMLICO

760

WHY THE ALLIES WON

Richard Overy is Professor of History at the University of Exeter. He is the author of many books on the Second World War and the Third Reich including *Russia's War* (1998), *The Battle of Britain* (2000) and *Interrogations: The Nazi Elite in Allied Hands, 1945* (2001). His latest book *The Dictators: Hitler's Germany, Stalin's Russia* (2004) was joint winner of the Wolfson Prize and the Hessel-Tiltman Prize in 2005. He is a Fellow of the British Academy and a Fellow of King's College, where he taught for twenty-five years. He is currently writing a book on inter-war cultures of decline.

WHY THE ALLIES WON

—

RICHARD OVERY

PIMLICO

Published by Pimlico 2006

6 8 10 9 7 5

Copyright © Richard Overy 1995
Second Edition copyright © Richard Overy 2006

Richard Overy has asserted his right under the Copyright, Designs
and Patents Act 1988 to be identified as the author of this work

First published in Great Britain by Jonathan Cape in 1995

Second Pimlico Edition, fully revised, 2006

Pimlico
Random House, 20 Vauxhall Bridge Road,
London SW1V 2SA

Random House Australia (Pty) Limited
20 Alfred Street, Milsons Point, Sydney,
New South Wales 2061, Australia

Random House New Zealand Limited
18 Poland Road, Glenfield,
Auckland 10, New Zealand

Random House South Africa (Pty) Limited
Isle of Houghton, Corner of Boundary Road & Carse O'Gowrie
Houghton 2198, South Africa

Random House Publishers India Private Limited
301 World Trade Tower, Hotel Intercontinental Grand Complex
Barakhamba Lane, New Delhi 110 001, India

Random House UK Limited Reg. No. 954009
www.randomhouse.co.uk

A CIP catalogue record for this book
is available from the British Library

ISBN 9781845950651

Papers used by Random House UK are natural,
recyclable products made from wood grown in sustainable forests;
the manufacturing processes conform to the environmental
regulations of the country of origin

Typeset by SX Composing DTP, Rayleigh, Essex
Printed and bound in Great Britain by
Cox & Wyman Ltd, Reading, Berks

Contents

Illustrations

Maps

Acknowledgments

The author and publishers are grateful to the following for permission to reproduce illustrations: Library of Congress, Washington DC – 17, 18, 19, 20; Ullstein Bilderdienst – 5; Imperial War Museum, London – 2, 3, 4, 6, 7, 8, 9, 11, 12, 13, 14, 15, 16, 22, 23, 24, 25, 30, 31.

Every effort has been made to obtain the necessary permissions to reproduce the remaining images (taken from the author's personal collection of World War II pictures, posters and cartoons); however, should there be any omissions in this respect we apologise and shall be pleased to make the appropriate acknowledgments in any future edition.

The maps were drawn by Graham Malkin, Advanced Illustration.

Preface to the Second Edition

I T IS NOW almost twelve years since the first edition of *Why the Allies Won* was written. In the interval a great deal of additional material has been published and new perspectives opened up on the history of the Second World War. Some touch on the questions raised here, but others have opened up fresh avenues of approach to the issue of war as a cultural or social phenomenon rather than on the issues of war-making and strategic choice with which explanations of victory and defeat are bound up. In most respects the arguments about this narrower question of Allied victory stand or fall today in much the same way as they did a decade ago. As a result it has not seemed necessary to write a different book, but instead primarily to take the opportunity to correct mistakes and make clearer what was not clear in the earlier version.

There is one area, however, where the historiography and the base of evidence are changing very rapidly. The history of the Soviet war effort has been the beneficiary of a decade of close research, which has not altered the basic shape of the narrative but has deepened understanding about the big issues raised and modified or overturned versions of events that seemed standard in 1995. The major question – how did the Red Army succeed in holding back and then defeating the overwhelming bulk of the German armed forces – now has fuller answers than it once had. The research on detailed aspects of the reform of Soviet military practice and thinking during the war has shown conclusively that mere numbers did not suffice to explain the difference between the two sides. The sheer depth of Soviet preparation for war, for all its drawbacks and miscalculations, has been shown to be more wide-ranging and significant than was once thought, thanks to the work of Lennard Samuelson and David Glantz. For all the catastrophic losses of 1941, the Soviet Union was never starting from ground zero in its attempts to rebuild its armed and economic strength in 1942 and 1943. The vexed question of Lend-Lease, about which a generation of Soviet scholars were so deprecating, has now been transformed by the evidence that the Soviet leadership understood very well how vital these resources were in a context where their rump economy could not produce both armaments and the raw materials and equipment necessary to sustain the war effort. Finally the whole idea that the Soviet side sought an armistice or separate peace in 1941 and again in 1943 has been shown to be without the serious foundation attributed to it a decade ago.

On the wider arguments about the reasons behind Allied victory there will always be disagreement. *Why the Allies Won* did not differ from much of the literature of the early 1990s in emphasising just how important the Soviet contribution was. Perhaps in reaction to the predominantly Soviet-centred analysis of the outcome, there has been a drift back to a more balanced view. The impact of bombing, for example, though the subject of increasing moral condemnation has come to be regarded as more significant in limiting German (and Italian and Japanese) options than was once thought. It is worthwhile here to make clear, since it has been a subject of some confusion in the arguments surrounding the book, that explaining the role of bombing in the defeat of Germany is not the same as endorsing the legitimacy of the campaign. There were pressing military and political reasons which explain the British and American choice, but the consequent death, disablement and dispossession of millions of civilians, was not in any sense consistent with the liberal values of the two states that embarked on the campaign or with their pre-conflict view of what was and what was not permissible in international law. Mass civilian deaths are not something of which anyone could approve or feel morally indifferent towards; but it is necessary in this context to understand the nature of the impact that such bombardment produced.

There were two areas in the book where I displayed a woeful failure to grasp the technical and scientific complexity of what I was describing. The first was in the naval war, the second in the development of nuclear weapons and technology. I am very grateful for the advice I received on both areas, on whose basis I have made the necessary changes. The details tend to strengthen rather than weaken the overall thrust of the argument, but the details are important, particularly as so much of the discussion hinges on small but significant improvements in tactics and weaponry, which still tend to be underestimated in analysis of military conflict.

I was fortunate that *Why the Allies Won* was chosen as the theme for the annual conference of the German Committee for the History of the Second World War when it met in Hamburg in 2002. The conference was an interesting reflection of the two very differing approaches to the history of the war still current among those interested in its outcome as a historical problem. The idea that this was a war that Hitler lost rather than one that the Allies won has remained embedded in much of the analysis of the conflict. The chief justification for writing this book in the first place was to challenge the assumption that the outcome was determined by the nature of the strategic gamble and persistent errors, political and otherwise, attributed to Axis leaders. That is still a pertinent ambition, all the more so since with the passage of time the nature of the stakes between the two sides has become clearer and more fantastic. The (often deluded) assumption that world history had reached a dead end in the years after the end of the First World War produced a growing popular sense of civilisation in crisis for which a new civilisation or a new order were deemed to be the only remedies. The bitter divide between the Allies and Axis can better be understood in the context of the collective anxieties and paranoia that fuelled an age of political extremes, encouraged desperate and vicious recipes for survival and resulted in Europe's and Asia's descent into a decade of barbarism and atrocity. An Axis victory would have brought radical change dressed up as the triumph of a new version of modernity. The intensity of the efforts to prevent that happening, which mobilised populations otherwise unaware of or indifferent to the world-historical forces they confronted, explains why this war became a war to the death – and why it took the lives of more than sixty million worldwide. While it is possible to explain (and argue about) how one side prevailed over the other, there lurks a still larger question about why the developed world in the inter-war years, having experienced and been horrified by the Great War, should descend in twenty years into a crisis more deadly and damaging than the first and in a milieu of political oppression, civil conflict and ideological hatred without precedent in the modern age. In a century's time this may well seem a more significant question than the narrower, if important, issue with which this book began.

I would finally like to acknowledge the helpful advice I have had from the following: Corelli Barnett, M.G. Brewer, George Bornet, Reg Curtis, Joseph Forbes, Evan Mawdsley, Henry Ploeestra, E.A. Rawes, Lt. Comm. David Waters and Alfred Weber. I am also grateful to Will Sulkin for giving me the opportunity to update and overhaul the first edition.

Richard Overy
University of Exeter
May 2006

Preface

WHEN PEOPLE HEARD that the title of my next book was to be 'Why the Allies Won', it often provoked the retort: 'Did they?' There are many ways of winning. With the passage of time it has become possible to argue that none of the three major Allies – Britain, the United States and the Soviet Union – won a great deal. Britain lost her empire and her leading world role; the United States found that they had traded one European enemy for another, an 'evil empire' apparently more dangerous and unfathomable than Hitler's; as for the Soviet Union, the cost of sustaining the super-power status won in 1945 eventually produced a crisis in Soviet society which led to its collapse in 1991. The three Axis states – Germany, Italy and Japan – have made no attempt to become major military powers again, but they have all produced economic success stories instead. Germany and Japan became the superpowers of the world market, and their citizens a good deal richer than the British, whose war effort almost bankrupted what had been one of the wealthiest economies in the world in 1939. When people ask 'Did they?', these are the things they have in mind.

The Allies unquestionably won the military contest in 1945, and it is with victory in this narrower sense that this book is concerned. I have not tried to provide a general history of the war – there are plenty of those already. The focus of the book is to explain the outcome, rather than to describe its course. I have restricted the narrative to those parts of the conflict I regard as decisive, first the areas of combat, then the other elements of the war – production, technology, politics, and morale. As a result, many familiar aspects of the story are dealt with only briefly. The eastern front has been-given a prominence it surely deserves, but the battles in the Pacific and the war between Japan and China here must take a back seat. It is fashionable to see the use of intelligence as a critical difference between the two sides, but I am not sufficiently persuaded of this to give the subject a chapter of its own. Where intelligence clearly had special significance, its story has been woven into the narrative. All of this has been done in order to answer very directly the question of 'why the Allies won'.

There are conventional answers to this question. There is a commonly-held assumption that the Axis states were beaten by sheer weight of material strength, which ought inevitably to prevail in an age of industrialised warfare. To this might be added a related assumption, that Germany, Japan and Italy made fundamental mistakes in the war, not the least of which was biting off more than they could chew in fighting Britain, the United States and the Soviet Union together. Neither of these assumptions is very satisfactory, and it would be wrong to pretend that what follows is not in some sense a response to them. The more I have worked on the history of the Second World War, the more I have become convinced that the outcome had not just a material explanation but also important moral and political causes. I am also sceptical of the view that the Axis powers lost the war through their own efforts rather than those of the Allies. Mistakes were obviously made on both sides, but the outcome on the battlefield ultimately depended on a very great improvement in the military effectiveness of Allied forces. The Allies did not have victory handed to them on a plate; they had to fight for it.

This might seem an obvious point to those who lived through the war, but it is one that is seldom made with much force. I owe a debt to all those veterans of the conflict I have talked with and listened to over the years. Their testimony has prompted me to think more critically about Allied success. I have accumulated many other debts along the way too numerous to mention. I would like to give particular thanks to Ken Follett who has read more of the manuscript than anyone else; also to Andrew Heritage for help with the maps; and to Geoffrey Roberts, Peter Gatrell and Mark Harrison for help on aspects of the Soviet war effort. I also owe a great deal to my publisher, Neil Belton, and to my editor Liz Smith, both of whom have helped enormously to make this a much better product than I could have made it. My agent, Gill Coleridge, has been more patient than I deserve.

Richard Overy
March 1995

Author's Note

THROUGHOUT THE TEXT the terms 'Allies' and 'Axis' have been used. These terms need to be qualified. The 'Allies' covers a set of shifting coalitions: Britain, France and Poland from 1939 to 1940, Britain and the Soviet Union in 1941, and Britain, the Soviet Union, the United States and a host of other smaller states from 1942; from 1944, with the fall of the French Vichy regime, France again became one of the major Allied powers. Both the terms 'Britain' and 'France' have to be understood also to cover their respective empires. In the British case this included the Dominion states Canada, Australia, New Zealand and South Africa, all of which made very substantial contributions to the Allied war effort. Neither the Allied nor Axis powers were united as a whole in a formal military or political alliance. Only Britain and the Soviet Union had a firm alliance, sealed in 1942. The Axis states were united only by informal agreements. Italy broke with the Axis in 1943, though Italians continued to fight for both sides. I have persisted in using the conventional shorthand fully aware of its lack of historical precision. The alternatives are simply too cumbersome to sustain readability, but the defects of the existing terms must be borne in mind.

Measurements provide difficulties too. I have in general kept to imperial weights and distances, i.e. pounds, tons and miles. But in cases where the metric system is commonly used (for example in expressing certain gun calibres and wavelengths) I have kept the metric measurement. The use of tons needs to be clarified. I have used the word interchangeably for imperial, American and metric tons. In general tons applied to Soviet or German production are metric tonnes, 2,204 pounds instead of the imperial 2,240 pounds.

In America the ton is generally 2,000 pounds. Measures of Japanese and American oil production have generally been quoted in barrels: approximately 7.5 barrels equal 1 ton of oil. Again, it proved too cumbersome to make all these differences explicit throughout the text.

UNPREDICTABLE VICTORY
Explaining World War II

WHY DID THE allies win World War II? This is such a straight-forward question that we assume it has an obvious answer. Indeed the question itself is hardly ever asked. Allied victory is taken for granted. Was their cause not manifestly just? Despite all the dangers, was the progress of their vast forces not irresistible? Explanations of Allied success contain a strong element of determinism. We now know the story so well that we do not consider the uncomfortable prospect that other outcomes might have been possible. To ask why the Allies won is to presuppose that they might have lost or, for understandable reasons, that they would have accepted an outcome short of total victory. These were in fact strong possibilities. There was nothing preordained about Allied success.

This is a difficult view to accept. The long period of peace and prosperity for the west that set in with the Allies' triumph in 1945 could be said to show that Progress was once again in the saddle of history from which the momentary aberration of world war had unseated her. It has always been comforting in the west to see victory in 1945 as a natural or inevitable outcome, the assertion of right over might, of moral order over nihilistic chaos. For western liberals victory was a necessary outcome, a very public demonstration that in the scales of historical justice democracy counted for more than dictatorship, liberty for more than servitude. To hammer the point home, enemy leaders were put on trial at Nuremberg and Tokyo for all the world to see what happened to regimes that thrived on crime.

Of course no one pretends that the triumph of freedom over despotism is an entirely sufficient explanation. The western powers defeated the Axis only in alliance with the Soviet dictatorship, which before 1941 they had shunned and vilified with only a little less vehemence than they reserved for Nazi Germany. The Soviet Union bore the brunt of the German onslaught and broke the back of German power. For years the western version of the war played down this uncomfortable fact, while exaggerating the successes of democratic war-making. Yet if the moral image of the war is muddied, the material explanation for victory seems unambiguous. Alliance between the British Empire, the Soviet Union and the United States created overwhelming superiority in manpower and resources. If there is any consensus about why the Allies won, it rests on the unassailable evidence that one side vastly outnumbered and outproduced the other.

There is danger here of determinism of a different sort. As one historian of the conflict recently put it, the Allies 'were certain to defeat Germany' once German and Japanese aggression brought them together in December 1941.[1] By 1943 the material gulf was huge. That year the Axis produced 43,000 aircraft; the Allies produced 151,000. The temptation has always been to assume that the figures speak for themselves. The balance of populations and raw materials greatly favoured the Allies: hence whatever the Axis powers did they would always come up against the strategic dead end of material inferiority. This is at best an unsophisticated argument. It begs endless questions about the comparative quality of weapons, or the gap between potential and real resources, or about how well weapons were used once the forces actually got them. It also ignores the very considerable efforts made by both sides to deny resources to the other, for example the submarine offensive against British trade, or the Anglo–American bombing of German industry. There was a wide gap between potential and actual output on both sides. The statistics do not simply speak for themselves; they require interpreters.

It can be seen at once that there are no simple answers to the question of 'why the Allies won'. Much popular belief about the war is illusion. Take, for example, the view that the war represented the triumph of democracy over tyranny. In reality democracy was narrowly confined in 1939 – to Britain, France, the United States and a handful of smaller European and British Commonwealth states – and was even more restricted once the conflict got under way. Far from being a war fought by a democratic world to bring errant dictators to heel, the war was about the very survival of democracy in its besieged heartlands. Victory in 1945 made democracy more secure in western Europe, America and the British Dominions, but outside these regions this form of government has had at best a chequered career in the half-century since the defeat of the Axis.

If anything the war made the world safe for communism, which was as embattled as democracy in the 1930s and close to eclipse by 1942. One of the most significant consequences of the war was the spread of communism in Europe and Asia and its consolidation in the Soviet Union. This outcome reflected the significant role played by Soviet forces in defeating Germany. There is now widespread recognition that the decisive theatre of operations lay on the eastern front. Without Soviet resistance it is difficult to see how the democratic world would have defeated the new German empire, except by sitting tight and waiting until atomic weapons had been developed. The great paradox of the Second World War is that democracy was saved by the exertions of communism.

This unpalatable fact is usually explained in terms of the common cause against Hitlerism, which helped to bridge the ideological gulf between capitalist democracy and communist authoritarianism. The pooling of resources and military effort was clearly a better way to secure survival than going it alone. But here, too, a word of caution is needed. Collaboration between the three Allied states could not be taken for granted. Until 1941 the Soviet Union was regarded by the west as a virtual ally of Nazi Germany following the signature of the German-Soviet Pact in August 1939, which ensured a steady supply of resources and food for Germany from Soviet production. When the Soviet Union was attacked by Germany in June 1941, there was a residue of profound distrust between the Soviet leadership and the west which had to be dispelled before any alliance could be built. There was still powerful anti-communist sentiment in Britain and the United States. 'How can anyone swallow the idea that Russia is battling for democratic principles?' asked Senator Taft in Congress. 'In the name of Democracy are we to make an alliance with the most ruthless dictator in the world?'[2]

The alliance was forged in the end from the bare metal of national self-interest. It survived as long as each side needed the other to help achieve victory, and no longer. 'We are much better off', wrote Roosevelt's Assistant Secretary of State, Adolph Berle, in his diary, 'if we treat the Russian situation for what it is, namely, a temporary confluence of interest.'[3] Throughout the war each side worried that the other might reach a separate agreement with the common enemy, and the naked pursuit of national interest led to endless squabbles between the three partners. In the end they fought separate wars, the Soviet Union on the eastern front, the United States in the Pacific, Britain and America in the Mediterranean and western Europe.

The coalition certainly did produce a great weight of resources in the Allies' favour as long as they fought together. But there are illusions to dispel again. God does not always march with the big battalions. In World War I Britain, France and Russia mustered 520 divisions in the middle of 1917, but could not prevail over 230 German divisions and 80 Austrian. But in March 1918, with Russia out of the war, 365 German and Austrian divisions could not defeat the 281 of the Allies. Instead the German powers admitted defeat six months later with the balance of divisions 325 apiece.[4] Of course in World War I there were other factors at play: the Allies had more tanks and aircraft, and in 1918 there came a flow of vigorous American armies across the Atlantic; German and Austrian forces were hampered by collapsing home economies and declining enthusiasm for war; and so on. The basic figures are presented here only to illustrate the dubious value of relying on plain numbers to explain the outcome of wars.

Such simple analysis is particularly inappropriate in the case of the Second World War, when the material balance changed sharply several times during the course of the conflict. Up to 1942 the balance favoured the aggressors and might well have allowed them to win before American economic power could be placed in the scales. German victories brought the vast spoils of Continental Europe and of western Russia, turning the German empire in two years into an economic super-power, capable of turning out twice the quantity of steel that Britain and the Soviet Union together could muster. Japan's seizure of the rich resources of northern

China and of south-east Asia vastly improved her strategic position by 1942, while denying vital supplies of tin, rubber, oil and bauxite to the Allies. Even when America entered the war it took time before her enormous capacity could produce significant quantities of weapons, or, for that matter, the trained forces to use them. If the eventual outcome of the war owed something to the great Allied preponderance in material produced by American industrial strength, we still have to explain why the Axis states failed to use their economic advantages when they had them.

What gradually swung the balance of resources back in the Allies' favour were two factors: the sheer speed and scale of American rearmament, which dwarfed anything that the Germans and Japanese, or even the British, had thought possible; and the swift revival of the Soviet economy after the mauling it received in 1941, when it sustained losses so severe that most pundits assumed the worst. Without the extraordinary exodus of Soviet machinery and labour from the war zones of the western Soviet Union to the harsh plains of Siberia, Stalin's armies would have been like the Tsar's in 1916, the soldiers in the second rank picking up the guns and boots of their dead vanguard as they scurried into battle. Against every expectation Soviet society worked feverishly to turn out the tanks and aircraft their soldiers needed. With only a quarter of the steel available to Germany, Soviet industry turned out more tanks, guns and planes than her German enemy throughout the war.

This surprising outcome also tells us something about the organisational capacities of the Axis powers. The huge disparity in weapons was due not only to American rearmament and Soviet revival, but also to the inability of their enemies to make the most of the resources they had. Some of this deficiency could be put down to the circumstances of war. Japan never got what she wanted from the oil-rich southern islands because American submarines lay across her supply lines, reaping a destructive harvest from Japan's poorly defended merchant shipping. Germany extracted far less than Hitler wanted from the captured areas of the Soviet Union because Soviet forces torched anything they could not carry away before the Germans arrived. The rain of bombs on German, Italian and Japanese cities slowly eroded once flourishing industries. By 1944 bombing reduced German output of aircraft by

31 per cent, and of tanks by 35 per cent.[5] To all these external circumstances were added homegrown failures. Italian war production was riddled with corruption and administrative incompetence; Japan's economic effort was stifled by tensions between her soldiers and her businessmen, and a debilitating rivalry between the navy and the army; German industry and technology were the victims of ceaseless rivalry between the Nazi satraps and a military establishment whose technological fastidiousness made mass-production almost impossible. If these internal weaknesses had been resolved, the Axis by 1942 might well have proved the irresistible force.

Even when the balance-sheet of resources is broken down, the crude quantities tell us nothing about the quality of weapons produced from them. In fact there existed wide differences in levels of technical achievement between the two sides. The large forces of the Red Army in 1941 were a less formidable asset than they looked on paper because they lagged in the main behind German standards. But in 1942 the famous T-34 tank began to pour in large numbers from Soviet factories, and the quality of Soviet fighter aircraft improved sufficiently to make combat less one-sided. There were significant gaps on the German side too. The new generation of aircraft designed to replace the ageing models developed in the mid-1930s failed to materialise in the first years of war for a whole host of reasons, and the Luftwaffe was stuck for the duration of the conflict flying the models with which it started the war. The balance of air technology, in which Germany enjoyed an enviable lead in 1939, swung the Allies' way largely for reasons of Germany's own making. Even more striking was the reality of Germany's famous mobile forces. Though her scientists could produce rockets and jets by the end of the war, Germany failed to build the simpler trucks and jeeps needed to keep her armies on the move. By 1944 American and British forces were fully motorised, but the German army was still using one and a quarter million horses. When Hitler's massive invasion force stood poised on the Soviet frontier in June 1941, it deployed 3,350 tanks but also 650,000 horses.[6]

The material balance also tells us little about *how* the weapons were used once the forces got them. This was not a question of economic power or technical ingenuity, but of fighting skills.

During the war there were plenty of instances of poorly-armed troops fighting superbly; a super-abundance of weapons and equipment was no guarantee that forces could use them effectively. Fighting power was determined not just by weapons, but by training, organisation, morale and military elan. Both German and Japanese forces showed this as they were pushed back in 1944 and 1945: fighting against very large odds, with a deteriorating supply of weapons, they maintained their combat skills and a resolute willingness to fight almost to the end. Hitler was convinced throughout the war that the German soldier was a superior fighter, both more competent and more spiritually fortified than his opponent, and that this could in some sense compensate decisively for the greater numbers of the enemy. In Japanese society the military were supposed to be imbued with *bushido,* a spiritual shield that gave each soldier the strength to fight to the limits of endurance and self-sacrifice, even against overwhelming material odds. The emphasis placed on sheer fighting skills in Germany and Japan brought them remarkable victories between 1939 and 1942; it also forced their enemies to think more about the quality of their own forces, rather than their evident quantity.

The balance of fighting power, like that of resources, did not remain constant during the war. As might be expected, after the initial shock of defeat the Allies looked hard at the way their forces were trained, deployed and led. Lessons were quickly learned from Axis successes. The Allies were forced by the nature of their enemy to stretch their strategic imaginations to embrace ways of warfare that were more ingenious and effective. By contrast the early victories lent a certain complacency to Axis strategy and operations; there were few fundamental changes made to the military recipe that had worked so well the first time around. The gap in fighting power between the two sides narrowed remarkably quickly. There are echoes here of the Napoleonic wars. In both cases the aggressor initially demonstrated superior fighting skills and leadership, against forces that were divided and operationally ineffectual. In both cases the gulf between the two sides narrowed over time as the lessons of early defeats were evaluated and the weaker forces were expanded and reformed. By the end of the war Allied forces performed much more effectively than they had done at first. Axis forces, on the other hand, like the armies of Napoleon, stagnated – both were remarkably skilled in retreat, but it was retreat none the less.

Clearly, then, the possession of bigger battalions fighting in a just cause is not a sufficient explanation of victory. We need to be much more precise about the reasons for Allied victory to find an explanation that is historically convincing, all the more so given the nature of the crisis that generated the war in the first place. In the 1930s the world order had been subjected to seismic shifts, rocked by forces which western liberal statesmen could scarcely understand. Democracy was in retreat everywhere, feebly keeping at bay the tide of violent imperialism, racial conflict and popular dictator-led nationalism. When war broke out there were widespread fears in the west that the march of Hitler's brutal armies signalled the end of liberal civilisation. Well into the war the outcome hung in the balance, even after America joined the fray. The war was not some deviation from the natural development of the world towards a democratic utopia, but was a hard-fought and unpredictable conflict about which of a number of very different directions the world was going to take.

Viewed from this perspective, Allied victory represented a remarkable reversal of fortunes. In the 1930s the western liberal world seemed at the point of eclipse. The Great War of 1914–18 had called into question the values of the self-confident and wealthy states that fought it. It plunged the world from civilisation into a new barbarism, giving lie to the Europeans' claim to be the bearers of peace and plenty. By the time it was over, Europe was impoverished, a whole generation of young Europeans brutalised and embittered. In Russia the war provoked social revolution as the Tsarist empire collapsed under the strain. The triumph of Lenin's Bolsheviks in 1917 gave birth to an age of bitter social and ideological conflict. Communism threatened to undermine the very foundations of capitalist society, and was loathed by anyone with an interest in the status quo. Yet in 1929 capitalism almost destroyed itself. The Great Crash brought the worst recession of the modern age. Millions were thrown out of work, and millions more plunged into poverty. The old political system could not cope. Communism was one alternative. In the 1930s men and women from all walks of society were drawn to the communist claim that socialist order must be preferable to capitalist chaos. Terrified at such a prospect, the anti-communist right swung towards militarism and fascism and the promise of social harmony and a strong nation.

The rise of extremist mass movements in the 1930s offered a profound and dangerous challenge to liberal democracy and conventional capitalism. On all sides could be heard a chorus of voices calling for a 'New Order' of planned economies and totalitarian societies. Communism and fascism offered a way out of what a great many had come to regard as a bankrupt political and economic system whose days were numbered. The fear that the existing order was in terminal decline produced a deep sense of moral anxiety among those who sought to defend it. The violent conflicts between liberalism, fascism and socialism in the 1930s were expressed by their champions in terms of fundamental human values, in a language of moral decline and moral renewal. The spirit of the age was of crisis, decay, transformation.

This spirit was expressed internationally in the growing division between a small number of states anxious to preserve the existing balance of power and a larger number who sought to revise it. The defenders of the existing system were few indeed. At their forefront stood Britain and France. Both states lay at the centre of global empires which together covered one-third of the world's surface. Neither state had enough money or military resources to defend its worldwide interests, and each knew it. No other major power had much interest in the survival of the Franco-British world order. Other nations could see a gap widening between the apparent and the real strength of Britain and France. Less than a generation separated the two major world powers of the 1930s from the humiliating *débâcle* at Suez in 1956. The United States shared the liberal politics of western Europe, but was hostile to old-fashioned colonialism, and deeply distrusted what American leaders saw as a reactionary and decadent Europe. The Soviet leadership saw the old imperial states as historically doomed, and, though the Soviet Union did little in the 1930s to hasten their demise, Stalin looked in the long run for what he called a 'new equilibrium'.[7] There was no hint in the 1930s of the later wartime coalition.

The chief challenge to the existing equilibrium came from

Germany, Italy and Japan. All three had been democratic states after the Great War, but their democracy turned sour in the heat of economic crisis and popular authoritarian nationalism. In 1922 Mussolini came to office in Italy at the head of the first fascist party to gain power; in 1931 Japan came gradually under the domination of the military; in 1933 Hitler's National Socialist Party, profiting from the social and political chaos generated by the slump in Germany, stormed the ramparts of respectable politics and imposed single-man, single-party dictatorship with a little over one-third of the popular vote. All three states were united by resentment. They were, Mussolini argued, the 'proletarian states', dominated by the wealthy plutocracies, Britain, France and America. The fashionable view that empire was a source of political strength and economic nourishment, particularly for states that were over-populated and weak in natural resources, led all three to the conclusion that in the crisis-ridden 1930s their only hope of salvation lay in acquiring empires of their own. The term everyone used was 'living-space'; since the globe's territorial resources were finite, such space could be acquired only at the expense of someone else, and violently. The evident weakness of Britain and France and the unwillingness of either the United States or the Soviet Union to take their place in world politics, exposed a temporary window of opportunity, through which the three 'New Order' states hesitantly climbed.

In the space of half a dozen years the fragility of the old order was made transparently clear. In 1931 Japan invaded and occupied Manchuria, the industrially rich province of northern China. An uneasy truce followed. Japan refused to abandon her conquest in the face of western disapproval and no one had the strength or willingness to expel her by force. In 1937 a full-scale war broke out between Japan and China. Japanese leaders declared an Asian New Order and set out to conquer China's Pacific provinces. The attempt lasted to 1945 and Japan's defeat at American and Soviet hands. In 1935 Mussolini attacked Abyssinia to extend Italy's colonial empire, and to lay the foundation for Italian domination of the Mediterranean and northern Africa. Again western protests did nothing to reverse the aggressor. By May 1936 Abyssinia was conquered. In July that year Italy sent forces to help Franco in the Spanish Civil War. Long before the onset of world war both Japan and Italy had abandoned any idea of peaceful settlement. They fought for their living space, and calculated correctly that none of the other powers would be willing to obstruct them by force.

Germany was the most dangerous component of the Axis, though German forces, unlike Japanese or Italian, did not fire a shot in anger until the invasion of Poland in September 1939 which launched world war. The source of the German threat was Hitler. Other German nationalists wanted Germany to reassert herself as a major state in the 1930s following years of enforced subservience to the victor states of 1918. Few Germans of any political shade had accepted the Allied demands for reparations and German disarmament, or been reconciled to the loss of territory to Poland and France. But very few Germans wanted to run the risk of war again. Hitler's outlook was quite different. Any account of the origins and course of the Second World War must give Hitler the leading part. Without him a major war in the early 1940s between all the world's great powers was unthinkable.

Hitler began life in 1889 in the small Austrian town of Braunau-am-Inn, born into the vast multi-racial Habsburg empire, ruled by the grand old man of the dynasty, Emperor Franz-Joseph. He was the third son of an Austrian customs official. He was indulged by his parents, perhaps to compensate for the death of all but one of their other children. Hitler's parents both died by the time he was eighteen. The year his mother died, in 1907, Hitler failed to gain entry to the Vienna Academy where he had hoped to pursue an artistic career, the one thing in which he was interested. Over the next six years he did little, living off a legacy from his mother, painting when he could, escaping from the Austrian authorities who wanted him to do military service. He had no desire to fight for the largely Slavic empire. When he was finally caught he had moved to Munich. Early in 1914 he was examined by Austrian military doctors and they found him, to his relief, unfit for service. He stayed on in Munich in 1914, where on 1 August he stood in a large crowd on the Odeonplatz listening to news of the outbreak of war. Now he was eager for military service, and the German authorities were less circumspect about his health. He joined the Bavarian Reserve Infantry and went to the front. He was exhilarated by war. In December 1914 he wrote to his Munich landlady: 'I have been risking my life every day, looking death straight in the eye.' That same month he was awarded the Iron Cross, Second Class. 'It was', he wrote, 'the happiest day of my life.'

Against all the odds Hitler survived the whole war. After his early exuberance, he later confessed to a growing fear. He did not like the constant artillery barrage, the daily confrontation with death. He stopped himself from going mad or from shirking his duties by what he described as a long inner struggle. By 1916, he later wrote, 'my will was undisputed master . . . Now fate could bring on the ultimate tests without my nerves shattering or my reason failing.' His later determination to show his willpower at moments of crisis, despite inner misgivings, surely has its roots in the horrific experiences of the Western Front. When the war ended with Hitler in hospital following a gas attack, the will temporarily snapped.

He was consumed in 1918 by a deep despair at German defeat. He found the psychological shock almost unbearable, the humiliation personal. In his struggle to come to terms with reality he reached two conclusions: defeat was the fault of Jews and Bolsheviks on whom a terrible revenge would one day be wrought; and he, Adolf Hitler, was the chosen instrument of destiny to lead the German people in the coming era of racial struggle and conquest. The messianic complex married to the savage prejudices made Hitler a far more dangerous and explosive force in world politics than Italy or Japan, or even Stalin's Soviet Union. Hitler's appetite for vengeance, his unquenchable hatreds, his powerful self-belief, all proved irrepressible. Once he had his hands on the controls of the German state after 1933 there was nothing else he wanted to do but to keep his tryst with fate, first through the construction of a powerfully armed, racially-aware community embracing all the German peoples, and then through the unavoidable struggle against Bolshevism and the Jews.[8]

He revealed little of these fantastic ambitions to those around him. Historians have been inclined to take them with a pinch of salt. There was no clear plan, no detailed blueprint for German world power. But there did exist an incontestable continuity from Hitler's first angry scribbled notes in 1919, through *Mein Kampf*, written in 1924, to the memoranda, speeches and secret meetings of the 1930s. His warped view of the world was his reason for living. By marrying his fortunes to German destiny he ceased to be the humble, insecure veteran, the man who looked, according to one of his close friends, 'like a waiter in a railway station restaurant', and instead rubbed shoulders with Charlemagne, with Frederick Barbarossa, with Bismarck, the heroes of the German pantheon. Hitler was Jekyll and Hyde; at rest he appeared pallid and puffy, personally nondescript, socially inept, full of nervous small talk; but in his stride, eyes ablaze, theatrical, unstoppably articulate, intemperately self-absorbed, he imagined himself a real-life Siegfried.[9]

Hitler was astute enough politically to recognise that he could not transform the world order overnight. Germany had to revive its shattered economy; the National Socialist Party needed to secure itself domestically; above all Germany needed to rearm. By 1938 much of this was achieved. Hitler's personal power was beyond challenge and he now exerted it to turn the German state towards war. The first object was to build a united Germany. In March 1938 Austria was fused with the German Reich. In May Hitler ordered the armed forces to prepare an attack on Czechoslovakia to bring the three million German-speakers living there under German rule. Only the unexpected intervention of Britain and France, unable to countenance an entirely free hand for Hitler in eastern Europe, frustrated his desire for violence. The German areas of Czechoslovakia were won instead by negotiation. In March 1939 the rest of Czechoslovakia was swallowed up, this time with no intervention from abroad. In April 1939 he ordered the armed forces to prepare for another small war, this time against Poland, for the reconquest of the German areas lost in 1919. By 1939 Germany was poised, like Japan and Italy before her, to use war to tear up the existing order.

Those states bent on preserving this order had a number of uninviting options left by the late 1930s. They could give in and acquiesce in the redrawing of the world in the hope that something might be salvaged; they could side with the aggressors, as many smaller states did; or they could fight to preserve the old system and their own way of life. There were circles in London and Paris in support of each of these options. Most opinion outside Britain and

France assumed that the western failure to stop either Japan or Italy was a tacit recognition that acquiescence made most political sense. In truth, neither Japanese nor Italian expansion was perceived to be so vital to western interests that they would risk open war. Germany was a different matter. The German threat was closer to home, and much larger. Germany's economic size and military potential marked her out as the only revisionist state that could seriously menace the national interest of the other great powers. Though Nazism was disliked by much of British and French opinion, it was not the nature of Hitler's system that worried western leaders as much as the degree of power it could exert abroad, and the warlike aspect of its unpredictable leader. By the late 1930s both the British and French governments opted for the most difficult course, containing German ambition by the threat of military force.

The threat was real enough. Both states began to arm heavily from the mid-1930s and by 1939 outproduced Germany in tanks and aircraft. In 1938 the threat was just sufficient to force Hitler to abandon his war on the Czechs and at the Munich Conference to submit to the compromise insisted on by Britain and France. But it was difficult to persuade Hitler long term that the military threat was convincing, and that the western states, after a decade in which they had failed to obstruct any major act of aggression, had the will to fight for the old political order. In 1939 Hitler gambled that they would not fight; to ensure this he pulled off what he regarded as a spectacular diplomatic coup in August 1939 by reaching agreement with Stalin on Soviet non-aggression and the division of Poland between the two states. In his view the old balance of power was in ruins by the autumn of 1939. He ruled out the possibility of intervention and attacked Poland on 1 September. Neither Britain nor France actively wanted war in 1939, but neither could face the consequences of accepting international humiliation and German domination of Europe. At home a broad patriotic consensus emerged in 1939 in favour of confronting Hitler. The British and French armed forces agreed in March on a common war plan to wear Germany down in a long campaign of blockade and attrition as they had done in the Great War. The two states allied with each other and with Poland in the spring of 1939, a commitment from which none of them could easily escape when Hitler launched his war on the Poles. On 3 September Germany, Britain, France and Poland were at war. It was not the war Hitler had expected, but he fought it willingly enough. It was the war the western allies expected, but they fought it with great reluctance.

When the crumbling world order plunged into violence in September 1939, the eclipse of democracy and international stability threatened to become total. On news of the war, Harold Ickes, Roosevelt's Secretary for the Interior, confided in his diary the gloomy view that civilisation was now doomed, 'headed for a decline of fifty or one hundred years, or even longer, during which our descendants will lose many of the gains we have made'.[10] There was grim evidence in the eighteen months that followed to support the bleakest prognostications. The New Order triumphed everywhere. German forces swept aside all resistance in democratic Europe. In the spring of 1940 Denmark and Norway were overrun. In May and June Belgium, the Netherlands, Luxembourg and France were defeated and a German army of occupation installed. The French army was the only major centre of military power on the Continent outside Germany. Its defeat in six weeks – a product not of numerical inferiority, but of poor organisation and fighting skills – effectively destroyed the order established after the Great War. Britain escaped the same fate by the narrowest of margins. Her army defeated in France, her air force worn down in the weeks of intervention in the European struggle, her navy at risk from German air power, Britain was saved by the poor level of German preparation for a cross-Channel assault, and the failure of German air forces to wrest control of the skies of southern England from the RAF. The Battle of Britain saved the island from invasion, but it did not do more than that. By the summer of 1940 Britain was isolated in Europe, with no means to re-enter the European main-land, and little prospect of alliance.

The situation in the east was more delicate for Germany. Poland was easily defeated in the autumn of 1939; the remaining states in eastern Europe and the Balkans began to align themselves with Germany, centre of the New Order. The Soviet Union, which Britain regarded as virtually an enemy state, supplied Germany with food, oil and raw materials. For Hitler the opportunity now opened up invitingly to complete the programme of empire-building by seizing the coveted living-space of Eurasia, the rich steppe areas of the Ukraine, the oil of the Caucasus, the sprawling iron and steel basin of southern Russia. Only the Red Army now stood between him and the dream of world power, and there was every indication that it was poorly armed and led. In the autumn of 1940 he turned his back on Britain, who could, he argued, be finished off by the Luftwaffe in good time, and looked eastward. By December he had drawn up a plan, 'Barbarossa', for an assault along the whole length of the Soviet frontier, deep into Soviet territory, to seize Moscow, Leningrad and the industrial south. It was a plan of exceptional scope and risk, for Hitler believed nothing less than that the Soviet Union could be defeated in four months.

From Britain's point of view the German threat was only one worry among many. On the back of German victory over France and Britain in the Battle of France the other revisionist states, Italy and Japan, began to flex their own muscles. Italy declared war on 10 June and confronted Britain with a sizeable army and navy astride her imperial sea route to Suez and India. During 1941 Italy, with German assistance, pushed back British Commonwealth forces towards Suez and threatened the whole Middle East. Italy attacked Greece too, and in the process provoked a German-Yugoslav conflict, which led German forces to occupy much of the Balkan peninsula. Britain was faced with the very real threat that the Axis states would seize Gibraltar and Suez and shatter what remained of Britain's strategic position overseas. Only Hitler's obsession with the assault on the Soviet Union prevented Britain's almost certain defeat. In the Far East Japan moved forces down to the French colony of Indo-China (Vietnam) and threatened British and Dutch possessions in the East Indies, with their rich supplies of oil. In September 1940, in recognition of their successes, the three aggressors met at Berlin where they signed a Tri-partite Pact to divide the world between them, 'to establish and maintain a new order of things'.[11]

The ultimate confrontation for Hitler was the contest with Bolshevism, which he had longed for ever since the agonising days of defeat in 1918. On 22 June 1941 four million soldiers poured across the Soviet frontier, including contingents from Germany's allies and co-belligerents, Finland, Hungary and Romania. Despite numerous warnings from sources even the Soviet intelligence authorities could have regarded as unimpeachable, Stalin insisted to the very last moment that Hitler would not attack. He thought he had the measure of his fellow dictator. The shock was complete. Soviet forces were quite unprepared for the scale of attack. Thrown into complete confusion the Soviet front broke open, and German armies streamed towards their targets. On 3 October Hitler broke a three-month silence by returning from his headquarters to Berlin to announce final victory to the German public. His excitement was evident to all who saw him. To his carefully selected audience at the Sportpalast he announced that he had returned from 'the greatest battle in the history of the world', which Germany had won. The Bolshevik dragon was slain 'and would never rise again'.[12] There was a tumult of applause. A few days later the news was formally released to the German public: there were two large Soviet army groups encircled by German forces and on the point of surrender, but after that the war in the east was over. Hitler planned, like some mediaeval Mongol khan, to destroy Stalin's capital. Moscow was not to be occupied but 'completely wiped from the earth'. Neutral journalists were invited to the Propaganda Ministry where, in front of a colossal map of the Soviet Union, German spokesmen outlined to the anxious newsmen the dimensions of the German victory and the shape of the New Order.[13]

Everyone who visited Hitler's headquarters that autumn could sense the euphoria. In just two years the political map of the world was torn up. In July Hitler had already authorised new armament programmes to build up a large battlefleet and overwhelming air power to destroy British resistance and to keep America at arm's length. A clumsy attempt by the Soviet side to buy time in October by hinting at a negotiated end to the conflict was swept aside by Hitler, still confident that he was close to destroying Soviet power once and for all. A few weeks later, prompted in part by the scale of German success in blunting any Soviet threat, Japan turned south to attack America and Britain in the Pacific and carve out a new empire in south-east Asia. On 7 December Japanese aircraft attacked the American naval base at Pearl Harbor. Four days later

Hitler declared war on the United States as well. German leaders did not consider America a serious military threat. Two months before, Hitler's Foreign Minister, Joachim von Ribbentrop, had told the Japanese ambassador that 'American policy represented one great bluff'. Germany planned to complete the establishment of the New Order before America could intervene.[14] Over the weeks following Pearl Harbor Japan's forces took the south by storm; by February they had captured Malaya, Singapore, most of the Philippines, the Dutch East Indies and much of Burma, and had threatened Australia and India. On all fronts desperate defensive efforts kept alive both Soviet and western hopes, in the outskirts of Moscow, which German forces failed to take over the winter of 1941, in the approaches to the Suez Canal, on the borders of northern India, and in the northernmost territories of Australia. These battlefields proved to be the limits of Axis advance, but who in those catastrophic months would have believed it?

On the face of things, no rational man in early 1942 would have guessed at the eventual outcome of the war. In the jargon of modern strategy, the Allies faced the worst-case scenario. The United States was not yet armed, and would have large trained forces only by 1943 at the earliest; the Soviet Union had lost the heart of its industrial structure and German forces were poised to seize the oil of the Caucasus and the Middle East. The situation for the Allies – and the coalition only emerged in December 1941, not sooner – was desperate, demoralising. In January 1941 Harry Hopkins, Roosevelt's personal emissary to Churchill, conveyed the President's conviction 'that if England lost, America too would be encircled and beaten . . .'[15] Even the belligerent Churchill had moments of bleak despair when he sketched to his staff his picture of 'a world in which Hitler dominated all Europe, Asia and Africa' and left to Britain and America 'no option but an unwilling peace'.[16]

It was from this sorry foundation that the Allied powers first halted, then reversed, the apparently inexorable drive to conquest of their enemies, Germany, Italy and Japan. Between 1942 and 1944 the initiative passed to the Allies, and Axis forces experienced their first serious reverses – at Stalingrad and Kursk on the eastern front, at the battles of the Coral Sea and Midway in the Far East,

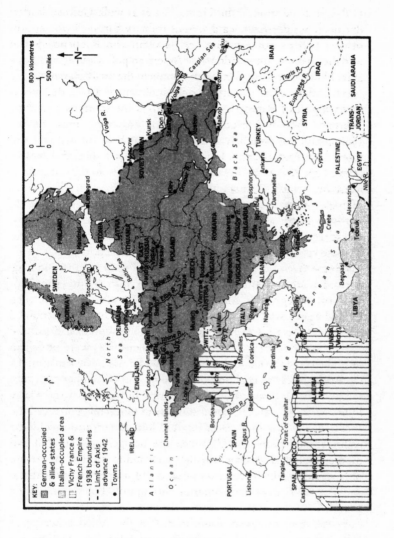

1 Axis expansion in Europe 1938–42

and El Alamein in the Middle East. By 1944 the demoralisation of the Allies was dispelled; contemporaries could see that the odds were now overwhelmingly on Allied victory. The neutral states that hedged their bets earlier in the war – Turkey, Spain, Sweden – now looked for association with the winning side. The countries of Latin America came, one by one, to declare war on Germany. Argentina was the last, on 27 March 1945, six weeks before German defeat. Persia declared war on Germany in September 1944; Saudi Arabia and Syria in February 1945; Romania changed sides in August 1944, from the Axis to the Allied cause. The embattled democracies of 1939 led a world crusade six years later.

*   *   *

Somewhere in the changing fortunes of war between 1942 and 1944 lies the heart of the answer to our question of why the Allies won. So dramatic was the transformation that it is hard not to assume that there was a particular turning-point, some uniquely significant battle like Waterloo, a decisive error of judgement, a moment of strategic hesitation which cost the Axis the war. Of course there were important battles, and human error explains much on both sides. But the war was fought worldwide for six long years. The chances of a single battle or decision seriously explaining its outcome are remote. For much of the war the chief campaigns were based on attrition, for months or years on end – in the Atlantic Battle, in the air war, on the eastern front, in the slow erosion of the German foothold in western and southern Europe or the Japanese hold on the islands of the south Pacific.

The explanation of Allied victory requires a broad canvas and a wide brush. The war was unique in its scale and geographical extent. Colossal resources were mobilised over vast distances. The battlefield was a world battlefield in a very literal sense. For the Allies there was no question of winning the war in some defined area of engagement – it had to be won in every theatre and in every dimension, land, sea and air. This made the pursuit of victory costly, extensive, and above all time-consuming. The war made extravagant demands on the warring states of both sides. They each pitched a third (or more) of their manpower into battle, and converted up to two-thirds of their economy to feed the insatiable demands of the front-line. This was warfare on a scale the nineteenth century could not have contemplated, indeed on a scale hardly possible today, and it drew its justification from the desperate, Darwinian view of the world peddled by the doom-mongers of the 1930s. All states, fascist, communist, democratic, shared the common but terrifying assumption that war had to be 'total', what Mussolini called a 'war of exhaustion', to win the struggle for survival.[17] The outcome of war depended as much on the successful mobilisation of the economic, scientific and moral resources of the nation as it did on the fighting itself. This may not be as glamorous an explanation as one of simple battlefield performance, but it was a civilian's war as well as a soldier's. Allied success in the long campaigns of attrition can be convincingly explained only by incorporating the role of production and invention.

In the discussion that follows a rough balance has been kept between two different kinds of historical approach, between the war as a series of decisive military campaigns and the war as a set of distinct themes, between how the Allies won and why the Allies won. The first half of the book examines the four main zones of conflict in which the Allies prevailed between 1942 and 1945 – the war at sea, the land struggle on the eastern front, the offensive from the air, and the reconquest of Europe. The second part explores the elements that conditioned and caused those military successes: the balance of resources, combat effectiveness, leadership and strategic judgement, the mobilisation of the home front, and last, but not least, the moral contrasts between the two warring camps.

The zones of conflict are self-defining, for they were the arenas in which the Allies chose to exert their maximum effort. It may well be argued with hindsight that they should, and could, have made other choices, but that is hardly the issue here. To understand why the Allies prevailed in these zones is to understand the outcome of the war. Though each zone was fought for independently, the outcome in any one affected the outcome in others. If the submarine menace had not been contained in 1943, the invasion of Europe the following year would have been infinitely more hazardous; if the bombing offensive had not diverted large quantities of men and materials away from the eastern front, the Soviet advance might well have been slower and less secure; and so forth. There is, in short, a strong line of connection between each zone, which explains the Allies' determination to prevail in them all.

The war at sea was a critical one for the western Allies, for the simple reason that all their major arteries of communication and supply were across water. Sea power was the only means by which they could bring other kinds of military force to bear on the enemy, and the only means by which they could fight a genuinely global war. For most of the Second World War Britain and the United States fought a predominantly naval conflict, and relied more heavily on naval power than anything else, until they shipped the armies to Europe in June 1944 to start the reconquest of the Continent. In the Pacific the United States navy bore the brunt of the conflict with Japan right up to the point of Japanese defeat. The naval war linked all the zones of conflict. Supplies for the Soviet Union were dangerously convoyed to Archangel or Vladivostock or the Persian Gulf. The vast American war effort – men, tanks, planes and trucks – was shipped in great armadas across the Atlantic Ocean. The British war effort was unsustainable without the flow of materials, food and equipment from all over the world. 'It is in shipping', Churchill wrote to Roosevelt, 'and in the power to transport across the oceans . . . that the crunch of the whole war will be found.'[18]

The Axis states knew how much the oceans mattered, which is why they made such strenuous efforts to sever the arteries one way or another. By 1942 German submarines were sinking British ships faster than they could be replaced, while the Japanese Imperial Navy won for itself a brief period of ascendancy in the Pacific and Indian Oceans. For the next two years the western Allies struggled to defeat the submarine and to contain Japanese naval power. The reasons for their eventual success in both theatres are elaborated in what follows. Axis defeat on the high seas paved the way for the more effective reinforcement of the Soviet Union and western Europe, and the complete defeat of Italy and Japan.

While the western Allies tried to secure the seas, the Soviet Union was locked in the world's largest land battle, stretched across the heart of Eurasia, from the Baltic to the Black Sea. Germany was by far the most powerful of the three Axis states; the core of that power was her huge army, eight million strong. The overwhelming bulk of it was directed at Soviet Russia. In 1942 Germany deployed 178 divisions on the Soviet front; her allies and co-belligerents, Hungary, Italy, Finland and Romania, provided another 39. In North Africa Rommel had only four divisions. When the Soviet Union was invaded in June 1941 Hitler expected to conquer it in four months. Few western observers gave the Soviet Union much chance. Roosevelt's Secretary of War, Henry Stimson, gave it even less than Hitler: 'a minimum of one month and a possible maximum of three months'.[19] Within a matter of weeks Soviet forces lost two million men and five thousand aircraft, dwarfing the losses in World War I; within months German armies besieged Leningrad and Moscow.

The key to the eventual victory of the Allied states lies here, in the remarkable revival of Soviet military and economic power to a point where the Red Army could first contain, then drive back the German invader: remarkable, because it followed the loss by December 1941 of 4 million men, 8,000 aircraft and 17,000 tanks, equivalent to almost the entire strength of the Soviet forces in June;[20] remarkable, because it followed the German capture of more than half the Soviet steel and coal output, and the entire Soviet 'breadbasket', the fertile black earth regions of the Ukraine and the western steppe, where the vital food surplus for the cities was produced. So severe was the mauling that it is hard to imagine any modern state under these circumstances continuing to fight. It has been claimed that in October Stalin tried to negotiate with the Germans through a Bulgarian intermediary, but this was most likely a ploy to buy time. When the Germans were nearing Moscow most of the Soviet government apparatus evacuated, but Stalin decided to stay in his capital to help rally its defence, spurred on, it has been argued, by the evidence of patriotic defiance among elements of the Moscow population.[21] In 1942 there were local successes, but also long retreats, as the German armies pressed on the southern flank through the Crimea and beyond. But in 1943 the Soviet forces defeated their enemy at Stalingrad, and then at Kursk, and the long drive back into Europe began. How and why this happened, against every reasonable expectation, remains the central question of the war.

Stalin, quite naturally, hoped that in the critical years the western Allies would find some way of relieving the pressure on the Soviet war effort. They did so in two ways, neither of which entirely satisfied the Russians. First, they launched a bombing offensive against the Axis on a scale without precedent. Second, and after much inter-Allied argument and hesitation, they launched two vast amphibious operations, one against Italy in July 1943, the other a year later, in June 1944, against German forces in northern France. Bombing has always been a contentious issue. Aside from the strong moral objections which it rightly raised, and continues to raise, there have always been serious doubts about its strategic worth. The bombing offensives absorbed very large resources and their results were regarded as ambiguous at best, even by the politicians who ordered them in the first place. Nevertheless bombing was the first choice of Anglo-American planners in their efforts to get directly at Germany in 1942, and it remained a central element in western war-making until the defeat of Germany and Japan in 1945. For this reason, if for no other, its contribution to Allied victory deserves serious assessment.

There exists a great deal of confusion about what bombing was supposed to achieve in the war. The popular expectation that bombing alone would cripple the German economy, destroy the morale of its people and bring about German surrender was never seriously entertained by British and American leaders. The achievements of bombing were more modest, but were nevertheless substantial. Bombing speeded up the re-entry to Europe of western forces; it helped to open up a 'Second Front' in 1942 and 1943 by diverting large quantities of manpower and equipment away from the Russo-German conflict to the defence of the Reich. Finally, the choice to confront Germany through an air campaign created the conditions for the defeat of the German air force. In Italy and Japan bombing undermined the home economy and home morale critically; in Germany bombing prevented the effective development of an economic super-power. In any argument about why the Allies won, this is an impressive catalogue of reasons.

Bombing, like the war at sea, created circumstances that made possible the main thrust of Allied armies in 1944 to defeat Germany on three European fronts, east, west and south. This was rightly seen as the only way to secure victory, but for the western states it involved an operation of very considerable risk. The assault on the Normandy coast on the morning of 6 June 1944 was the largest amphibious attack ever launched. History was replete with examples of failure: Gallipoli (which almost finished Churchill as a serious politician in the First World War); the Spanish Armada; Napoleon in 1805 (when his strength on land was overwhelming); and more recently Hitler's own failure to invade Britain in the autumn of 1940, when the opposition on the beaches of Kent and Sussex was nothing to the network of defences facing the Allies four years later. So difficult was the enterprise that the Germans thought the Allies might try a more indirect route through the Balkans, or Portugal, or Scandinavia rather than risk the frontal assault on Fortress Europe. The success of the D-Day landings sealed Hitler's fate, as the landings in Italy a year before sealed Mussolini's. Like the bombing, this was an enterprise whose outcome deserves careful explanation.

The history of each of these four zones of conflict is central to the overall explanation of Allied military triumph. They are linked together by the wider themes which follow. The success in combat was determined in great measure by issues of production, scientific discovery, military reform and social enthusiasm. In all these spheres there are marked contrasts between the Allied and Axis sides which require elaboration. It has already been observed that the balance of human and material resources between the two sides can be reduced to three critical questions: how did the Soviet Union recover its industrial resilience? how did the United States turn itself in a year into a military super-power, when every other state had taken years to rearm? why did Germany, with so rich and industrially developed a continent at her disposal, produce so much less than the Allies? On the theme of fighting power it is tempting to reduce the issue to the simple question of why the Red Army managed to transform its effectiveness in a matter of months, when it looked a clumsy, spent force in 1941. This is almost certainly the most important question; but should we not also ask why the two military superstars, Germany and Japan, failed to sustain their momentum in the second half of the war? If the graph of Allied combat effectiveness rose steeply upwards, that of the Axis levelled off, and finally declined.

Some of these contrasts in fighting power can be explained by better use of intelligence or by superior technology, but it is impossible to ignore the human factor. Leadership counted for a great deal. So, too, did popular enthusiasm for war. The leading personalities contributed in all kinds of ways to the final outcome. Churchill, with his dogged hatred of Hitlerism, Roosevelt, with his defence of embattled democratic values, and Stalin, who roused a furious people to the defence of Mother Russia, all emerged under the stress of war as leaders of quality; but they tempered their own contribution by listening to advice and leaving much of the day-to-day task of running the war to others with more time and competence. In Germany the opposite happened. The easy victories persuaded Hitler that he had an inspirational grasp of strategy and operations. As the war went on he concentrated the war effort more and more in his own hands and trusted almost no one to give him advice. The German war became a remarkable one-man show in which intuition displaced rational evaluation, and megalomaniac conviction ousted common sense. Hitler was able to achieve much more than might have been expected given the manifest limitations of his education and experience, but in the end he took on too much.

Not surprisingly, such wide differences in the style of leadership produced contrasts between the warring societies as a whole. In the Allied states there developed a powerful bond between leaders and led which helped to sustain populations through the bad times, and brought societies closer together. This was true even of the Soviet Union, where the home population was regimented and oppressed to a greater degree than that of Hitler's Germany, but nevertheless exhibited a fervent, crusading patriotism that transcended the risks and miseries of everyday life. The response of Axis populations was more ambiguous. Important sections of society became disillusioned with the consequences of wayward, one-man leadership – so much so in the German case that some of Hitler's senior officers tried unsuccessfully to assassinate him in July 1944. The costs of Mussolini's overblown ambitions became too much for the

Italian King and army, who kicked him out in July 1943 and sought a peace with the Allies three months later. In Japan there were large numbers who thought from the start that the war was a mistake. Though they fought with fanatical tenacity from fear of what they thought the vengeful Allies might do to them, there was always an ambivalence about the Axis war effort. As the war slowly turned against the Axis states, they were forced to rely more on naked terror and crude propaganda to keep their populations fighting. On the eastern front the German authorities shot the equivalent of a whole division of Germans, more than fifteen thousand men, for indiscipline, defeatism or dereliction of duty.[22] In Japan in 1945 gangs of militarist thugs toured the home front bullying and murdering anyone who talked of peace, while young recruits were browbeaten into adopting suicidal tactics.

There is a striking moral contrast here. Whatever the rights or wrongs of the case, the Allies were successful in winning the moral high ground throughout the war. There are clear advantages in moral certainty and moral superiority. The Allied populations fought what they saw as a just war against aggression. They were able to appeal to neutral states to collaborate in a good cause; enthusiasm for war was straightforward; much was justified in the name of a higher ideal, such as the bombing, which provoked a real heart-searching only after the conflict was over. It was impossibly difficult, on the other hand, for the aggressor states to slough off their merited reputation for oppression and violation, although Axis leaders saw their own cause as just in their terms. In every theatre of war the language of liberation and resistance was directed against the Axis. The Japanese were regarded by the west as little more than barbarians. The Gestapo and the SS, even before the lurid revelations at Nuremberg, were bywords for inhumanity. The war was never, in reality, a simple war of good against evil, of civilisation against Dark Age, but the Allies' ability to make it seem so simplified their war aims and cemented a domestic and international consensus in their favour.

At the core of this moral certitude was a shared hatred of Hitler and Hitlerism. The Allies were never in any doubt that Japan and Italy were the lesser threat to their way of life. They were united in a moral revulsion at everything that the new German Reich and its leader stood for. As a moral crusade the war aims were reduced to one single ambition, to rid the world of Hitler. Even before the war and the Holocaust, Roosevelt regarded Hitler as 'pure, unadulterated evil'.[23] More than any other contemporary leader, Hitler was personified as the dark force that threatened to take civilisation by storm and drag it into the abyss. Why he aroused such strong passions, and still does, a half-century later, is not as easy a question to answer as it appears. But it is worth asking because they were passions that held together the Alliance when it threatened to dissolve in the face of bitter military or political disputes. They were passions that stimulated the Allies to the greatest of efforts (including the search for atomic weapons), and they explain the Allies unyielding commitment to unconditional surrender.

If we are to understand why the Allies won we must recognise that material explanations, of resources, of technology, of fighting men, are not enough on their own. There is a moral dimension to warfare inseparable from any understanding of the outcome. Allied populations were sustained by the simple morality of defending themselves against unprovoked assault; Axis populations knew in their hearts that they had been led into campaigns of violence which the rest of the world deplored. If the Axis states had won, the qualms of their populations would not have mattered. But the moral ambiguity underlying that violence surely explains something about why they did not. When Franz von Papen, one time Chancellor of Germany before Hitler came to power, heard of the outbreak of war in September 1939 he told his secretary: 'Mark my words; this war is the worst crime and the greatest madness that Hitler and his clique have ever committed. Germany can never win this war. Nothing will be left but ruins.'[24] Even Hermann Göring, Hitler's right-hand man and confidant in 1939, greeted the news of war with agitated alarm: 'then God help Germany!' Confidence in victory, misplaced though it proved in the short-term, was much in evidence on the other side. As the conflict loomed late in August 1939, General Pownall, head of British Army Intelligence, scribbled exultantly in his diary: '. . . we must have a war. We can't lose it!'[25]

*   *   *

In the end both these views were borne out, though not before the
world dissolved in flames; 55 million people lost their lives, and
destruction was wrought on a scale almost unimaginable fifty years
later. Out of the ruins of war a new political and economic order
was forged which is now, in its turn, in the throes of a painful
transformation. The threats to peace are far less than they were in
the 1940s, when Hitler was poised with his allies on the very brink
of world conquest, but none the less there are dangers to confront
in a nuclear world which have not disappeared with the collapse of
the Soviet Union at the end of the Cold War. Asking 'why the
Allies won' is done not in some spirit of modern triumphalism –
though the explanation *is* a tale of triumph over adversity – but in
a spirit of genuine historical inquiry, in order to be precise about
the explanations that matter, and may matter once again in the
century to come.

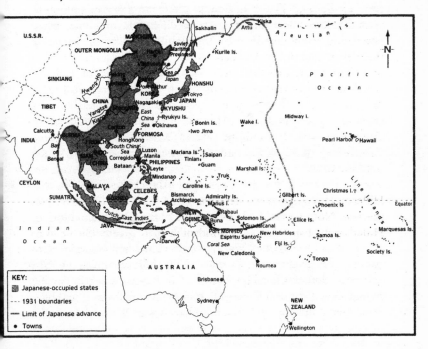

2   Japanese expansion 1931–42

LITTLE SHIPS AND LONELY AIRCRAFT . . .
The Battle for the Seas

'It was the job of the little ships and lonely aircraft, a hard, long and patient job, dreary and unpublicised, against two cunning enemies – the U-boat and the cruel sea.'

Captain G. H. Roberts, Cambridge, 1950

O N   A   D U L L, misty August morning in 1941 the British destroyer *Oribi* steamed into the great naval base at Scapa Flow in the Orkney Islands, at the northern extreme of the British Isles. Her unusual cargo, disgorged into ferry boats, was transferred to the very latest British battleship, the *Prince of Wales*, a mountain of a ship by comparison, sporting ten 14-inch guns. The consignment began with the British Prime Minister, Winston Churchill, who had left London in great secrecy the day before on board a sealed train. Behind him came a cavalcade of senior officers and officials, followed by a box of grouse, a world globe, and enviably unrationed quantities of sugar, beef and butter. The battleship moved slowly out of the harbour, shepherded by three escort destroyers. In faltering sunlight the ships set off north-west, taking Churchill to the first Anglo-American summit talks on the far side of the Atlantic.[1]

The invitation to meet had been Roosevelt's, but Churchill jumped at the opportunity. The two men had only met briefly years before; Churchill was uncharacteristically nervous about this second encounter. 'I wonder if he will like me,' Churchill asked the American envoy Averell Harriman before the first session.[2] There were solid grounds for anxiety. Churchill desperately needed American help for Britain's war effort; a close personal relationship would do a great deal to cement an unspoken alliance between the two states. All Churchill could do was wait in excited anticipation during the five-day passage. For security reasons there could be no contact with the outside world. The battleship dipped and zig-zagged through unseasonally heavy weather, ever-conscious of the threat of German submarines. Churchill behaved 'like a boy let out of school', temporarily absolved of the heavy burdens of office. He struck a nautical note. Abandoning the luxurious but noisy admiral's cabin prepared for him, he spent the whole voyage up on the bridge, sleeping in the admiral's sea-cabin, close to the action. He read *Captain Hornblower R.N.*; in the evening the mess cinema showed *Lady Hamilton,* the story of Nelson's mistress.[3]

Though trained as a cavalryman, Churchill's links with the sea had deep roots. In the Great War he was First Lord of the Admiralty until forced to resign over the fiasco at Gallipoli. When war broke out again in September 1939 he was recalled to the same office and searched restlessly for naval action. He supervised yet another catastrophe, the failed Norwegian campaign in April 1940, though this time disaster was followed by unexpected promotion when he succeeded Chamberlain early in May. As Prime Minister the naval war absorbed a great deal of his time. When he began his five-year correspondence with Roosevelt he coyly signed himself 'Former Naval Person'. Roosevelt shared his enthusiasm for sea warfare. He too was a former naval person, Assistant Secretary for the Navy from 1913 to 1920. As President he kept in close, almost daily, touch with naval affairs. Sea power exercised a special fascination for him throughout the war. He too rode to their meeting in a modern warship.

On the morning of 9 August the *Prince of Wales* reached her destination concealed from the world, at Placentia Bay, on the south Newfoundland coast. The British flotilla was early: clocks had been adjusted to the wrong time. Churchill rose before the ship's company, and nervously paced the deck. At 7.30 in the morning the American cruiser *Augusta* came into view, with the President seated on the deck. Amidst the cheering and band-playing Churchill was ferried across the bay to the American vessel. Protocol dictated that Churchill, a chief minister, should first pay his respects to Roosevelt, a head of state. The two men greeted each other with great cordiality, and then got down to business.

There was much to discuss. The situation for both states was critical. German armies were now deep in Russian territory, within striking distance of Leningrad and Moscow. In the Far East Japan had recently occupied French Indo-China and now threatened the whole of south-east Asia and the southern Pacific. Since 1939 Britain had lost over two thousand ships totalling almost 8 million tons to enemy submarines, aircraft and merchant raiders. The Atlantic run, which Churchill had just experienced first-hand, was Britain's lifeline. By 1941 a stream of vital foodstuffs, machinery and raw materials flowed from the New World, including most of Britain's oil and aluminium. Without these supplies Britain's war effort in 1941 could not have been sustained. Both men knew how much control of the sea mattered to the democracies. Churchill told Roosevelt in December 1940 that shipping was 'the crunch of the whole war'; in May 1941 Roosevelt suggested to Churchill that the war 'would be decided in the Atlantic' and if Hitler could not win there 'he cannot win anywhere in the world in the end'.[4] At Placentia Bay naval power was uppermost in both their minds. Roosevelt agreed to give any help short of war in the Atlantic struggle. Both leaders agreed to warn Japan in the sternest tones to encroach no further into the Pacific. Both agreed to do what they could to ship supplies to the beleaguered Soviet Union. Roosevelt could report the onset of American rearmament on the largest scale, but he could not promise belligerency.

For public consumption the two leaders concocted a common declaration of principle, the Atlantic Charter, for which the meeting became famous. The document was not a treaty; neither party was bound by its terms. It was a very public statement of democratic solidarity, expressed in recognisably Churchillian prose, defining the hopes of both men for a 'better future for the world' through democratic politics, national self-determination and open trade. The Charter was testament to how well the two men got on despite their many differences in political outlook or interests. Churchill need have had no worries. 'The President is intrigued and likes him enormously,' wrote Harriman to his daughter in London.[5] The two men parted with genuine warmth. In grey damp weather the *Prince of Wales* slipped from her moorings late in the afternoon of 12 August for a second Atlantic run. Three days later the battleship overtook a convoy of 72 merchant ships steaming eastwards. Churchill had never seen a convoy and was overcome with excitement and sentiment. The signal flags for 'church' and 'hill' were flown from his ship. The crews aboard the armada of tankers, liners and tramp ships cheered with passion and Churchill waved and cheered in return. The thirteen columns of ships with funnels smoking 'looked almost like a town'. Churchill then asked the captain to go round again, and the battleship performed a wide arc back through the convoy before pulling away. Churchill stood on the deck and watched until the last smoke disappeared behind the horizon.[6]

The *Prince of Wales* made only one more voyage. Japan ignored the warning sent after the Atlantic meeting, and on 7 December attacked the American Pacific Fleet at Pearl Harbor and British possessions in the Far East. Churchill, confident that battleships could still defend themselves against air attack, sent the *Prince of Wales* and *Repulse* to Singapore. Both were caught by Japanese torpedo bombers in the South China Sea on 9 December and sunk. Admiral Phillips and Captain Leach stood at salute on the bridge of the *Prince of Wales* as it disappeared into the sea. 'In all the war', Churchill later wrote, 'I never received a more direct shock.' After the Japanese attacks there were no major British or American warships in the whole of the Pacific or Indian Oceans. 'Over all this vast expanse of waters', he continued, 'Japan was supreme, and we everywhere were weak and naked.'[7]

*   *   *

The sea mattered to Churchill and Roosevelt because the states they led were, first and foremost, naval powers. The United States possessed the largest navy in the world in 1941, but her army was ranked eighteenth, a tiny skeleton force. The British had a navy in 1939 second to none among the powers then at war, but could muster no more than two fully equipped divisions to send to France when war came. Though both states built up large well-equipped armies by the war's end, they fought the large part of the conflict as a naval war – blockading their enemies, securing foreign supplies, and fighting small local engagements, supplied and protected from the sea. Axis victories ensured that Britain and America could bring their enemy to battle in only one way, by securing the sea-lanes for amphibious assault.

For both Britain and the United States naval power was a

geographical necessity, though for different reasons. British naval strategy was chiefly concerned with protecting overseas trade and safeguarding a far-flung colonial empire. Britain purchased half her food and two-thirds of her raw materials from abroad. Without this flow of supplies homewards, and the flood of manufactured exports sent in return, Britain might have remained what she was until the eighteenth century, an impoverished, underdeveloped island off the European mainland. Without a powerful navy Britain would not have been able to protect her trade, or build the network of imperial possessions that nourished and guarded it. Without the navy Britain would not for centuries have been blessed with the means to defend the home islands from assault and the luxury of being able to fight on other people's soil at times and places of her own choosing. There were drawbacks. British strategy was global rather than local. The seamless cloth of ocean made for a vast, unmasterable field of conflict. Britain was manifestly vulnerable to the interruption of her sea traffic by states with pretensions to regional maritime power, and even, in the worst case, to the blockade of the home islands. In the fifty years before the Second World War these dangers had become first apparent, then real. The rise of American sea power challenged British maritime primacy, though not dangerously so. The growth of German naval power, though blunted by defeat in the Great War, was regarded as much more dangerous, particularly when it revived again in the 1930s, and even more so when Germany began to drift closer to Italy and Japan, both of whom had large modern navies in menacing proximity to Britain's chief imperial interests. By 1939 Britain simply could not afford a navy of a size sufficient to ensure global security. Yet she entered the war in September 1939 reliant for her very survival on the ability of her navy to defend the arteries of trade and to shepherd troops and supplies worldwide.[8]

America was much less vulnerable than Britain. Geographically remote from potential enemies, the United States possessed abundant food supplies and extravagant natural resources. Foreign trade made her richer, but it was not a lifeline. Neither did America have a far-flung empire on which she relied for men and supplies. Her few overseas possessions, in the Pacific and the Caribbean, were sentry posts to the western hemisphere and not stepping stones to world empire. The American navy was not a means to other ends but was mustered purely for the armed defence of the New World. Ever since 1822, when President James Monroe announced his famous Doctrine for the Americas, the United States navy assumed unsolicited responsibility for the defence of the New World against outside military interference. Its vital strategic function brought it the lion's share of Congressional appropriations for defence. In the crisis-ridden international order of the 1930s it was the navy's task to insulate America from the dangers of war, east and west. If it ever became necessary to commit American forces to battle, as Roosevelt realised might be the case sooner than most, the existence of a powerful navy ensured that the fighting would be overseas in distant theatres, provisioned and protected by American ships. Without naval power American intervention in World War II could not have been contemplated, even less attempted.

When war broke out in Europe in September 1939 America was still more than two years away from belligerency. For most of that period the British Royal Navy confronted the European Axis states across the global battlefield with its own limited resources. The outlook at first was set fair. Britain had French naval power to call upon in the Mediterranean, the critical supply line to the imperial east. The German navy was tiny. Britain and France between them had 22 battleships and 83 cruisers; Germany possessed 3 small 'pocket' battleships and 8 cruisers.[9] The German navy knew that there was no prospect of fighting any kind of face-to-face battle with the enemy fleet. Neither were prospects for submarine warfare against British trade much better. The German submarine arm had only eighteen operational boats in the Atlantic, against an enemy who began at once to convoy all shipping and to provide escort vessels equipped and trained for anti-submarine warfare. The German navy's Commander-in-Chief, Grossadmiral Erich Raeder, regarded the outcome as a foregone conclusion, hoping only that his forces would know 'how to die gallantly' when the time came.[10]

Yet within a few months the balance in the sea war began to move away from the Royal Navy. The German conquest of Denmark and Norway transformed the geography of the conflict by securing a long coastal flank for the movement of German vessels into the Atlantic. The German defeat of Belgium and France in June 1940 resulted in the nightmare that had haunted British governments since the days of Napoleon: control of the Channel coast and ports by a hostile power. Now German surface forces and submarines were able to operate from an Atlantic coastline against British trade. Hitler's directive to his navy to 'deal an annihilating blow to the English economy', which looked hollow when it was published the previous November, began to fill up with early successes.[11] The defeat of France was a double blow for the Royal Navy, for not only did it mean the loss of the French navy for the Allied cause, but it also invited Mussolini's Italy to join in the war, hungry for spoils. Straddled across Britain's vital imperial route through the Mediterranean, from Gibraltar at the western end of the sea to Suez in the east, was the Italian navy, half a million tons of hostile vessels, including more than a hundred submarines.[12]

There was worse to come, for the real enemy of the British fleet was not on the sea but in the air. When war broke out, navies on both sides were still wedded to the traditional view of sea power, exercised by ships on or under the surface of the ocean. Few people foresaw how swiftly air power would render redundant naval strategies still rooted in the battleship age of Fisher and Tirpitz, who led the Anglo-German naval race before the First World War. The very first naval contest of the war, off the coast of Norway, was a rude shock. With secure air bases on the Channel coast by the summer of 1940, German bombers began to take a terrible toll of British shipping. When the German air force supplied the navy with a small group of converted long-range passenger aircraft, the Focke-Wulf Condor, attacks on shipping were extended far out into the Atlantic shipping lanes. In 1940 aircraft alone sank 580,000 tons of British shipping; the following year over a million tons, more than British dockyards could make good.[13] In the Mediterranean, with its narrow channels and clear skies, ships made easy targets, first for the Italian air force, then from 1941 for the bombers and dive-bombers of the Luftwaffe, transferred there to secure the Balkans and to help Italian endeavours in North Africa. The persistent illusion, shared even by Churchill, that the big battleships could defend themselves against air attack ended with the destruction of the Italian fleet at Taranto on 11 November 1940 by a mere twenty Swordfish biplanes of the British Fleet Air Arm,

the sinking of the German battleship *Bismarck* in May 1941, and the
devastating destruction of the *Prince of Wales* and *Repulse* seven
months later.[14] Great ships that took years to build and commission
were sent to the sea-bed in a matter of minutes, destroyed by a
handful of bombs and torpedoes. The most remarkable mismatch
of the whole war was the trial of combat between these great
dinosaurs of the sea and the tiny aircraft that circled them like
venomous insects, waiting to sting.

The Royal Navy was soon stretched to the limit, fighting with
shrinking resources in waters from the North Sea to Egypt, from
the Arctic to the far reaches of the South Atlantic. The enemy
avoided fighting face to face, for he had no main fleet. Instead he
fought a bitter war of attrition, sapping away the lifeblood of the
British merchant marine by sea and air, seizing any opportunity to
sink naval vessels that came to its defence. During the course of
1940 the tables were turned on Britain: the blockader was
blockaded. Though Hitler lacked the naval strength to invade
Britain, he was easily persuaded by his naval chiefs that Britain's war
effort could be crippled, perhaps decisively, by cutting through the
main artery of trade across the Atlantic sea-lanes. Though the
German navy was able to send no more than ten or fifteen
submarines into the area for the first eighteen months of war, these
were sufficient to create a debilitating haemorrhage.

British preparations were undone. Though the navy took over
the organisation of all British shipping in a remarkable feat of global
planning, German cryptographers succeeded in unmasking the
British naval codes that carried the directives. German submarines
were able to compensate for their small number by the fore-
knowledge of convoy movements. The convoy escorts relied on the
use of ASDIC, a sound-detecting instrument first developed in the
Great War, which betrayed the presence of submerged submarines.
To counter this threat the German submarines adopted the simplest
of counter-measures: they attacked at night on the surface, where
they could neither be seen by the escort look-outs, nor detected by
ASDIC. Deprived of their eyes and ears, their position on the ocean
betrayed, the ships were easy prey. The submarines were helped by
the British decision not to convoy ships with speeds of less than 9
knots or more than 13; during 1940 sixty per cent of ships sunk were

not sailing in convoy. In that same year 992 ships were sunk, totalling 3.4 million tons, a quarter of British merchant shipping.[15] By careful use of brief radio transmissions it proved possible for the Germans to gather large numbers of submarines together in 'wolf-packs' to snap at the heels of intercepted shipping. In the first four months of 1941 almost 2 million tons of shipping were destroyed, over half in the North Atlantic. For German submariners these were '*die glückliche Zeiten*', fortunate times.[16]

German forces succeeded against British sea power beyond their wildest expectations. British trade was slowly bled white: 68 million tons were imported in 1938, 26 million in 1941.[17] When figures of sinkings for February were released Churchill was finally moved to give anti-submarine warfare priority over everything else. On 6 March he announced that Britain was now fighting 'the Battle of the Atlantic' which he regarded with justice as 'the real issue of the war' for Britain.[18] The concentration of British minds on the naval war produced some positive results. British cryptanalysts at the code and cypher school at Bletchley Park succeeded in breaking enough of the German naval code to be able to steer convoys away from the gathering wolves for the summer months. Desperate efforts were made to provide better air cover for convoys and escort vessels for the whole of the North Atlantic run, and a submarine Tracking Room set up in London succeeded through a combination of secret intelligence and intelligent guesswork in steering large numbers of convoys clear of danger altogether.[19] But losses continued to mount elsewhere. By late 1941 the British merchant fleet began to decline steadily, while the long hauls necessary to avoid danger-spots wasted vital shipping capacity. Over the year 1,299 ships were sunk, and over half their crewmen were lost.[20] Still bereft of naval allies, Britain faced the prospect of isolation from American supplies and the loss of the Mediterranean. Already all that linked her with her Empire in the east was the long 8,000-mile haul around the Cape of Good Hope, into seas increasingly threatened by Japanese aggression.

It was with this gloomy outlook that Churchill sat down to supper with Averell Harriman and the American ambassador, John Winant, on the evening of 7 December 1941. Harriman found the Prime Minister 'tired and depressed'. Churchill said little throughout dinner, deep in thought, 'with his head in his hands'. Just before nine o'clock his butler brought in a small radio for the evening news from the BBC. He was slow to switch it on and missed the opening headlines. Then minutes later the news reader returned to the opening story, which had reached him shortly before going on air: 'Japanese aircraft have raided Pearl Harbor.' Churchill leapt out of his chair and banged the radio. He telephoned at once to Roosevelt: 'What's this about Japan?' Roosevelt confirmed that Japan and America were now at war: 'We are all in the same boat now.'[21] Churchill promised to declare war on Japan the next day. The two largest navies in the world were now ranged side by side, an alliance for which Churchill had worked hard for two years. It was impossible for British leaders not to feel a sense of relief at the news. General Ismay, secretary to the British Chiefs-of-Staff, told a companion the following day that the alliance made 'ultimate victory certain'. But in the short term Britain and the United States faced what Ismay later called a 'cataract of disaster'.[22]

There was no disguising what a disaster it was. Within a matter of weeks Japan crippled the American Pacific Fleet and eliminated the British and Dutch navies in the Far East. British, Dutch and American possessions were seized one after the other, their fragile defences swept aside with contemptuous ease. The Indian and Pacific Oceans lay wide open to Japanese sea power. None of the world's broad seaways could be safely sailed, or easily defended. 'If we lose the war at sea', observed Britain's naval chief Admiral Sir Dudley Pound in March 1942, 'we lose the war.'[23]

★   ★   ★

No one was more surprised by the speed and completeness of Japanese success than Japanese leaders themselves. They had expected a campaign of six months or more, not twelve weeks; they had anticipated losing a quarter of their fleet, but lost only three destroyers. It took only a dozen divisions to capture the entire southern region.[24] Whatever qualms existed before Pearl Harbor, they evaporated; most Japanese became intoxicated with triumph, 'victory-drunk' as they called it. The temptation to capitalise on that success was overwhelming. Japanese self-confidence was riding high. On New Year's Day 1942 Admiral Matome Ugaki, Chief-of-

Staff of the Japanese Combined Fleet, saw a year ahead brimming with promise: 'The future is filled with brightness. The course of events this year will determine the fate of the war . . . The main thing is to win, and we surely will win.'[25]

There were now so many options open that Japanese army and navy leaders began to squabble about which one to take. Some wanted to drive through to Australia to complete the conquest of the whole south-west Pacific. There was talk of moving westwards to join up with the Germans in the Middle East, driving the British from Asia. The army rejected them both because it lacked the manpower for operations on such a bold scale. The Combined Fleet led by Admiral Isoroku Yamamoto favoured a more modest but strategically significant move: the extension of Japan's new Pacific perimeter to embrace the western Aleutians in the north, the island of Midway in the central Pacific, and a circle of bases north of Australia – Port Moresby in Papua, Ocean Island and Nauru. The object of these operations was to cut sea communications between the United States and the western Pacific, and to tempt the crippled American Pacific Fleet to a final showdown where it would be annihilated by the larger Japanese force.[26] The Japanese admirals were obsessed with the traditional rules of sea warfare, the pursuit of a great fleet engagement like the one they had won against the Russian navy in the Straits of Tsushima 37 years before, when Yamamoto was a young midshipman. Only the decisive defeat of the American main fleet could make the Pacific a Japanese lake.

The plan was first to seize the south-western islands in May, while the bulk of the fleet and its carrier forces rested, regrouped and prepared for the main operation against Midway in June. The attack on the Aleutian islands, seizure of which would safeguard Japan's northern flank, was scheduled for the same time, in the hope of dividing American forces at a critical point in the confrontation. The core of the Japanese strike force was provided once again by the large aircraft carriers and the small force of highly trained naval pilots which had inflicted so much damage at Pearl Harbor. Armed with the highly effective 'Zeke' or Zero fighter, faster and more manoeuvrable than anything yet available to the Allies, and with 'Long Lance' torpedoes, fuelled with liquid oxygen to give them a speed and range unrivalled by any other navy, the Japanese approached the coming battles with a dangerously arrogant sense of invulnerability.

On the other side of the Pacific Japanese victories produced a frenzied response, part fury, part fear. The American public, used to regarding the Japanese as their racial and cultural inferiors, was suddenly faced with an enemy apparently unstoppable, whose military skills bordered on the magical. The thirst for vengeance was unquenchable.[27] This posed difficulties for Roosevelt and the military chiefs. Ever since 1940 the assumption in Washington was that Germany was the chief enemy; Roosevelt had reiterated to Churchill at Placentia Bay the commitment to fight 'Germany first' if America found herself at war. Japanese aggression challenged this commitment. It was not simply a question of satisfying the bloodlust of ordinary Americans, powerful lobby though it was. The real issue was the imminent collapse of America's position throughout the Pacific basin. This required an urgent response. Admiral Ernest King, appointed Commander-in-Chief of the US navy in 1941, pleaded for the priorities to be reversed: Japan first, then Germany. In the end Roosevelt settled on a compromise: Britain was given a guarantee that the defeat of Hitler was still the primary ambition, but the Pacific got the lion's share of naval and army resources. By the middle of 1942 there were almost 400,000 American soldiers in the Pacific theatre; against Germany and Italy there were only sixty thousand.[28] King had another bee in his bonnet: he did not want to share the Pacific war with the British, against whose naval ambitions he displayed a powerful prejudice. There was little the Royal Navy could do in the Pacific with unlimited commitments elsewhere, and in March Churchill agreed to divide oceanic responsibilities, America to operate in the Pacific, Britain in the Indian Ocean, and both states in uneasy condominium in the Atlantic.

In the early months of 1942 American strategy crystallised into a single objective, to keep some kind of military foothold in the southern Pacific as a springboard for a future offensive. Australia was the obvious choice and it was here that the American commander General Douglas Macarthur, rudely expelled from the Philippines, gathered together the ragged remnants of the retreating armies for the desperate task of halting the Japanese onslaught. On 30 March Roosevelt appointed him Supreme Commander, South-west Pacific. At the same time King appointed Admiral Chester Nimitz Commander-in-Chief, Pacific Ocean Area. On these two men fell the responsibility for keeping open the lifeline from America; Macarthur, flamboyant, ambitious, a jealous guardian of army responsibilities, and Nimitz, a more modest personality, a man of method and good sense, a solid organiser.

Both men could see that Japan was unlikely to stop after the first flush of victory. By mid-April Nimitz had reliable intelligence that enemy forces were once again on the move south towards Australia, almost certainly intending to seize Port Moresby on the south-east coast of New Guinea, a short flight from Queensland, and probably to take the remaining necklace of islands bordering the Coral Sea. On 16 April Nimitz despatched the aircraft carriers *Lexington* and *Yorktown* from Hawaii to rendezvous with a motley assembly of smaller warships to form a task force to oppose the Japanese assault. The Japanese moved down in four groups, organised according to a complex schedule of operations. There were two invasion forces, one for Port Moresby, the other for the ring of islands, stretching from the Solomons to Ocean Island. They were accompanied by a covering force of larger warships. The main striking force, built around the large aircraft carriers *Shokaku* and *Zuikaku,* steered farther to the south-east, to try to catch American ships in a pincer movement. The whole assault was to take four days, from 3 May to 7 May.[29]

Japanese soldiers had already landed on Tulagi in the Solomon Islands on 3 May when the American task force, commanded by Rear Admiral Frank Fletcher, sailed into the Coral Sea. No stretch of water more deserved its name. Calmer and much bluer than the ocean, the sea was ringed with coral islands dotted with tropical greenery. Along the southern coast lay the 1,500-mile Great Barrier Reef, which produced a fringe of white foam around the azure waters. Neither side had accurate intelligence on the whereabouts of the other, and squally, grey weather made aircraft reconnaissance difficult. Both forces circled round, anxious to make contact. When they finally did so there followed a catalogue of errors on both sides. The Japanese carrier force under Admiral Takagi

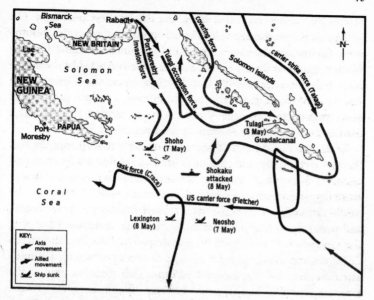

3   Battle of the Coral Sea, 5–7 May 1942

mistook the American fuelling ship *Neosho* for the enemy task
force, pounding the unfortunate vessel with the full weight of an
aerial assault. American reconnaissance aircraft mistook the light
naval force supporting the Port Moresby invasion for the Japanese
carriers. Fletcher sent off his own aircraft, which by mere chance
found the larger covering force on their way. They attacked and
sank the light carrier *Shoho*, the first of its kind lost by Japan. This
was enough to halt the attack on Port Moresby. The assault force
retreated north-west, its air cover lost. On 7 May the weather was
too poor for successful air attack, but Takagi sent off his aircraft
regardless. Reaching the end of their fuel supply, his fighter
bombers were continuing to hunt for the American ships when
they were attacked by enemy fighter aircraft. Nine were shot
down. In gloomy weather the remainder turned for home. Some
tried to land on the *Yorktown,* mistaking it for their own ship. Short
of fuel the rest were forced to land in the sea. Only one-fifth of the
force returned safely.

The following day both fleets knew that they must be close enough to each other to make contact. American aircraft found the two large Japanese carriers in cloudy weather. They attacked in two waves, beginning at just before 11 a.m. The result was disappointing: neither carrier was sunk. The *Zuikaku* avoided any serious damage; the *Shokaku* was hit by just three bombs – this was however sufficient to destroy her engine repair shops and damage her flight deck severely enough to send her on a long haul back to Japan for repairs. At almost the same time Japanese airmen found Fletcher's ships. Both American carriers sustained severe damage, but were not sunk. Only later in the day did the *Lexington*, listing badly, suffer a massive internal explosion. At eight in the evening she was put out of her misery by an American destroyer. Her crew watched from nearby rescue ships as she finally went down, 'crying and weeping like young girls'.[30]

The Battle of the Coral Sea was the point at which Japanese expansion in the Pacific was halted. The Allied force hardly distinguished itself, but enough was done to turn back the invasion of Port Moresby, and to blunt the assault on the islands. On 12 May an American submarine torpedoed the *Okinoshima,* flagship of the invasion force for Naura and Ocean Island. 'A dream of great success has been shattered,' wrote Ugaki in his diary, although in Japan the authorities announced yet another startling victory.[31] Any attempt to renew the attack on the southern islands was shelved until later in the year. Every effort was made to prepare for the next operation, codenamed 'MI', the invasion of Midway. In planning it, Japanese commanders drew almost no lessons from the failures in the Coral Sea. Yet it was a conflict dictated by air power, a naval battle in which no ship fired its guns at another ship in anger, indeed no ship even sighted the enemy fleet. It showed as clearly as the sea war in the Atlantic and the Mediterranean that command of the sea also required command of the air.

Midway Island was an unprepossessing target. A tiny atoll no more than 6 miles across, much of it under water, it formed the farthest point of the Hawaiian archipelago, and the most westerly point of United States territory. Claimed for America in 1867, it was not occupied until 1903 when the navy chased off its itinerant population of Japanese feather-hunters. Only in 1940 was a ship channel built to allow larger vessels to enter the harbour, and Midway became a full military base, with aircraft, flying boats and marines. Japan wanted Midway as a gateway to the shipping lanes of the Pacific, and as a base from which to threaten Hawaii itself. On 5 May preparations began. The Japanese forces were once again widely dispersed in five main groups: a carrier strike force under Admiral Nagumo consisting of four large carriers; an occupation force for Midway itself; the main fleet of seven battleships, including the flagship *Yamato,* intended as the instrument to destroy the remains of the American Pacific Fleet; a diversionary force to seize the two westerly islands of the Aleutians; and finally an advanced force of submarines to scout ahead and to shield the main fleets. Yamamoto's plan was simple. Midway would be occupied, American ships would sail from Pearl Harbor to save the island, where they would be softened up by the carrier force before being annihilated by the main battleship fleet following in its wake. There was little flexibility in the plan, and Japanese commanders could see small reason to be flexible. They had scanty intelligence on the American position, but they were confident that the *Lexington* and *Yorktown* had been sunk in the Coral Sea. By skilful misinformation American radio intelligence persuaded the Japanese that the remaining two carriers of the American Pacific Fleet were in the south-west Pacific protecting Australia, too far away to interfere in time with the battle of Midway.[32] The date was fixed for 5 June, Japanese time, 4 June in the United States.

The American position was precarious. Reinforcement was slow, and the calls on American shipping in every marine theatre compromised efforts to supply Hawaii. American dockyards were turning out large numbers of new ships, but there were no new carriers or battleships to confront the current threat. American naval aircraft were sound, but the torpedoes were slow and inaccurate. Fighter protection for the dive-bombers and torpedo-bombers was still insufficient to cope with the numerous shipboard fighters of the Japanese carrier force. American ships had radar, which their enemy did not, but against mass air attack even radar warning was of limited value. All Nimitz had for Midway were two carriers and partial but accurate intelligence of Japanese intentions; that, and the knowledge that here was a battle he did not dare to lose.

Two carriers were better than none. The *Hornet* and the *Enterprise* had sailed briefly for the Coral Sea, but returned when it was clear that they could not arrive in time. The pugnacious commander of the carriers, Admiral William Halsey, who rallied his men with the slogan 'Kill Japs, kill Japs, kill more Japs', had to be hospitalised with a serious skin disorder, and his place was taken by Rear-Admiral Raymond 'Electric Brain' Spruance.[33] He earned his nickname because he could think on his feet in tough situations with a calm, remorseless logic. He was very different from Halsey, but was almost certainly the better man for the difficult task in hand. It was no mere chance that placed aircraft carriers at the centre of the American task force. Ever since General Billy Mitchell had demonstrated twenty years before that warships could be bombed successfully from the air, the US navy had been alive to the significance of naval aviation. In the 1920s the navy commissioned the carriers *Lexington* and *Saratoga,* the largest ships afloat until the war. Under Admiral King's leadership in the 1930s naval aviation made great strides in tactics and training. King's own career was linked with naval aviation. He taught himself to fly when he was well over forty, and was commander of the carrier forces in the late 1930s. He was not a big battleship sailor; certainly not the man to pick up Yamamoto's challenge to a fleet duel.[34]

The one solid advantage enjoyed by the American navy over its opponent was intelligence. The Japanese were able to intercept American radio traffic, but could not decode it. Fortunately for Nimitz a telegraph cable had been laid early in the century from Hawaii to Midway and a great deal of his communications could be sent down the line, entirely secure from the enemy. Japanese radio communications were leaky by comparison. The Fleet Radio Unit Pacific, stationed at Pearl Harbor, was able to read about one-third of the Japanese naval code, JN25. The task of decrypting it was doubly difficult, as each message was not only placed in code, but was also then enciphered on a table of 100,000 five-digit numbers mixed at random – the cipher had to be stripped away even before the task of decoding the message. The Unit, led by Joseph Rochefort, was based in a cramped, disorganised bunker. With a small staff, Rochefort worked day and night, sleeping for brief spells, living off sandwiches and coffee. The Unit was the very opposite of naval spick-and-span, but it did what Nimitz wanted by dint of exhausting effort. By early May he knew that a major operation was planned for the Hawaiian area. By mid-May the Unit identified Midway as the most likely target, with the Aleutians as a diversionary operation. The target code was the symbol 'AF', but the Unit needed to be sure that this was Midway, and not Hawaii itself. A radio trap was laid. A transmission was sent *en clair* from Midway to Pearl Harbor announcing that the freshwater distilling plant on the island had broken down. This was duly intercepted by Japanese intelligence and relayed in code back to Tokyo. The symbol they used for Midway was indeed 'AF'.[35]

This was 21 May. A few days later the Unit told Nimitz the exact date for the invasion: 3 June in the Aleutians, 4 June at Midway. The intelligence picture was vital for the Americans, for they knew that their forces were much smaller than those of the enemy. The only way this margin could be reduced was by deploying the carriers in exactly the right place for maximum effect. On 27 May Nimitz issued his battle plan. The two carriers and a small screen of cruisers and destroyers were to sortie from Pearl Harbor to a point north-east of Midway, out of range of Japanese search aircraft and beyond any submarine screen that the Japanese might establish around Hawaii. Here the carriers were to wait until land-based planes from Midway could tell them exactly where the Japanese carriers were, and then American aircraft were to wage a war of attrition against enemy vessels. On no account were Spruance and his fellow commander, Admiral Fletcher, to expose their small force to the full weight of Japanese fleet attack. On 28 May the American force sailed. Sitting in the harbour was the damaged carrier *Yorktown,* scheduled for a ninety-day refit after her exertions in the Coral Sea. Against all expectations an army of shipworkers, 1,400 strong, swarmed over the ship making good the damage in 48 hours, working round-the-clock, improvising where they had to. It was a triumph of American technical skill; on 31 May she was ready to sail back into combat. Against Yamamoto's 4 fleet carriers, 7 battleships, 12 cruisers and 44 destroyers Nimitz now had 3 carriers, 8 cruisers and 15 destroyers.[36]

On 29 May Yamamoto's forces, a vast flotilla of warships, fuellers, supply ships, seaplane carriers and minesweepers, sailed out of the main Japanese naval base at Hashira Jima, through the Bungo Strait between the islands of Kyushu and Shikoku and out into the Pacific. At the centre of the fleet was the giant battleship *Yamato,* a 62,000-ton monster launched two years before, bearing the sacred name of the Japanese race itself. From here Yamamoto directed the operation, surrounded by his staff. The whole fleet was under orders to maintain the strictest radio silence. Ships communicated by flag or morse code. The weather was foul, with high winds and heavy seas. North of the island of Iwo Jima the force split up, the covering fleet for the Midway invasion steering due east towards the island, the carrier task force and the main body of the fleet going north-east, with the carriers in front, waiting to pounce on any American forces sent out from Pearl Harbor. Japanese commanders could hear a heavy increase in American radio traffic but, believing their codes to be secure, assumed that this indicated routine precautionary measures in Hawaii. Throughout the seven-day voyage every effort was made to find out if the enemy had detected Japanese intentions. So poor was the weather that sighting from the air was unlikely. Heavy mist on 1 June and thick fog the next day shielded the force from detection by enemy submarines. But Japanese reconnaissance was equally hampered. Seaplanes sent to French Frigate Shoals, a small group of coral islands west of Hawaii which it was hoped would provide a base for air reconnaissance, found them already occupied by American forces. The submarine screen was late in arriving at its destination, and by the time it was in place Nimitz's forces had long left Pearl Harbor and were well to the west. On 3 June the commander of the carrier force, Admiral Nagumo, recorded in his log that the enemy did not know of the Japanese plan, and that there was no evidence of an enemy task force anywhere in the vicinity. Confident of success, the carriers now veered south-east on a course for Midway, entirely ignorant of an American force 200 miles away, forewarned days before of their arrival.[37]

The first sighting by either side came on 3 June when a Catalina flying boat from Midway spotted the small island invasion force approaching from the west. The pilot reported that he had seen the main body of the Japanese fleet, but the American carrier force knew from intelligence that the main Japanese fleet was farther north. Bombers were despatched to attack the invasion force, while Fletcher and Spruance moved their carriers to a point where they would lie exactly on Nagumo's north flank, within easy striking distance of his carriers. The sun rose the following morning at four o'clock. Visibility was excellent, a mixed blessing for a force in hiding. The Japanese ships launched a limited reconnaissance. One aircraft even flew over the American task force but its negligent observer failed entirely to see it. A second was held up on deck with a faulty catapult. A third plane turned back with engine trouble. Confident that there was no enemy force to be found, Nagumo ordered his carrier aircraft to strike the island of Midway. Over a hundred aircraft were launched from the decks of *Akagi*, *Kaga*, *Hiryu* and *Soryu* and at 4.45 a.m. they flew off south-eastwards.

Half an hour later the first American aircraft sighted the carrier force and relayed its approximate position. By seven o'clock the first attack was made from aircraft based on Midway. For over an hour the carriers swerved and slewed about to avoid bombs and torpedoes; American aircraft scored not a single hit. But that hour

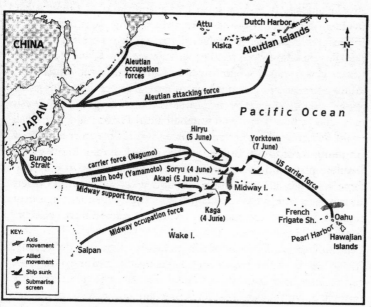

4   Battle of Midway, 4–5 June 1942

was a decisive one. At 7.28 the one remaining Japanese reconnaissance aircraft at last sighted the American carrier force, though it reported only cruisers and destroyers. Nagumo dithered. He had ordered his remaining torpedo-bombers to be converted to carrying land bombs for a second attack on Midway; most were below decks in the process of rearming. In an hour he would need to land the first wave of aircraft returning from Midway. He asked the reconnaissance aircraft to be more precise, while he ordered the bombs to be replaced by torpedoes again. At 8.20 he learned that there might be one carrier with the enemy force. He became uncharacteristically hesitant. Rather than attack the enemy at once he decided to land and rearm his aircraft now arriving back from Midway and then reorganise his forces for an attack later in the morning. For the next hour his carriers were at their most vulnerable, full of refuelling and rearming aircraft, with only light fighter cover, their position known to the enemy.[38]

Fletcher and Spruance could not have planned things better. They held back their carrier attack aircraft so that they would hit the Japanese ships during the crucial changeover. The first wave of torpedo-bombers attacked at about 9.30, a second shortly after. Short of effective fighter cover, the slow planes were shot to pieces by the defending Zero fighters. Out of 41 torpedo-bombers, only six returned; not a single torpedo reached its target. But circling above the doomed torpedo-bombers were 54 Dauntless dive-bombers. Undetected by the Japanese fighters they dived out of the sun 'like a beautiful silver waterfall'.[39] Nagumo's flagship *Akagi*, with forty refuelling aircraft on deck, was hit by three bombs and in minutes became a floating inferno, wracked by one devastating explosion after another. Nagumo stood disbelieving on the bridge amidst a sea of flames. His staff pleaded with him to abandon ship; finally he was physically dragged from the bridge, leaving by the only remaining exit, a rope over the side. All around him unmanned guns were firing as the flames licked at their ammunition. As Nagumo made his way to a waiting destroyer he could see the carriers *Kaga* and *Soryu* burning furiously. *Kaga* was hit by four bombs out of nine aimed at it, *Soryu* by three. On decks crowded with fuelled or refuelling aircraft, petrol tankers and bombs, the explosive effects were magnified. *Kaga* sank at 7.35 that evening, *Soryu* a few minutes earlier, after blazing fiercely all day. *Akagi* stayed afloat but was scuttled as beyond repair the following morning.

Within ten minutes the heart of the Japanese navy's strike force was destroyed by a mere ten bombs on target. Aboard the *Yamato* far to the rear of the carrier force, the mood of jubilation which had greeted news that an American fleet was near enough to engage, turned to utter dismay. When Yamamoto was handed the message that three carriers were lost, with all their aircraft, he stood groaning, too stunned to speak. His first reaction was to order all available warships to make full speed ahead towards the American force for a pitched battle. They sailed on through fog so thick it was impossible to see the next ship. Encouraging news reached Yamamoto during the course of the day. Aircraft from the remaining carrier, the *Hiryu,* had succeeded in damaging the recently repaired *Yorktown* (which was to be torpedoed by a Japanese submarine three days later). But then at 5.30 in the afternoon came news that American aircraft had destroyed the *Hiryu* too. Half an hour earlier, 24 dive-bombers from the *Enterprise* had dropped four bombs on to the carrier, causing such damage that she was to sink the following morning, with Admiral Yamaguchi, widely tipped to be Yamamoto's successor, standing forlorn, sword in hand, on the bridge. That evening Japanese warships circled round preparing to search out the American force for a night engagement, but with the loss of all air cover Yamamoto bowed to reality and cancelled the entire operation at two o'clock in the morning. 'How can we apologise to His Majesty for this defeat?' asked one of his staff. 'I am the only one who must apologise,' was the reply.[40]

The Battle of Midway was the closest the war at sea came to a Trafalgar. It was the most significant fleet engagement of the war. 'After Midway', recalled Navy Minister Mitsumasa Yonai, 'I was certain there was no chance of success.' All the Japanese naval officers interrogated by the United States navy at the end of the war picked Midway as the decisive turning-point, 'the beginning of total failure'.[41] It achieved the American objective of keeping the Pacific open to sea traffic, but more than that Midway saw the destruction of the outstanding instrument of Japanese success since

Pearl Harbor. Nagumo's force was the navy's elite. The losses of ships and aircraft at Midway were hard enough to sustain, but the pilots were almost irreplaceable. The six hundred carrier officers were outstanding aviators, flying *samurai,* trained to the highest standard for arduous flight over water and pinpoint attacks on enemy ships. At Midway one-third lost their lives and 40 per cent sustained injury.[42] Many of the rest were swallowed up in the wars of attrition fought in the autumn in the Solomons, where American air forces won local superiority. Little thought had been given by the Japanese commanders about what to do if the carriers and pilots were ever lost. During 1942 losses of Japanese carrier aircraft were almost double the numbers produced to replace them.[43] In 1943 Japanese shipyards supplied only three more aircraft carriers, and four in 1944; the United States navy in these two years procured another ninety. Yamamoto's ambition to destroy American power at one blow was frustrated once and for all at Midway. A few months later he was killed when his plane was shot down by American fighters, his flight details decoded by American cryptanalysts. In 1945 his flagship, *Yamato,* condemned to see out the war without a single naval engagement, made one desperate suicide run to relieve the Japanese garrison holed up on Okinawa and was battered into the sea by a swarm of American aircraft.[44]

Midway did not end the war in the Pacific, but it threw Japan on to the defensive, and allowed the United States to divert men and materials to the German war. Over the next three years Nimitz and Macarthur wore down Japanese resistance island by island, from Guadalcanal to Okinawa, against fierce, suicidal defence. Japan's fleet was decimated, her merchant shipping obliterated. When bases near enough to Japan's home islands could be secured, a relentless aerial bombardment of her cities eroded what economic strength remained. Without the Coral Sea and Midway this bitter erosion of Japanese fighting power would have been both longer and more costly. The United States might well have had to fall back on California as its front line in the Pacific. Midway was won by the narrowest of margins – ten bombs in ten minutes – but it was not an accidental victory. It was rooted in sound intelligence and the effective deployment of air power at sea for which the United States navy had prepared for twenty years. Victory in the mid-

Pacific was the first leg in the long haul back to Allied command of the seas.

*   *   *

The desperate efforts to stem the tide in the Pacific had immediate repercussions for the war against Germany and Italy. Although the American service chiefs wanted to send an army into Europe as soon as possible in order to get to grips with the enemy they regarded as the more dangerous, they were compelled to accept that not everything could be done at once. There were shortages of ships, trained men and equipment of every kind. So generously provisioned was each American division with food, vehicles and services that it required 144,000 tons of shipping space to move just one infantry division, and almost a quarter of a million tons for one that was armoured.[45] By October 1942 only one and a half American divisions had reached Britain. With great reluctance Admiral King, and his army colleague General George Marshall, had to accept that no invasion of continental Europe was possible in 1942, though they pressed all year for a firm commitment that some kind of amphibious assault would take place in 1943 against German-held territory.[46]

The British had recognised sooner than their new ally that it was foolhardy to attempt an assault on Europe without adequate numbers and prolonged preparation. Their preference was for limited seaborne operations in the Mediterranean theatre, where British Commonwealth troops were already engaged in a holding operation to prevent Italian and German forces from reaching the Suez Canal and the oilfields of the Middle East. American leaders found British preoccupation with the Mediterranean hard to understand. American military doctrine was straightforward: take the offensive with all available force against the main force of your adversary. British strategy was less direct, more subtle, honed by years of experience in fighting larger powers with small resources. British leaders knew that they faced disaster if they pitted an inadequate land army against Germany in 1942; they had been defeated in just such an attempt two years before, with the whole French army at their side. Naval power allowed them the luxury of picking easier fields of combat, against the weak links in the enemy's armour. There was nothing particularly special about the Mediterranean. Britain had no large settler community there; from Gibraltar to Aden she had only a few thousand colonial officials and traders. There were no vital economic interests at stake. Although she hoped to deny Middle Eastern oil to the Axis powers, almost all the oil used in Britain came from the New World by 1942.[47] The chief argument for Allied presence in the Mediterranean was that here they were fighting a corner of the war that they could win in 1942, against weak Italian forces spiced up with a handful of German divisions and air squadrons.

It was not a glamorous alternative, but it was a realistic one. Against the strong advice of his military chiefs Roosevelt agreed in July 1942 to a diversion of scarce shipping and equipment to supply an invasion of North Africa, codenamed 'Torch'. Churchill hoped that the decision would lead to great things: the defeat of Italy, the reopening of the Mediterranean to Allied sea traffic, and the cutting of a route into the 'soft underbelly' of Axis Europe, through the Balkan peninsula, or Yugoslavia or southern France. For the Americans it was the least they would consider doing. They refused to regard Torch as an alternative to the invasion of northern Europe, and accepted it only on the understanding that the larger enterprise would not be long postponed.[48]

The joker in the pack was the German submarine, or U-boat ('U' being short for *Untersee*). No Allied strategy was possible across the Atlantic Ocean without the defeat of the submarine. The supply of goods and oil to Britain and the Soviet Union was threatened on every route, from the Arctic convoys to Murmansk and Archangel, to the long hauls from Trinidad or around the Cape of Good Hope. Any operation, whether in the Libyan desert, in north-west Africa, or across the Channel, relied on securing the sea-lanes first. Almost immediately on American entry into the war the advantage in the Battle of the Atlantic swung towards the submarines, and Allied naval power faced its most severe test.

After the successes of 1941 the German navy was no longer the Cinderella of the German armed services. From only a handful of U-boats in 1939, the submarine arm was now being supplied with thirty new boats a month; it had almost three hundred in total at the beginning of 1942 and almost four hundred by the end. The German Commander-in-Chief, Grossadmiral Erich Raeder, now saw the submarine as the key to 'decisive victory'. Even Hitler, whose grasp of naval affairs was rudimentary, came round to the view in April 1942 that 'victory depends on destroying the greatest amount of Allied tonnage'. By June he was completely converted: 'the submarine will in the end decide the outcome of the war.'[49] The strategy was unsophisticated. U-boats were ordered to sink Allied shipping when and where they could, merchant ships for preference, naval vessels if they found themselves under attack. The commander of submarines, Admiral Karl Dönitz, estimated that a monthly sinking of 700,000 tons would be more than enough to erode combined Anglo-American shipping capacity and slow down or even stop Allied operations. The aim was to maintain sinkings at at least 4-500,000 tons a month, the level achieved in December 1941. The submarine war was a war of closely calculated attrition, its success measured not in any one engagement but in the calendar of stricken tonnage.[50]

Dönitz shared his chief's conviction that the Anglo-Saxon powers would be defeated 'only at sea'.[51] The son of a Berlin engineer, Dönitz joined the navy straight from school in 1910, and rose during the Great War to command a submarine until captured by the Royal Navy shortly before the war's end. He was regarded by all his superiors as an outstanding officer, dedicated, cool-headed and sociable. He bitterly resented Germany's defeat and welcomed the revival of German strength under Hitler. In 1936 he was appointed Führer of the German U-boat arm and dedicated himself singlemindedly to the task of building up a sufficient force of submarines to alter German prospects decisively in a renewed naval war with Britain.[52]

He always regarded three hundred as the optimum number of submarines to make Britain's sea-lanes unnavigable, and by 1942 he had all but achieved this figure. This did not mean there were three hundred submarines in action at any one time, for large fractions were always in transit, refitting or training; but it did mean eighty or ninety were operational at any one time, most of them in the Atlantic. The bulk of the boats were the Type VII, weighing approximately 750 tons, with a maximum speed of 17 knots, a radius of 8,000 miles and an armament of eleven torpedoes. In 1942 they were supplemented by the Type IX boat, of 1,100 tons with a cruising range of 13,450 miles and 22 torpedoes.[53] The boats were organised in groups of three, and were formed into the wolf-packs only when a definite convoy sighting had been made. A control room in France tracked the progress of both U-boats and merchant ships, directing the boats to the contact point by brief radio messages. Some U-boats were sent to the coast of Africa or to the Indian Ocean, returning after eighteen-month voyages with crews parched by the sun and bearded almost beyond recognition. But Dönitz preferred the principle of concentration of force against the richest pickings on the Atlantic trade lanes. Submarines followed the ships.

In the early months of 1942 there was no doubt about the target. With American entry into the war the submarines were sent to the far side of the Atlantic to attack the concentration of shipping on America's vulnerable eastern seaboard. The submarine packs were given suitably predatory names – *Leopard, Panther, Puma* etc – and ordered to begin Operation *'Paukenschlag'* ('Drumbeat'). For the submariners it was the start of 'fortunate times' once again. So easy was the hunting that the submarines operated on their own rather than in groups. From the middle of January they began to reap a remarkable harvest. The Americans sent merchant ships and oil tankers without escort, unconvoyed, using radio so openly that the submarines had no difficulty in closing on the isolated vessels as they each betrayed their position. When the submarines surfaced near the coast they met an extraordinary sight. The seaboard cities were a blaze of light, illuminating the silhouettes of slow-moving ships passing along the familiar peacetime routes, their own lights still shining. The submarines fired at will. In four months 1.2 million tons of shipping was sunk off the American coast alone. The Allies lost 2.6 million tons of shipping between January and April, more than had been lost in the Atlantic in the whole of 1941. U-boat losses in January were only three, in February only two.[54]

This situation could not last. The American navy had encouraged cargo ships to sail independently, without escorts, because of what it had seen of damage to the convoys in 1941, when British and Canadian escort vessels had been unable to cope with the submarine once it had sighted its prey. The navy's offensive traditions inclined it to try to hunt down the submarines using specialised naval groups along the coast, but the patrols were ineffectual, looking for a needle in a haystack. Not surprisingly, few contacts were made, while the wolf-packs gobbled up the defenceless merchantmen. Only a shortage of submarines prevented an even worse disaster, for at just the point that Dönitz hoped to tighten the screw Hitler ordered U-boats to Norway to prevent what proved to be a phantom invasion. By the time the submarines were available in strength there had occurred a revolution in American tactics. Convoying was instituted, together with a blackout and radio silence. Continuous air patrols forced the submarines away from the coast, down to the Caribbean and farther out into the Atlantic.

Here there were still plenty of opportunities. During 1942 the submarine force had the benefit of a vital edge in the intelligence war. The German naval intelligence service, the B-Dienst, successfully broke the British Naval Cypher No. 2 in September 1941, and three months later was able to read the new Cypher No. 3, which was used by the British, Americans and Canadians for controlling the movement of all trans-Atlantic convoys. In February 1942 the Allies lost their newfound ability to read the German Enigma traffic – machine-enciphered signals thought by the Germans to be impenetrable (known on the Allied side as Ultra) – when the naval Triton cipher was altered. This intelligence black-out lasted until the end of the year. It was now possible for German submarines to be directed accurately into the path of oncoming convoys, but extremely difficult for the Allied navies to know exactly where the wolf-packs lay in wait.[55] Other advantages piled up. New supply submarines of 2,000 tons – nicknamed 'milch-cows' – were sent out to enable the smaller boats to refuel and rearm without having to make the long trip back to the submarine bases in Europe, during which time they were completely ineffective. The radio messages which held the wolf-pack together and directed it to the target were converted to a concentrated form to make their duration as brief as possible, in order to avoid detection.

Yet for all the tactical sophistication the submarines remained vulnerable to the well-managed, heavily-escorted convoy, or to air patrol by converted bombers and flying boats, armed for anti-submarine warfare. During 1942 the radius of air action was extended, with great arcs of the Atlantic becoming accessible from Ireland, Iceland and Newfoundland. What was left was an area 600 miles wide in the mid-Atlantic, the so-called 'Atlantic Gap' or 'Black Pit'. The gap stretched from Greenland in the north through to the area around the Azores, and it was here that Dönitz concentrated his submarines from the middle of the year. The wolf-packs lived in the gap, attacking convoys as they left air cover and breaking off the attack when they regained it. The packs would then refuel from the milch-cows and await the next convoy. The attacks carried great risks, but the rewards were substantial. In 1942 sinkings in the Atlantic reached 5.4 million tons; total Allied losses reached 7.8 million tons, 1,662 ships in all. This was a rate of loss difficult to sustain. British imports sank to one-third of their peacetime level; by January 1943 the British navy had only two months' supply of oil left. 'The only thing that ever really frightened me during the war', Churchill later wrote, 'was the U-boat peril.'[56]

*   *   *

In an area as vast as the Atlantic the submarine was an inherently difficult enemy to defeat. Cruising on the surface it could outrun most ships; submerged it was invisible without the right scientific equipment. The initiative lay with the submarines who could choose where and when to engage enemy traffic. The Allies by 1942 were faced with a daunting but not insoluble mission. They adopted two very distinct approaches. The first was to find ways of avoiding the submarine altogether. Indeed the chief priority was not to sail pell-mell into a submarine trap in the hope that a handful of small escort vessels could sink the attacker in time, but to get across the Atlantic entirely unmolested. If this was galling to naval commanders who relished combat, it was none the less the best way of getting supplies (and thankful sailors!) across the ocean. The second was to find a way of accurately locating submarines so that effective anti-submarine forces, naval and air, could be brought to bear without wasting endless time on long sweeping searches. It was never enough simply to avoid the submarines: at some point they had to be fought.

The common denominator required for both strategies was precise knowledge of the movement of submarines. Convoys could then take evasive action; aircraft and combat vessels could close in on the betrayed location. Accurate intelligence about the whereabouts of U-boats was the first building block in the Atlantic Battle, but it was frustratingly difficult to secure. For much of 1942, with Allied cryptanalysis blinded by Triton, the navy relied on the Submarine Tracking Room set up in the Admiralty at the outbreak of war. The Tracking Room used a wide variety of intelligence sources, from agents' reports in the submarine bases on the French coast, to bearings reported by the units for radio interception scattered around the Atlantic rim. It was deliberately housed next to the Trade Plot Room, from which the global movement of merchant vessels was controlled. Intelligence from the Tracking Room was immediately available to alter the movements of merchant shipping. The room was run not by desk-bound sailors but by civilians – academics, economists, lawyers – drafted in to bring intelligence, in its conventional sense, to bear on the sea war. In charge was Rodger Winn, a barrister of exceptional talent, who devoted himself, to the point of physical collapse, to providing the best estimate day by day of the number and position of submarines in the Atlantic. Even without cryptographic intelligence the Tracking Room was able to warn the American navy of the impending assault on east coast shipping lanes in January. It did not take Admiral King long to realise the value of accurate plotting, and a similar room was set up in Washington, which cooperated closely with its British counterpart for the rest of the war.[57] The intelligence effort paid rich dividends. Between May 1942 and May 1943 105 out of 174 convoys sailed across the Atlantic without interference from submarines; out of the 69 sighted by the wolf-packs 23 escaped without attack and 30 suffered only minor losses. Only the remaining 10 per cent were severely depleted.[58]

Good intelligence was the first step. It was up to the convoy commander and his naval escort to do the rest. However, there were obvious weaknesses in the escort system. Escort vessels were in short supply, even more so when the crisis in the Pacific drew American vessels away from the Atlantic. Many of the escorts were obsolescent boats, converted to a role for which they were

inadequately prepared and armed. The task of the escort commander was hazardous and demanding. In all weathers, with only
imprecise knowledge of the exact whereabouts of the submarine
packs, he had to remain constantly vigilant, shepherding his loosely-
formed flock of merchantmen scattered over miles of ocean, and
rescuing stragglers. Air cover was almost non-existent. The large
aircraft carriers were needed elsewhere, and the conversion of
tankers and cargo vessels to a carrier role was a slow process, their
potential realised only late in the day. Until 1942 the training for
escort duty was rudimentary. Commanders were expected to
devise their tactics on the ocean. When confronted with a U-boat
some commanders ordered the escorts to give chase, leaving the
merchantmen weakly defended. Others concentrated on herding
their vessels out of danger, using the escorts like so many sheep-
dogs, snapping at the heels of slower transports to save them from
the wolves. But gradually a common set of tactics emerged: when
numbers permitted, the escorting ships would divide into two
groups, the larger part protecting the cargo vessels, a smaller group
hunting the convoy perimeter for submarines.

Direct attack on the submarines required information of a
different kind. There was little chance of a kill without some way
of locating to within feet the position of the enemy. At night-time,
against submarines operating on the surface, this knowledge could
not be acquired with conventional technology and naval tactics,
and chance played a part in any densely packed convoy battle.
During 1941 and for much of 1942 the number of submarines
destroyed in all theatres remained small, only 35 in 1941, and only
21 in the first six months of the following year. This was a rate of
attrition that could comfortably be supported by the U-boat arm.
The only way to increase that rate was to apply to the naval conflict
the new weapons of war – long-range aircraft, radio and radar.

The importance of aircraft was plain to see. Wherever air patrols
were in operation submarine activity was much reduced. Aircraft
were 'the great menace for the submarines' Dönitz told Hitler in
September 1942. But until well into 1942 the aircraft was a limited
weapon in the actual destruction of U-boats. The number of
aircraft allocated to over-sea duty was small in relation to the task,
and it proved difficult to persuade the RAF to divert larger

bombing aircraft from the attack on Europe to the war at sea. The original anti-submarine bomb proved to have almost no destructive power, and was replaced by an aerial depth-charge, though only in the course of 1942 were weapons of enhanced destructive power, charged with a new explosive (Torpex), finally introduced.[59] The central problem for the anti-submarine aircraft was to find the submarine and, having found it, to sustain an attack before it submerged to the safety of waters beyond the range of depth-charges. The answer lay with Radio-Direction Finding, later known by its familiar name of radar.

The development of effective Air to Surface Vessel (ASV) radar played a key part in the Battle of the Atlantic. A radar using wavelengths of 1.7 metres was developed, capable of detecting a surfaced submarine at 5 miles. The radar allowed aircraft to find submarines by day or night, but about a mile from the target their exact location disappeared on the screen because of distortion from the sea. By day visual contact could be maintained, but at night, when submarines cruised on the surface confident of invisibility, they were almost impossible to find. The solution emerged almost by accident. A junior RAF officer, Sir Humphrey de Vere Leigh, privately developed the idea of a powerful marine searchlight, located in the nose or under the wing of anti-submarine aircraft, which could be switched on at the point where the radar picture faded. The device was developed in defiance of orders from above, but its advantages became evident and it was accepted as a tactical experiment. During 1942 the Leigh Light was turned rapidly into an operational instrument. Flying dangerously close to the surface of the sea, the attacking aircraft tracked the target with ASV radar, until approximately 1 mile away from it an operator floodlit the area ahead. The aircraft then dived to 50 feet to deliver a pattern of depth-charges. The first successful attack was carried out on the startled crew of U-502 cruising in the Bay of Biscay on 5 July 1942. U-boat sinkings in July were four times as many as those in June. By August U-boats were forced to travel submerged and at night when they crossed the Bay to and from their bases; or they would move during daylight hours, with anxious look-outs scanning the skies for RAF patrols. But two months later the pendulum swung back the other way. German scientists rapidly developed a receiver set, 'Metox', which gave warning of aircraft approaching using the 1.7-metre frequency. The ball was back in the Allies' court, to find a radar with greater accuracy and beyond submarine detection.[60]

Ships suffered from the same deficiency as aircraft in the accurate location of submarines. Here too radar was the only answer. The standard long-wave 1.7-metre radar was of little use for tracking surface vessels, for there was too much activity from the sea itself, even in relatively calm weather. In 1941 the British navy sponsored the search for much narrower wavelengths. At Birmingham University the breakthrough was made. The invention of the cavity magnetron permitted the development of radio valves that could operate on much shorter frequencies, of 10 centimetres. When a radar set was built around the new valve it proved able to detect even a submarine periscope. The new Type 271 sets were installed on escort vessels from the middle of 1941. They did not yet work perfectly. The valves had a small power output until the development later in the war of the 'strapped' magnetron, the CV56 valve, which powered the radar sets of all naval vessels by 1943.[61] In rough seas even this more sensitive radar was deceived by the waves. But centimetric radar gave a solid scientific foundation to the war at sea. Its constant refinement enabled more and more escort vessels to see in the dark around their threatened convoys.

By 1943 all British and American vessels were fitted with some basic form of centimetric radar. Once the target was in range it could be destroyed either with gunfire or depth charges. During 1942 this technology was also improved. Instead of depth charges being dropped from the back of the vessel at the point where contact with the submarine through radar or ASDIC was lost, a new method was devised by the British Department of Miscellaneous Weapons Development. A multiple mortar, nicknamed 'Hedgehog', was installed at the front of the ship. It fired 24 small bombs, which fanned out in front of the ship and dived swiftly into the sea 200 yards ahead. The bombs were fitted with contact fuses, so that only a submarine hit would produce an explosion. The remaining bombs sank to the bottom of the sea. For sailors used to the climactic roar of the depth charges the new weapon was a disappointment; but it claimed fifty submarine kills by the end of the war.[62]

Despite all these innovations the rate of sinkings hardly abated. It might well be objected that the painstaking organisation of sound intelligence, and the slow conquest of the scientific frontier made little difference to the Atlantic Battle. But the statistics deceive. During the latter half of 1942 there were many more submarines available to Dönitz, but their sphere of action was restricted more and more to the poorly defended areas in the mid-Atlantic, or the coast of southern Africa. Each U-boat in the Atlantic sank the equivalent of 920 tons per day in October 1940; in August 1942 the figure was only 149 tons, and in November 1942, even though the aggregate figure for tonnage sunk was the highest of the war in the Atlantic Gap, no more than 220 tons.[63] The ratio of operational submarines to successful sinkings declined sharply over 1942. Allied efforts did not halt the submarine offensive, but in 1942 they prevented the carnage at sea from being very much worse. Merchant losses were still unacceptably high but the rate of loss was slowed down and contained, while the losses of U-boats rose steadily. During the second half of 1942 65 were sunk, four or five times the number lost in the first months of the year.

There were other solid achievements for Allied naval power. In the Mediterranean, too, the careful application of air power, and the use of radar and radio intelligence, turned the tide. The convoys began to move through the areas of greatest danger with more success. Supplies were run to Malta, which became a front-line base in the fight against Axis aviation and shipping. During the course of the year the Royal Navy used the submarine weapon against the Italian fleet, cutting vital supplies to Axis land forces stretched out along the sandy supply lines to the Egyptian frontier. Two-thirds of Italy's merchant marine ended up at the bottom of the sea, a quarter sunk by British submarines, 37 per cent by Allied aircraft. The Axis forces in North Africa were denied almost half their supplies, and over two-thirds of their oil.[64] Pushed to their logistical limit against an enemy reinforced by supplies brought via the long sea route around the Cape, Italian armies, supported by Field Marshal Rommel's small Afrika Korps, were finally pushed back at El Alamein in early November. Between November and May 1943, when Axis forces finally surrendered in Tunisia, Italian vessels plied the seas between Sicily and Tunis through a hail of Allied bombs, torpedoes and mines. Sailors called it the Death Route, for there was almost no chance of survival. Between November and May, in rough winter seas, beset by squalls and fog and pursued by the relentless drone of enemy aircraft, 243 ships were sunk and 242 damaged passing through precarious narrow channels cut across the minefields, each one 40 miles long and at some points less than a mile wide.[65]

The U-boat threat was also contained sufficiently to permit the transport of troops and equipment across the Atlantic for operation Torch. The effort disrupted the normal flow of traffic but the naval build-up was successfully concealed from the enemy. The passenger liners *Queen Mary* and *Queen Elizabeth* scurried at high speed and in isolation across the Atlantic carrying fifteen thousand troops apiece. In October a vast convoy of 102 ships sailed from America bound for Casablanca, where 25,000 troops were to disembark. From the Clyde in Scotland sailed six assault convoys and six slower supply convoys, a total in all of 334 ships. The effectiveness of the convoy when it had adequate air protection and the benefits of modern technology was demonstrated beyond dispute. Only one small supply vessel was lost out of the whole armada. On 8 November, supported by heavy fire from the big battleships, Allied troops landed at Casablanca, Oran and Algiers. Local French troops soon abandoned the struggle and the combined Anglo-American force joined up with British Commonwealth armies chasing the Axis westwards from Libya. Although the final Allied victory was much slower and more costly than expected, German and Italian troops were bottled up by western naval and air power, fighting until there was nothing left with which to fight.[66] On 13 May 150,000 Italians and Germans surrendered.

The war in the Mediterranean did nothing to solve the U-boat war. At the end of 1942 the Battle of the Atlantic was delicately poised. Allied shipping was still dangerously vulnerable; the morale of the merchant mariners was low and casualties high. But U-boat losses were rising too, and German submariners were faced with the stark reality that they would now survive no more than two or three sorties at most. Between January and May the Battle of the Atlantic reached its climax.

★   ★   ★

Nature itself supplied a backdrop to the crescendo of battle that was straight out of Wagner. The weather in the winter was never good; over the winter months of 1942–3 it was atrocious. Gale force winds for weeks on end drove the sea into icy mountains and valleys that could barely be navigated by submarine or victim. Driving blizzards made the sighting of vessels almost impossible. On the bridge of the submarine the men on watch were strapped down with strong security belts as the waves swept up and over the boat time after time. It was difficult to hold the wolf-packs together in pursuit of the convoys. In January 1943 only nine ships were sunk in the North Atlantic.

The respite was only temporary. Hitler, determined to press home the submarine campaign at all costs, sacked Raeder and appointed Dönitz Commander-in-Chief of the navy. All resources for the naval war were concentrated on the production of sub-marines and the training of their crews. The B-Dienst continued to supply a stream of decoded messages alerting the submarine groups to the convoy routes. More U-boats than ever before were available to cover the Altantic Gap, an average of a hundred or more each day.

On the Allied side there was the same sense of imminent climax. Meeting in January the Combined Chiefs-of-Staff agreed that the defeat of the U-boat was still their strategic priority, 'a first charge on the resources of the United Nations'. More escorts and aircraft were sent to shelter the convoys. In London the U-boat war was placed under the direction of a remarkable British submariner, Admiral Sir Max Horton. In November 1942 he was appointed Commander-in-Chief of the Western Approaches. His brief was to find some way to reverse the downward slide in the anti-submarine war. He was a pugnacious, hard-working commander, chosen deliberately as a breath of fresh air in the campaign. He was respected rather than liked, 'quite ruthless and quite selfish' according to one of his staff. He kept unusual hours and expected his colleagues to do likewise. He worked in the morning, then played a leisurely round of golf every day between two and six, finally returning to the submarine plot room where he worked until well past midnight every night, gulping pints of barley water, barking at his subordinates and displaying what one witness described as an 'uncanny prevision of what the enemy would do next'.[67] As a former submarine commander he understood more clearly than his predecessors how Dönitz might react. He could also see clearly the weaknesses still evident in the Allied anti-submarine campaign.

With the arrival of Horton to duel with Dönitz came the changes that would eventually tilt the Battle of the Atlantic the Allies' way. He campaigned immediately for more aircraft, particularly long-range aircraft to fill the Atlantic Gap. He insisted that all aircraft hunting submarines at night should be fitted with Leigh Lights and centimetric radar. He set about reorganising the whole escort system, insisting on employing more up-to-date ships and escort carriers, and on the establishment of 'support groups' of submarine hunter-killers which could be moved quickly about the Atlantic battlefield to the point of greatest danger. He insisted on higher standards of training and preparation. During 1942 escort commanders and crews received only two weeks' training for convoy work, and often entered the fray quite unequal to the task assigned to them. Horton appointed an officer to the task of training escort crews at greater length in the skills they needed, emphasising above all the ability to cooperate as a naval team, coordinating both attacks on submarines and the sheltering of their vulnerable charges.[68] All of this took time; the initiatives of late 1942 flowered only with the passing months. Until they bore fruit Horton fought Dönitz with the weapons he had to hand.

From the end of January, when the worst of the weather was past, were fought the fiercest battles of the Atlantic campaign. The first blows were struck against convoys HX224 and SC118 on route from America laden with goods for the Soviet Union. Contact was made by U-456, whose captain called up other U-boats and then moved in and sank three ships in strong gale-force winds. Dönitz stationed a wolf-pack in the Gap codenamed 'Pfeil' (Arrow). HX224 was hounded, although for little additional loss, but SC118, following two days behind, ran full tilt into the waiting submarine trap. Though the convoy had an unusually heavy air and sea escort, the submarines, pressing home their torpedo attacks at night against fierce counter-attacks, succeeded in sinking thirteen ships for the loss of three U-boats and severe damage to four more.[69] At the end of February the submarines mauled three convoys going the other way. One of them, convoy ONI66, was betrayed by radio intelligence and ran into a waiting wolf-pack. The ensuing battle stretched out over 1,000 miles of ocean and lasted four days. Ultimately two U-boats were lost but fourteen merchantmen were sunk. In London and Washington there was consternation. Even with improved escorts and air cover the convoys were still vulnerable. More worrying still, not even the fact that the British cipher school had succeeded at last in breaking into the Triton cipher prevented convoys from running into disaster. Messages could not always be decoded either fully or quickly though sufficient could be gleaned to permit the convoys to be re-routed. But the new routes could be detected by German intelligence, and so numerous were the submarines hidden in the Atlantic Gap that even the best-informed of convoys ran the risk of making accidental contact.[70]

In March the worst fears were realised. The slow convoy SC121 crossing from west to east lost thirteen ships, with no loss to the submarine pack. Then on 5 March a large slow convoy of fifty vessels, SC122, escorted by three Canadian corvettes and a mine-sweeper, set off from New York. Three days later a faster convoy, HX229, left from the same port with 41 ships and an escort of two destroyers and two corvettes, following almost the same route as the convoy in front of it. In the mid-Atlantic the two convoys piled up one behind the other, presenting a large and tempting target. Forewarned of the convoy routes by German intercepts, Dönitz stationed three large wolf-packs, 'Raubgraf', 'Stürmer' and 'Dränger', across their path. Once more the weather deteriorated, with heavy seas, fog, and intermittent flurries of fierce snow and hail. Through these miserable elements the two convoys plunged until sighted by the 38 U-boats waiting for them. With no air cover in the heart of the gap, but with increased naval escort, the battle commenced. The first victim was HX229, which was sighted by U-653 on her way to refuel. The submarine dived as the whole convoy passed slowly above it. When it resurfaced the message went out to 21 of the waiting submarines to close and engage the convoy. During the night of 16–17 March, in high seas that hampered the use of shipborne radar, the attack was pressed home. Seven merchantmen were sunk or immobilised. During the course of the night the two convoys began to merge together, and in bright moonlight the U-boats now turned on SC122. Over the course of the day five more ships were sunk, but that morning for the first time very-long-range (VLR) aircraft arrived over the convoys. Only three aircraft were sent, all converted B-24 Liberator bombers, and only two managed to find the convoys. These two could only stay with the ships for short periods before their fuel ran dangerously low. But the submarines were exceptionally cautious in their presence. The two ships sunk that day from convoy HX229 were lost in the interval between the departure of the first Liberator and the arrival of the next. Over the next two days isolated kills were made, but by 19 March the convoys were out of the gap and under air cover. It was only then that the single submarine kill of the whole battle was made, U-384, sunk by a B-17 Flying Fortress from Benbecula Island in the Outer Hebrides.[71]

For the Allies the battle was a disaster: 21 ships totalling 141,000 tons were lost for the cost of just one U-boat. Even worse might have followed but for the renewed intervention of the elements. In late March a fierce hurricane blew across the mid-Atlantic, so severe that the U-boats could not manoeuvre or attack. Most were withdrawn back to base to refit and refuel after a month in which they had sunk in all waters 82 ships of almost 700,000 tons. Two more convoys were sighted and attacked in March, but the submarine commanders noted for the first time strong air cover provided by escort carriers. They could not prevent sinkings, but they kept most of the U-boats at arm's length. This was small consolation in the Admiralty after a disastrous month. The naval staff recorded later that the U-boats came 'very near to disrupting communications between the New World and the Old'.[72] In the prevailing gloom there was speculation that the convoy system itself might no longer be effective. If such were the case the prospects for mounting any kind of Allied assault on Europe were poor.

Admiral Horton was undaunted by the crisis. A considerable number of convoys had sailed in March and been steered clear of U-boats altogether. Moreover the kill rate was rising: nineteen U-boats were sunk in February, the highest figure for any month of the war so far, and fifteen in March. Radio intercepts indicated a distinct lowering of morale among the submarine crews. Attrition worked both ways. Rather than halt the convoys Horton planned in April to lure the U-boats into battle in the hope of inflicting insupportable rates of loss. He insisted, and insistence was Horton's hallmark, that the Royal Navy release naval vessels to form the nucleus of his new anti-submarine hit squads, the 'support groups'. By the end of March he had his way. Five support groups made up of destroyers and including existing escort carriers were readied for action. The support groups were to keep in close contact with the convoys, to be called at a moment's notice. Overhead Horton now had the air support he had asked for in December, including the VLR aircraft fitted with centimetric radar and Leigh Lights which could plug the Atlantic Gap. He ordered all anti-submarine aircraft to concentrate on destroying submarines either around the convoys or in the approaches to the U-boat bases in the Bay of Biscay.[73]

Rather than avoid the submarines, Horton drove his convoys into their midst. There was no repeat in April of the fate of HX229 and SC122. Each convoy was now surrounded by a carefully trained escort, able to cooperate with aircraft and with the lurking support groups. The ships had the latest centimetric radar, and, more significantly, so did the supporting long-range aircraft. The submarine's Metox receiver was now useless against air attack. New radio equipment was installed in ships for the purpose of High Frequency Direction Finding (HF/DF), which provided an accurate bearing on any submarine that used its radio.[74] Convoy HX231 from Newfoundland fought its way through four days of gale-force winds against a pack of seventeen submarines. Four U-boats were sunk for almost no loss. At the end of April convoy ONS5, outward bound from Britain, was sent into the gap with a powerful escort. Some 39 submarines attacked and sunk twelve ships in all. But with the aid of two support groups and the Liberators seven submarines were sunk, and five severely damaged. By the end of the month only half the merchant losses of March were recorded, for the loss of nineteen more U-boats.

Offensive tactics were resumed in May. On the 11th Horton sent 37 merchantmen, formed in ten columns and escorted by eight naval vessels under Commander Peter Gretton, from Halifax, Nova Scotia, to England. Convoy SC130 had continuous air cover. For the first six days nothing happened. By the seventh submarine contact was made in what was left of the gap. Gretton marshalled his merchantmen like well-drilled guardsmen, turning in formation first this way, then that, to avoid the waiting submarines. Overhead, Liberators prowled the skies. The first to appear sank a U-boat in its initial attack. The others drew the escorts to the submerging boats by radio. The nearest support group steamed to attack and arrived behind the hastily reorganising U-boats, who were now caught between the two forces. A second boat was sunk. In the afternoon the relief Liberators arrived and in close co-operation with the surface vessels sank four more submarines. The small cargo ships, slow and old-fashioned, watched the explosions of bombs and depth charges in a distant noisy ring around them. But not one torpedo reached them. After two days of battle the wolf-packs broke up to lick their wounds. For Dönitz the convoy battle was especially poignant for on board U-954, sunk by a Liberator on 19 May, was his son, Peter.[75]

Horton's offensive did all that he had hoped and more. During May sinkings in the North Atlantic fell to 160,000 tons, the lowest figure since the end of 1941. The new tactics and technology blunted the U-boat threat in a matter of weeks. During May the U-boat arm lost 41 vessels. This was a catastrophic rate of loss. By the end of the month U-boats were being sunk faster than cargo ships. With great reluctance Dönitz bowed to reality and on 24 May he ordered all submarines to retreat from the Atlantic until his forces could be reformed and rearmed. On 31 May he reported to Hitler that the Atlantic Battle was lost for the moment.[76] His retreating forces were harried home to their bases across the Bay of Biscay where they were forced to fight it out with enemy aircraft armed with rockets, depth charges and pin-point radar. In June and July his force lost 54 more boats. For submarine crews each mission was more and more likely to be their last.

It was some time before the British realised what had happened. After years of painful attrition the U-boat threat was liquidated in two months. Radio interception soon showed that the submarines had gone. A strange silence fell across the battlefield, almost too good to be true. In June the weather calmed. Thirty-two convoys ploughed through the ocean. Not one ship was lost, not one attack was recorded. In July 1,367 ships crossed both ways without incident. In the autumn Dönitz sent back a part of his force to test the water, but losses were once again beyond endurance. Between June and December 1943 only 57 ships were sunk in the whole Atlantic theatre, for the loss of 141 submarines. In October Portugal allowed Allied aircraft to operate from the Azores, and the last hole in the Atlantic where submarines could hide was finally plugged. In the whole of 1944 Allied shipping losses were no more than 170,000 tons, a mere 3 per cent of the losses endured in 1942. The battles of May and June 1943 represented, Horton told his staff, 'a clear-cut victory over the U-boat'.[77]

So sudden was the end of the campaign that had hung dangerously in the balance only weeks before that it is tempting to assume that some special factor, lately introduced into the battle, explained its abrupt conclusion. The explanation is more routine than this: victory was the product of all those elements of organisation and invention mobilised in months of patient, painstaking labour. Under Horton's inspirational command these elements reached a critical mass by the late spring of 1943. German directives could again be read by British cryptanalysts at almost exactly the time that new Allied codes blinded their opposite number for the rest of the war. Radar and anti-submarine weaponry had evolved at last the necessary level of reliability and sophistication. Better training and tactics turned the escort vessel into a genuine help-mate for the convoys and a true deterrent to its hunters. Finally, air power worked its transforming art in the Atlantic as it had done in the Pacific and the Mediterranean. This was the most important change of all. Not until April did Horton get the escort carriers long promised, but from then on the number of carriers increased more rapidly, and air cover could be provided throughout the seaborne theatres. The number and quality of aircraft plying backwards and forwards over the Bay of Biscay or far out into the Atlantic, in lengthy, often unrewarding, cold reconnaissance, increased sharply by the spring of 1943. In January 1942 there had been only 127 of which just ten were converted long-range bombers; a year later there were 371 aircraft, including almost

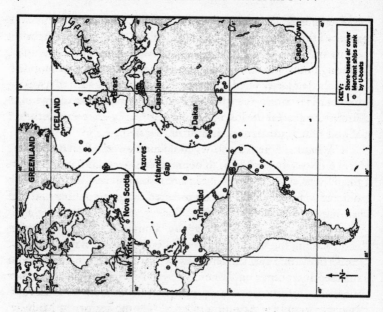

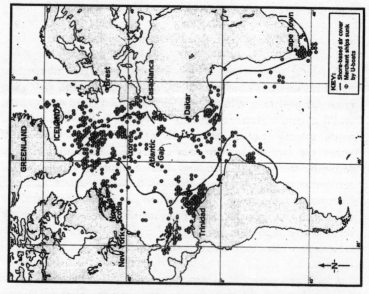

5   Merchant ships sunk from 1 August 1942 to 21 May 1943 (left)
6   Merchant ships sunk from 22 May 1943 to 31 December 1943 (right)

one hundred long-range aircraft. But the VLR aircraft, the Liberators with their extra fuel tanks, which could fly for eighteen hours or more, came in numbers only in April and May 1943, although only a fraction of what had been requested was supplied. Why the RAF remained resistant for so long to the idea of releasing bombers for work over the ocean defies explanation. A mere 37 aircraft succeeded in closing the Atlantic Gap, whose existence had almost brought the Allies' war plans to stalemate.[78]

The Battle of the Atlantic was not won like other naval conflicts, fleet face to face with fleet in decisive engagement. Its start and its finale were ragged and ill-defined, though real enough. The outcome of the struggle owed a great deal to organisations and staff far distant from the seamen and ships whose job it was to fight for the convoys' passage: the submarine Tracking Rooms in London and Washington, the Trade Plot Room that masterminded the global movement of merchant shipping, and the offices for radio interception and decipherment. It was a victory neither easy nor inevitable. Allied navies were stretched to the limit to achieve it. But victory in the Atlantic sea-lanes, like the victory at Midway, represented a decisive shift in the fortunes of war.

★   ★   ★

The war at sea was won not by the traditional instruments of maritime strategy, but by modern weapons – aircraft, radar and radio intelligence. Where ships sailed unprotected by aircraft, without benefit of modern means to detect other vessels and blind to the intentions of the enemy, they were all but defenceless. If either the British or American navies had been unwilling to recognise that maritime strategy had to undergo a fundamental revolution in the early years of the war, the sea-lanes might well have remained blocked. The Japanese navy was compromised for too long by the search for the big battleship confrontation, while the German navy was handicapped throughout the war by the reluctance of the German air force to develop a serious maritime strategy.

Under the impact of modern weaponry and tactics the sea war became a war of costly attrition conducted in the main by aircraft, submarines and small escort vessels. These spent the bulk of their time in monotonous routine, sweeping the skies and oceans for sight of the enemy, adding a tiny piece to the intelligence puzzle or supplying another statistic in the trade war. This kind of sea war, Churchill later observed, had none of the flavour of the past, no 'flaming battles and glittering achievements'. Instead its end product was 'statistics, diagrams, and curves unknown to the nation, incomprehensible to the public'.[79] Attrition warfare had none of the spectacular bloodletting of great land battles; there was no Somme, no Stalingrad. Yet the war at sea was exceptionally costly in men, who had to fight one small engagement after another, with all the skills of old-fashioned seamanship, against both the enemy and the harsh marine environment. Out of 39,000 German submariners, 28,000 were killed, or almost three-quarters of the force. From the 55,800 crewmen who went down with Britain's merchant ships, over 25,000 were drowned.[80] Chances of rescue on the high sea were slight. Life and death at sea were peculiarly harsh. 'There is no margin for mistakes in submarines,' Horton once told British submariners stationed at Malta, 'You are either alive or dead.'[81]

The victory of the Allied navies was the foundation for final victory in the west and in the Pacific. It permitted Britain and the United States to prepare seriously for the largest amphibious assault yet attempted, the re-entry to Hitler's Europe. It allowed the Allies to impose crippling sea blockades on Italy and Japan that destroyed 64 per cent of all Italy's shipping tonnage, and reduced the Japanese merchant fleet from over 5 million tons in 1942 to 670,000 tons in 1945.[82] The stranglehold on Italian and Japanese supply lines sapped fatally the industrial strength of these nations and made their reinforcement of distant fronts sporadic and costly. Finally, victory gave a growing immunity to Allied shipping so that the disparity in naval strengths and merchant tonnage between the two sides became unbridgeable. From the beginning of 1943 onwards the Royal Navy lost no more battleships or aircraft carriers; only six more cruisers were lost compared with 26 between 1939 and 1942; and only 36 destroyers against 112.[83] In the Atlantic the Allied merchant marine lost only 31 ships in 1944 against a figure of 1,006 in 1942, despite the fact that the U-boat arm actually had almost four hundred submarines in 1944, bottled up in their home ports or kept at arm's length by an almost impenetrable curtain of anti-submarine defences.[84]

The source of this enhanced naval presence was the vast production of American shipyards. The United States navy ended the war with 1,672 major naval vessels, the Royal Navy with 1,065 major warships, and 2,907 minor ones. The American merchant shipbuilding programme was a production miracle: 794,000 tons of shipping were built in 1941, 21 million tons in the next three years.[85] It is tempting to argue that sheer material strength won the sea war. Yet this is to ignore the question of timing. Midway was won against overwhelming Japanese superiority in warships. The convoy battles were won by a tiny number of maritime aircraft and limited numbers of escort vessels. The great imbalance in resources only began to tell later in the war, when it helped to prevent any re-entry by the Japanese navy or the U-boats as a serious peril in the sea war. In the critical year between Midway and the defeat of the submarine Allied resources were stretched to the limit. Tactical and technical innovation won the war at sea before sheer numbers came to matter.

The war at sea ended as untidily as it had been fought. When Germany finally accepted defeat on 8 May 1945, Grossadmiral Dönitz had become the most unlikely head of state – Hitler had designated him his successor as chancellor and war minister shortly before committing suicide in the Führer-bunker in Berlin. Dönitz ordered the remaining U-boats to surrender. Forty-nine of them were at sea and most complied with the order over the next two weeks. But two boats, U-530 and U-977, refused to do so. The first arrived off Long Island and attempted to torpedo shipping moving in and out of New York. Out of ammunition, the submarine made its way to Argentina where it arrived on 9 July, hoping for an enthusiastic welcome from Argentinian fascists. The crew was instead interned and their boat turned over to the American authorities. The second boat, U-977 made a slow, submerged crossing of the Atlantic and arrived in August on the Argentine coast. This was the submarine that was later alleged to have carried Hitler, his new wife Eva Braun, and his secretary Martin Bormann to safety in Latin America or Antarctica. The reality was internment for its fiercely pro-Hitler crew three months after the formal end of hostilities in Europe, and a matter of days before Japan's surrender.[86] But by the end of May Allied ships were reasonably clear of danger. On 28 May the United States navy and the British Admiralty issued a joint statement: 'No further trade convoys will be sailed. Merchant ships by night will burn navigation lights at full brilliancy and need not darken ships.'[87]

3

DEEP WAR
Stalingrad and Kursk

*'We speak of deep night, deep autumn;
when I think back to the year 1943
I feel like saying: "deep war".'*

Ilya Ehrenburg, *The War 1941–1945*

FOUR CENTURIES AGO on the broad river Volga, at the sharp elbow where it turns south-east to flow the last 300 marsh-fringed miles to Astrakhan on the Caspian Sea, the Cossacks built a small trading town, Tsaritsyn. Through this typically provincial 'three hotel town', dotted with wooden houses and jetties, flowed the rich produce of the Caspian and the Caucasus. It might have languished in this state but for the Russian Revolution in 1917, when Tsaritsyn found itself in the midst of a fierce civil war between the new Bolshevik forces and the 'White armies', a motley array of counter-revolutionaries and independent-minded Cossacks. The Whites laid siege to Tsaritsyn in the autumn of 1918, pushing back the Red Army until it held just a small horseshoe of territory on the west bank of the Volga, surrounding the town. The population began to evacuate. Local Bolshevik leaders cabled desperately to Moscow for reinforcements and arms of any kind. Nothing was forthcoming save a telegram urging the revolutionary forces to stand firm: 'Under no circumstances is Tsaritsyn to be given up.'[1]

The town was saved, according to Soviet legend, by the initiative of one man, the local chairman of the military committee, Josef Djugashvili, who in 1913 had adopted the name Stalin or 'steel'. Urging his comrades to fight to the death rather than abandon the town, he disobeyed orders from Moscow by recalling a Red Army division from the Caucasus; Zhloba's 'Steel Division', after a forced march of 300 miles, crashed into the rear of the Cossack host and saved the day. A month later Stalin was promoted to the national Council of Defence in Moscow. A year later he was back on the southern front in charge of a campaign that stretched from the steppe city of Kursk, through Tsaritsyn down to the

Caucasus. Once more, so it was said, Stalin was instrumental in saving the region for the revolution. For his success on the Volga, Tsaritsyn was renamed in 1925 his city, Stalingrad.

Twenty-four years later, by a strange quirk of history, Stalin found himself defending the city once again, in circumstances far grimmer. In the autumn of 1942 German forces reached to their farthest extent, to the snow-covered passes of the Caucasus mountains and the banks of the Volga on either side of Stalingrad. In the decades between these two assaults, the city had changed beyond recognition. It had become a major industrial centre, sprawling untidily along the river for a length of 40 miles. Its half a million inhabitants worked mainly in the new factories, turning out vast numbers of tractors to fuel the regime's agricultural revolution, and latterly large numbers of tanks. The city was a vital junction in Soviet trade. Industrial goods and machinery came down from the north; a steady flow of grain and oil was headed up the other way. Stalin had changed also. He was now the chief source of authority in the Soviet state, and the Supreme Commander of the armed forces. He wielded far more power than he had enjoyed in 1918, and vastly greater armies. On him alone rested responsibility once again for saving the city and the embattled Soviet system. On 28 July 1942 he issued a demanding order to the troops desperately trying to halt the German drive: 'Not a step back!' For four months they clung to the same horseshoe of land, while Stalin relived the nightmares of the civil war.

Then, slowly, the tide turned. Once again, Soviet armies crashed into the rear of the enemy host. The first major defeat of the war was inflicted. Over the next twelve months the Red Army drove German forces from much of western Russia, on a broad arc from Kursk to the Caucasus. These Soviet victories marked the turning-point of the whole war, as victory in 1919 had turned the tide of the civil war. In December 1942 Stalin promoted 360 officers to the rank of General for saving the city that bore his name. In March 1943 he awarded himself his first formal military title, Marshal of the Soviet Union.[2]

*   *   *

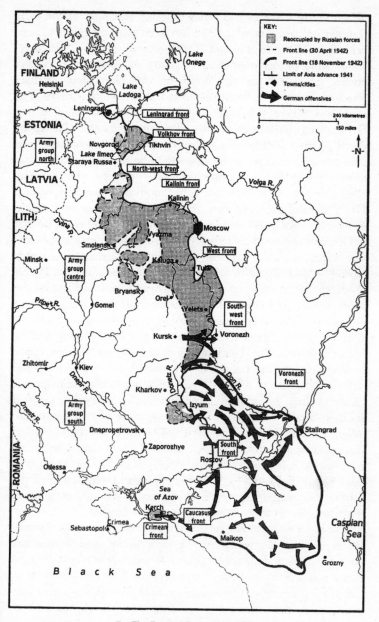

7 The Eastern Front 1941–1942

When German forces had renewed their onslaught in the early summer of 1942, Stalingrad had not been high on Hitler's list of priorities. His one thought had been to secure a decisive, annihilating victory over the Red Army and crush his Bolshevik enemy once and for all. With the east eliminated, German resources could then be turned to defeat the western Allies. The issue was where the blow should fall. German army leaders favoured an attack at the centre of the front, to seize the Soviet capital, Moscow. This was where the bulk of Soviet forces was concentrated; the loss of the city would be devastating for Soviet morale. Hitler thought otherwise. The conquest of the Soviet Union was ideologically inspired, but motivated by material greed. Hitler wanted the industries, the oil and the grain of southern Russia; here was real *Lebensraum,* or living-space. He reasoned that if Germany captured these resources from the enemy the Soviet war effort would be brought to a halt, while the Third Reich would become all but invincible. On 5 April he issued his Führer Directive for the new summer campaign: a general blow to the south against the Crimea, the Don steppe and the Caucasus.[3]

Soviet forces were caught off-guard, as they had been the summer before. Stalin had anticipated the strategy proposed by Hitler's advisers that German forces should renew their attack against Moscow and Leningrad with the object of encircling and defeating the core of the Red Army. When a German light aircraft crashed behind Soviet lines in June with full details of the planned southern attack, Stalin saw it as a clumsy piece of misinformation. When the British passed on details of German dispositions culled from decrypted German signals, Stalin had no more faith in British motives for doing so than he had displayed over their warnings in 1941 about Barbarossa.[4] On 28 June German forces launched the southern campaign, Operation 'Blue', against the weakest point of the whole Soviet front, achieving complete surprise.

The result was almost a repeat of the previous summer. German forces punched their way forward using a combination of large tank concentrations and massed air power. In a matter of weeks the Red Army was driven out of the whole area south of Kharkov and the Crimea was captured. Large numbers of Soviet soldiers were taken prisoner, or fled eastwards in disorder. The Black Sea port of

Sebastopol held out against overwhelming air and artillery attack until it fell on 4 July to General Erich von Manstein. His reward was promotion to Field Marshal. The crumbling of Soviet resistance threatened to turn into a rout. Rostov-on-Don, east of the Crimea, fell with little resistance on 23 July. Hitler was now quartered in the Ukraine, slightly north of the town of Vinnetsa. The summer base was given the codename '*Werwolf*'. It consisted of a small group of log huts concealed in woods. It was here that Hitler, ill at ease in the intense, muggy heat, prey to bouts of insomnia and soaring blood-pressure, first got news of the remarkable progress of German forces. Despite the climate, his spirits were restored. The victory that eluded him in 1941 was closer, the enemy 'considerably weaker' than a year before.[5]

Impatient for the final showdown Hitler decided at the end of July to speed up the campaign on the southern front. He divided his forces into two separate groups, Army Groups A and B. The first group was given the task of plunging across the Don river in pursuit of the retreating Soviet forces and capturing the whole of the Black Sea coast and the Caucasus region as far as the oil cities of Maikop, Baku and Grozny. This vast area was to be conquered in a matter of weeks by the 1st Panzer army, under Field Marshal von Kleist. The second group, the 6th army and 4th Panzer army, under the command of General Paulus, was ordered to advance to Stalingrad, destroy Soviet forces on the Volga, and proceed rapidly down the river to Astrakhan cutting all communication between the north and south. Stalingrad became for the first time a primary target. Its 'early destruction' Hitler considered 'especially important'.[6]

In almost all postwar accounts of the German failure in the Soviet Union, this decision to divide German forces in pursuit of distant economic targets in the late summer of 1942 is regarded as the decisive misjudgement on Hitler's part. It is easy to see why he was tempted to make it. German forces won large areas of territory very quickly in June and July. Soviet forces were clearly in disarray. In North Africa Rommel was pursuing the British forces back into Egypt. A swift strike through the Caucasus opened up the prospect of German domination of the whole Middle East. A quick victory at Stalingrad opened up the prospect of sweeping the other way, too, in a vast outflanking movement north-eastwards behind Moscow. Hitler seems not to have sensed that he was taking a gamble or risk in the summer of 1942, but was once again confident that his amateur strategy would win him more than he would get by following the stolid advice of the professionals.

His military staff saw things in a far more sober light. They could see little sense in simply sending forces off to occupy open steppe against an enemy that had so far committed only a fraction of his forces to the southern contest. On 23 July, the day Hitler issued his directive for the new offensive, his Chief-of-Staff, General Franz Haider, complained to his diary that Hitler's misreading of the enemy was 'both ludicrous and dangerous'.[7] The easy summer victories were achieved against weak forces, which eluded defeat by pulling back across the vast spaces of central Russia. The more German forces pushed forward on the southern front, the more dispersed they became. A long, vulnerable flank developed along the whole northern wing of the offensive, which was defended mainly by troops of Germany's allies, Hungary, Romania and Italy. The supply trains had to follow with the fuel and ammunition over long, hot trails, with poor roads and rail routes constantly harassed by Soviet partisans. Hitler would have none of it. He berated his commanders for timidity, even defeatism, and sacked those who disagreed. Events bore out his optimism. The 1st Panzer army burst on to the northern Caucasian plain, sweeping through its rich grainlands and abundant orchards; within twelve days it had penetrated 350 miles to the very foothills of the Caucasus mountains. Here German and Italian Alpine forces, trained for mountain warfare, began the slow ascent to the mountain passes. Meanwhile Paulus's armies moved steadily eastwards, clearing Soviet troops from the Don steppe and forcing them back to the approaches to Stalingrad. By 19 August the first assault on the city began; by the 23rd German soldiers had reached the suburbs. There was, recalled one witness, an 'exultant mood' at the *Werwolf* headquarters.[8]

In the first flush of German victories a mood of panic began to grip Soviet society. Populations in the south moved eastwards in a headlong flight from an enemy they had come to regard as unstoppable. At Rostov-on-Don even the troops were infected by fear of what would happen to them if they were caught by the Germans. At the height of this crisis of morale came the historic order 'No. 227' from Stalin himself: the Red Army was to stand firm against the invader or be treated as criminals and deserters. The notorious Soviet NKVD security forces rounded up alleged defeatists and saboteurs, while national propaganda roused the Soviet people to a final do-or-die effort on behalf of Mother Russia. The panic subsided, to be replaced by a mood of sombre determination. Audiences in Moscow flocked to patriotic Tchaikovsky concerts. Great heroes from the Russian past, whose socialist credentials were nil, were invoked to inspire heroic resistance and hatred of the enemy. The journalist Alexander Werth, who lived in Moscow throughout the darkest days of the war, recalled that hate reached 'a paroxysm of frenzy' during the difficult weeks in July and August. Alexei Surkov's poem 'I Hate' was published in the army journal *Red Star* on 12 August:

> My heart is as hard as stone.
> I hate them deeply.
> My house has been defiled by Prussians,
> Their drunken laughter dims my reason.
> And with these hands of mine,
> I want to strangle every one of them!

A few weeks earlier *Pravda* published an editorial on 'Hatred of the Enemy': 'May holy hatred become our chief, our only feeling'.[9]

It was not mere fear of the NKVD that kept the Soviet people fighting in 1942. There was a widespread and spontaneous patriotic revival, and a wave of revulsion against German brutality. The revival was deliberately stoked by the regime. Churches were reopened, and religious attendance encouraged after years of persecution. *Pravda* began to capitalise the word God for the first time.[10] Stalin looked to build bridges between the Red Army and the traditions of the Imperial past. New medals were struck for heroism in July 1942 named after the great Tsarist generals, Kutuzov, Suvorov, Nakhimov. More heterodox still, a medal was introduced only for officers, the Alexander Nevsky Order from Tsarist times; and at the height of the Stalingrad battle it was announced that officers would once again wear distinctive insignia and gold braid to instil a sense of pride and discipline into a hitherto classless army.

Medals and braid would not defeat the German armies. They were the outward trappings of a more fundamental effort mounted by the Soviet Supreme Command (Stavka) to restore to the Red Army a sense of self-confidence and the means for effective resistance. Despite the rapid success of German forces in the south, the overall balance between the two sides was more even than the contest in the south would suggest. At the start of the summer campaign the Soviet forces numbered five and a half million men, against six million Germans and their allies. Both sides had roughly the same number of aircraft, a little over 3,000; the Soviet armies had 4,000 tanks, German forces 3,200. Across the whole area of northern and central Russia the two sides built a vast defensive barricade, and it was here that Soviet forces were concentrated, protecting Moscow and the heartland of Russia itself. Only in the south were Soviet forces much weaker. By July the 187,000 Soviet soldiers here, with 360 tanks and 330 aircraft, faced an enemy force of 250,000 men, 740 tanks and 1,200 aircraft.[11] It was this temporary disparity that allowed German forces to move so rapidly across the Don basin and the northern Caucasus. Stalin was reluctant to send reinforcements from farther north until he could be certain that the southern front was the main area of campaign. Until July Soviet military leaders counted on a renewed assault on Moscow. As a result they were forced to respond in a hasty and improvised way once it became clear that the German goal was oil.

Stavka's first priority was to re-establish a clear Soviet front-line in the south. In the Caucasus a strong defensive line was set up under the veteran cavalryman Marshal Budenny, who had fought with Stalin in the civil war, defending Tsaritsyn. Once von Kleist's forces reached this enemy defence line on the Black Sea coast, in the foothills of the Caucasus mountains and on the approaches to the Caspian oilfields, their progress slowed down, and then halted. On 12 July a new Stalingrad front was formed on the outskirts of the city under General Gordov. Its aim was to try to slow down the German advance and to build a clear defensive line along the river Don, using the retreating armies in its construction. This proved a difficult task. Some of the soldiers in small groups, cut off from their units and their officers, never reached the front but were lost on the vast stretches of grassland, easy prey to roaming German aircraft. The army field headquarters had little idea how many troops they commanded, or, in many cases, even where the troops were. Stalin tried to salvage the situation by sacking commanders and bringing in men who had proved themselves in combat. Although this was unable to stem the tide of retreat, Soviet forces were at least falling back in better order on to the defensive line around the city, where they could make a more effective last stand. German forces reached the outskirts of the city, and by 23 August had even breached part of the Soviet front to reach the Volga north of the city, but their progress here slowed down in the face of fierce defensive fighting. The struggle now reached its critical stage. German forces, urged on by Hitler, expected to seize Stalingrad within days. For them the town enjoyed a symbolic significance that went well beyond its real strategic or economic importance. Stalin became day by day more alarmed. For him, too, Stalingrad was a symbol.

Late in August, Stalin played one last card. He called to his Kremlin headquarters General Georgi Zhukov, the man who a year before had organised the frantic defence of Moscow and brought the German assault there finally to a halt in December 1941. On 27 August Zhukov, commander of the Soviet western front, was appointed Deputy Supreme Commander to Stalin. That evening he arrived at the Kremlin, where he found Stalin and the State Defence Committee anxiously discussing the problem of the south. Stalin offered him tea and sandwiches. While Zhukov ate, Stalin outlined the situation: it was a matter of days before German forces would capture Stalingrad unless a proper defence could be organised. Zhukov was given the unenviable task of saving the city. The following day he spent in Moscow studying Soviet dispositions. On 29 August he flew to the Stalingrad headquarters to see for himself what could be done.[12]

Georgi Konstantinovich Zhukov, son of a shoemaker from a small village south of Moscow, had achieved a meteoric rise to become second to Stalin in the Soviet war effort. He was only 45 in 1942. A bright schoolboy, he was apprenticed to a Moscow furrier at the age of eleven. When he was nineteen he was drafted into the Tsarist cavalry where he rose to become an NCO by the time of the Revolution in 1917. During the Civil War he joined the new Red Army. He, too, fought for the defence of Tsaritsyn, where he was wounded by a grenade in fierce hand-to-hand combat. He remained a Red Army professional following the Civil War, a cavalry officer with a deep interest in military theory and the modern techniques of war. He was an able and tough commander, decisive and well organised, who paid close attention to detail and expected the utmost from his men. Stalin respected him, which perhaps explains why he survived the great purges of army staff in the mid-1930s. For Stalin, Zhukov epitomised the new young generation of communist soldiers, dedicated to the cause, anxious to move the army forward out of the cavalry age. He was an observer for Stalin of the Civil War in Spain in 1937; in 1939 Stalin chose him to command Soviet forces in a full-scale conflict with Japanese troops on the Mongolian-Manchurian border at Khalkin-Gol. For his success in routing the Japanese in a spectacular victory Stalin personally awarded him the title Hero of the Soviet Union. He had become Stalin's military troubleshooter.[13]

He was remembered by those who served with him as a hard and foul-mouthed leader, who sacked or punished officers who lacked sufficient will to win; but he was also regarded as a soldier who had a clear grasp of operational realities and could remain calm under the most adverse of circumstances. All these qualities were required at Stalingrad. When Zhukov reviewed the situation it soon became clear to him, as it was to Hitler's critics in the German army, that the attacking force was greatly over-extended. During August it became evident that German armies had very limited reserves in the south, though they enjoyed local air superiority and an advantage in tanks. Moreover the long flanks protecting the spearhead of the German attack were not composed of German troops, but of Italians, Romanians and Hungarians, who were less well armed and less committed to Hitler's life-and-death struggle against Marxism. Zhukov recognised almost immediately the opportunity to cut through the weakly defended sides of the salient and isolate German forces at Stalingrad in a giant pincer movement. But before that it was necessary to hold on at Stalingrad. A limited number of Soviet reserves were moved to the Stalingrad front early in September.

They failed to break the German encirclement but they did enough to prevent the Germans taking the city by storm. During the early days of September Stalin anxiously prodded his commanders to keep the enemy at bay at all costs. Though he wanted Zhukov to think of a way of saving the city, because he himself could not, he found it difficult to relinquish the habit of interference.

In the end Stalin allowed Zhukov to take the initiative. This was one of the most important decisions of the Stalingrad struggle, for it allowed the Deputy Supreme Commander to capitalise on his grasp of German weaknesses. On 12 September Stalin summoned him to the Kremlin. While he gloomily pored over maps of the front, Zhukov and the Chief of the General Staff, Alexander Vasilevsky, stood to one side whispering about the need to find another solution. Stalin's ears pricked up: 'What other way out?' He sent the two men away to work out in a day a solution to Stalingrad. At ten in the evening of 13 September they returned. Zhukov took the floor, patiently explaining to Stalin that as long as an active defence could be maintained at Stalingrad itself, large reserve forces could be assembled to the north and south-east which could mount a counter-offensive against the long-drawn-out flanks of the German attack, cutting the umbilical cord of supplies and reinforcements, and encircling enemy forces. Zhukov explained that the attack would have to be made against the weaker Romanian divisions, and it would have to be made only when adequate preparations were completed, at some point in mid-November. Stalin was sceptical at first. He objected that the counter-offensive was too far away from Stalingrad to relieve the city itself. Zhukov explained that the attack had to be carried out at some distance to prevent Germany's mobile Panzer forces from simply turning round and repelling the attack. Stalin said he would think about it, but by the end of September he was persuaded. Amidst the utmost secrecy a detailed plan was drawn up, officially approved by Stalin. Zhukov and the General Staff worked furiously for five weeks to get the counter-offensive organised. On 13 November Zhukov presented the completed version of Operation 'Uranus' to a cheerful Stalin. 'By the way he unhurriedly puffed his pipe, smoothed his moustache, and never interrupted once', Zhukov later recalled, 'we could see that he was pleased.'[14]

The whole success of a Soviet counter-blow rested on being able to hold Stalingrad at all costs, with very limited reinforcement. This called for the utmost sacrifice from the Soviet forces, who were to know nothing of the counter-offensive plan while they were bled white in the street-fighting. The defence of the city fell to two Soviet armies, the 62nd and the 64th. Determined German thrusts in late August pushed both armies back towards the centre of the city. By late August the front was split in two, the 62nd army besieged in the heart of Stalingrad, the 64th pushed to a small bridgehead on the Volga, south-east of the main battle. Most of the civilian population was now evacuated to the east bank of the river by a fleet of ferry boats that plied back and forth, carrying troops, guns and ammunition one way, the wounded and refugees the other. The 62nd army could continue only because of the frantic efforts of the military supply organisation and engineers in keeping open the Volga crossings in the face of heavy and repeated air and artillery attack. The long tail of the army – its artillery, air support and its rear services – was all on the far side of the Volga. From this less exposed position Soviet gunners and pilots kept up a ceaseless barrage against the approaching German forces.[15]

The fighting in Stalingrad was very different from the mobile cut and thrust that German armies had practised across the steppe. Because the city had been all but obliterated by the Luftwaffe during August, it was difficult terrain in which to manoeuvre. Tanks were vulnerable to ambushes, or were simply stuck in the rubble. Progress was no longer measured in miles but in yards each day. In addition, the geography of Stalingrad did not render it susceptible to a sudden swift seizure. The city sprawled for 40 miles along the river bank, from the large factories – the Red October Factory, the Barricades Factory, the Tractor Factory – in the north, backed by extensive workers' settlements, through the central residential and commercial area, dominated by the low hill Mamayev Kurgan, down to the houses and railway terminals of the south. Each area of the city became a battlefield, each building a fortress to be stormed. Slowly, German troops edged forward from the north and the west, using the surrounding hills as vantage ground from which to pour artillery fire into the city. By 3 September some of the units were only 2 miles from the river, pressing the 62nd army into a cluster of narrow footholds among the factories, around the central station and main jetties, and on to the remaining high ground in the city centre. The commander of the 62nd, General Alexander Lopatin, doubted that his army, which was suffering an appalling casualty rate, could hold the city. He began to move units across the Volga. For this he was promptly sacked. The commander of the Stalingrad front, General Andrei Yeremenko, a tough, thick-set Ukrainian, whose wife and four-year-old child had been killed in the German attack the year before, was forced to find a replacement at the very height of the battle.

The choice fell on Vasili Ivanovich Chuikov, the son of a peasant, and a veteran of the Civil War, who until a few weeks before had been in China as a military adviser to Chiang Kai-Shek. Joining the retreating Soviet army in July, he distinguished himself by launching limited attacks against the oncoming enemy to slow up his advance. He was lucky not to have died even before he reached Stalingrad. On the way to the front his drunken driver crashed the car at high speed, leaving him in hospital for a week with an injured back. A few weeks later, in late July, he had another fortunate escape when a reconnaissance flight over the Don steppe ended in disaster with his plane forced to crash by enemy aircraft. On hitting the ground it split in two, throwing Chuikov out. He survived with nothing worse than a bump on the head. By his own admission he was 'healthy and hardy by nature', a handsome, sturdy man, with a loud laugh that exposed rows of front teeth all crowned with gold. He learned lessons quickly in combat, and was a master at improvisation and surprise. He never questioned that his task was to stay in Stalingrad, and to die there if he had to. By all accounts he was an inspirational leader, and certainly one with nine lives.[16]

Over the next two months Chuikov fought a terrible duel with his opposite number, General Friedrich Paulus. The two opponents could not have been more different. At the time of Stalingrad Paulus was 52, a successful career officer, born in a modest middle-class home in Hesse, the son of a minor official. Though he is persistently described as 'von Paulus', he was in fact just Paulus, that rarity in the German military leadership: a successful bourgeois officer. An enthusiast for the new style of tank warfare, he earned his place on the army General Staff by virtue of his considerable

administrative skills. In 1940 he became Deputy Chief-of-Staff, but
was returned to the field in January 1942 to command the crack 6th
army following von Reichenau's sudden death from a heart attack.
He was an unlikely candidate for the Stalingrad confrontation. He
was a quiet, subdued man who loved Beethoven. Tall, almost
ascetic, he took an obsessive interest in his personal appearance and
would be found each morning with a bright white collar and highly
polished boots. During the approach to Stalingrad the heat made
him tired and listless. His health deteriorated with bouts of
dysentery, the 'Russian sickness'. As conditions in the city grew
worse, he became increasingly depressed about the toll on those
under his command.[17]

When Chuikov assumed command of the 62nd army, Stalingrad
was unrecognisable. 'The streets of the city are dead,' he wrote.
'There is not a single green twig left on the trees; everything has
perished in the flames. All that is left of the wooden houses is a pile
of ashes and stove chimneys sticking up out of them . . .'[18] Only the
concrete and iron structures of factories, and the larger stone
buildings of the city centre, remained standing above ground level,
roofless, the interior walls crumbled away. Every ruin was again
contested, until it, too, collapsed into rubble. On Chuikov's first
day of command, 13 September, German forces gathered for a final
concentrated effort to drive Soviet troops into the Volga. The
following day they captured the central Railway Station and the
heights of Mamayev Kurgan. Chuikov was forced to move his
headquarters from a rough mud dug-out on the hill to a safer
bunker on the banks of the river Tsaritsa, near its junction with the
Volga. From here he directed his dwindling forces to retake
strongpoints, or in small groups to base themselves in buildings in
front of the German advance, where they were to fight to the last
man and the last round. It is testimony to the extraordinary nature
of the contest that many did so. Over the following three days the
Central Station changed hands fifteen times; on Mamayev Kurgan
first one side, then the other, fought for the summit. Greatly
outnumbered, and subject to constant aerial bombardment, the
62nd was slowly pushed back. By 14 September the situation was
at its most desperate. At Stalin's headquarters the decision was taken
to throw in the only reinforcements to hand, the 13th Guards

division led by Hero of the Soviet Union Alexander Rodimtsev.

For Stalin, Rodimtsev's men were to play the part that the 'Steel Division' played at Tsaritsyn a quarter-century before. It was a race against time. Small groups of staff officers from Chuikov's headquarters were posted to hold the road leading to the jetty where the reinforcements were supposed to disembark. With only fifteen tanks between them they held the position until the first of the ten thousand guardsmen could be ferried across. They were thrown straight into combat from a gruelling forced march. One thousand of them had no rifles, and the rest were short of ammunition. Chuikov sent them straight into the heart of the battle for the central area of the city. They lacked knowledge of the terrain, and were inexperienced in city fighting, but they did succeed in stemming the German thrust, and gave valuable time for more reserves to be brought in, and for Chuikov to regroup his forces. The division took casualties of almost 100 per cent in the process, and had to be withdrawn. It took little serious part in the remainder of the battle, but Rodimtsev's men became part of the folklore of Stalingrad.[19]

Over the next few days the 6th army captured new stretches of the central district, including the giant Univermag department store on Heroes of the Soviet Union Square, where the Soviet defenders fought to the last in the basement. Paulus later made the building his headquarters. Farther to the south a fierce battle raged around a giant grain elevator, where a small detachment of Soviet soldiers held out for 58 days against tanks and artillery fire. A final push by the German forces secured the central landing stage on the river itself, but a wall of artillery and rocket fire from the far bank prevented them taking advantage of it. The Soviet front line now consisted of ragged pockets of resistance, with their main strength contained within the broad industrial area in the north of the city. Some units were isolated behind the German front, from where they stole out at night to harass the enemy. On 25 September Paulus turned his forces towards the factories. Three infantry and two Panzer divisions attacked on a 3-mile front. For over a month the same pattern was repeated, German attacks by day, Soviet counter-thrusts by night, hand-to-hand fighting for every workshop and every house. Soviet forces were overwhelmed by enemy firepower. One after another the giant factories fell, until the 62nd held only the Barricades plant, on the banks of the river.

Why did Paulus fail to take the city, which was held for two months by forces constantly short of supplies, suffering debilitatingly high losses, perched dangerously on the very edge of the Volga? German forces certainly declined in fighting power as they moved farther eastward; tanks and aircraft were difficult to maintain and all units suffered very high loss rates. The morale of German forces slumped as the tempo of battle increased. 'Stalingrad is hell on earth,' wrote a German NCO to his mother in September. 'It is Verdun, bloody Verdun, with new weapons. We attack every day. If we capture twenty yards in the morning the Russians throw us back again in the evening.'[20] German troops found it unsettling to move from large-scale, fast-moving operations to a narrow front of close fighting, where it was difficult to make sheer numbers tell so effectively. But above all, Soviet forces exploited the urban battlefield to full advantage.

Under Chuikov's command the Red Army developed a whole range of tactical innovations to hold the enemy at bay. Most postwar accounts of the fighting on the eastern front take an uncharitable view of Soviet tactical performance, but it is hard not to conclude that tactics played a key part in the survival of the 62nd army from August to November. Chuikov observed the German approach to warfare very carefully. He noted the German reliance on mass air and artillery attacks to push the infantry through, and the reluctance of German soldiers to engage in close combat without the protection of tanks. To reduce the impact of German firepower he insisted that the two fronts should be no farther apart than 'the throw of a grenade', making it difficult for German bombers to attack without fear of hitting their own side. He relied much more than his enemy on combat at night, and on hand-to-hand fighting, with daggers and bayonets, which suited the tough Siberians, Tartars and Kazakhs in his army more than tank and artillery warfare. Soviet forces became adept at camouflage and surprise, keeping their German enemy constantly on the alert and in fear. A sniper battalion found the ruined landscape a rich field for killing. Hidden in the rubble, with high-powered rifles and telescopic sights, they shot anything that moved. 'Bitter fighting,' wrote another German NCO in his diary. 'The enemy is firing from all sides, from every hole. You must not let yourself be seen . . .'[21] At night specially organised 'storm groups' would attack German bunkers well away from their own lines. As Soviet troops threw grenades and gave voice to the chilling 'Hurrah' that signalled an attack, bewildered German soldiers would wake to find automatic gunfire all around them. At dawn the storm groups would melt away, back to their dug-outs and foxholes. With them they brought up-to-date information on German strengths and dispositions. Unlike many Soviet generals, Chuikov insisted on good intelligence so that he could use his depleted forces as efficiently as possible.

As the battle went on the 62nd army received growing help from artillery and air power. On the far bank of the Volga three hundred guns were trained on the closing German forces. They were supplemented by a weapon even more feared by German infantry-men, the Katyusha multiple rocket-launcher. First developed in 1940, the B-13 rocket-launcher earned its nickname from a popular song at the time, 'Katerina'. It was capable of delivering a salvo of over 4 tons of explosives into a 10-acre area in seven to ten seconds. Its very inaccuracy was its strength, for enemy troops could barely hear the rockets coming, and could never predict where they would fall. Chuikov used rocket-launchers, mounted on the backs of ordinary trucks, right at the front line. When his forces were pressed back to the very edge of the river, the lorries were suspended over the bank, their back wheels above the water, in order to achieve the necessary trajectory.[22] To shells and rockets were added bombs. The 8th Soviet air army gave growing support to the defenders of Stalingrad, while German air activity, in deteriorating weather conditions, with mounting losses, slowly declined. There were definite improvements on the Soviet side in both equipment and tactics. By November there were 1,400 aircraft on the Stalingrad front, three-quarters of them modern combat planes, including numbers of the new Yak-9 and La-5 fighters, which could hold their own at last with German aircraft. An intensive programme of night-flying training permitted more sorties under cover of darkness against German positions. Soviet air forces were dogged with poor communications which made it

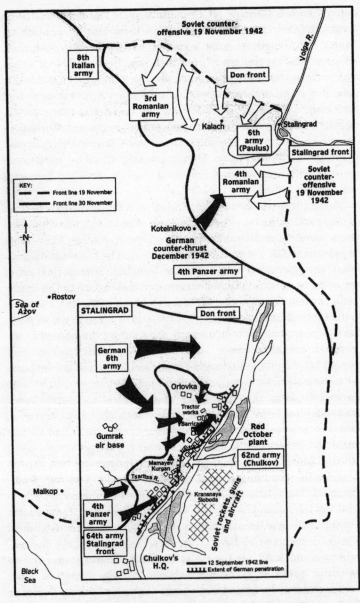

8   Battle of Stalingrad, September 1942 to January 1943

difficult to control large air operations, or to direct aerial strikes accurately. From September 1942 the commander of the Soviet air armies, A.A. Novikov, began an experiment with radio control of air forces on the southern front, which made it possible to marshal and direct aircraft just as the Luftwaffe did. Soviet air power had a great deal of ground to make up, but the gap between the two air forces began to narrow in the critical months covering the defence of Stalingrad. German air forces suffered from a constant drain on aircraft and pilots that could not easily be made good. Air Fleet 4 began the campaign on the southern front with 1,610 aircraft and ended it with 240 serviceable planes.[23]

A further bonus for the Soviet side was Chuikov himself, who continued to lead a charmed life. In September he narrowly missed death when a bomb scored a direct hit on his command bunker. On 2 October air attack destroyed large oil tanks above his new headquarters near the river bank, bringing a cascade of burning fuel flooding into the dug-out. Somehow he and his staff escaped from the midst of the flames, which gradually slid down on to the Volga itself, floating away on the current. For much of the last weeks of battle his command base was never more than a few hundred yards from the nearest German forces. From here he directed his scattered, exhausted troops, from one last-ditch encounter to the next. After the bitter defence of the factory district, which brought both sides to a standstill by late October, Paulus marshalled his forces for one final attack. Hitler was in Munich, attending the annual ceremony to commemorate the abortive Nazi coup of 9 November 1923. Before the Party faithful he pledged that he would shortly be 'master of Stalingrad'.[24] The same day, in the cold early hours of morning, five German infantry divisions and two Panzer divisions, all well under strength, launched the last German offensive in Stalingrad. By the end of the morning the 62nd army was split, and German forces seized a 500-yard stretch of the bank of the Volga. But elsewhere the attack made little progress. Chuikov organised an active defence, ordering local counter-attacks and ceaseless harassment. By the 12th the offensive petered out. A few days later Soviet storm groups began to win back what had just been conceded: the factory district, the slopes of Mamayev Kurgan, blackened by artillery fire in the otherwise snow-covered city, and the few large buildings in the city centre still standing. On 18 November Chuikov and his staff were gathered in his bunker to discuss further operations when they received a cryptic telephone call from the Front headquarters, telling them to expect a special order. At midnight it arrived: the following day armies on the neighbouring Don and South-western fronts were to attack and encircle German forces, snapping tight the trap that Zhukov had set in September.[25]

The Soviet counter-offensive worked like clockwork. Enemy forces were caught by complete surprise. Although local commanders had observed Soviet troop movements on the German flanks for some time, more senior commanders hesitated to pass the news on to Hitler, when victory at Stalingrad seemed so near. At Hitler's headquarters it was assumed that the battle for the city had consumed Soviet reserves, and that little was left for an extensive operation. It was later admitted by Hitler's Chief of Operations, General Jodl, that this was the 'biggest failure' of German intelligence. The day before the attack the head of army intelligence on the eastern front, General Reinhard Gehlen, predicted limited attacks, but would suggest no date or direction.[26] In fact under cover of darkness, cloud and snowfall, paying close attention to camouflage and the muffling of tank movements, the Soviet General Staff had built up a force of over 1 million men, with almost 14,000 heavy guns and mortars, 979 tanks (mostly the versatile, well-armed T-34) and 1,350 aircraft.[27] They were grouped on three fronts, the South-west and Don fronts to the north of German lines, and the Stalingrad front to the south and south-east. Facing them were the Romanian armies and weaker German reserve divisions. At 7.30 on the morning of 19 November the northern fronts began the artillery barrage. Two hours later Romanian forces were hurled back by a furious onslaught from three Soviet armies. For the first time Soviet commanders found themselves with the opportunity to do what their enemy did so well: to conduct a large-scale operation with mobile forces in open country, smashing forward with tanks and aircraft.

Soviet forces made rapid progress. On 20 November the southern armies, with the less difficult terrain, were unleashed against more hapless Romanians. Large numbers of prisoners, arms and supplies were seized in the first few days. 'In desperate plight since early morning,' wrote one unfortunate Romanian in his diary on the 21st. 'We are surrounded. Great confusion . . . We're putting on our best clothes, even two sets of underclothes. We figure very tragic end in store.'[28] He, like thousands of others, surrendered. Within three days Soviet mobile units reached the Don river, almost 150 miles from their starting point. So rapid was their progress that German sentries guarding a vital Don bridge, near Kalach, thought that approaching Soviet tanks were the next watch, come to relieve them. By the time they realised their mistake it was too late, and Soviet armoured forces were free to pour southwards to link up with armies coming the other way. On 22 November, at Sovetsky, south-east of Kalach, the two forces joined hands and Paulus was encircled. The whole steppe area in the German rear was in chaos. The snow-covered grassland was covered with the grotesquely frozen corpses of more than ten thousand horses some, like statues, hardened where they stood. Almost all the weak divisions holding the salient were eliminated. Soviet armies fanned out east and west, creating a heavily defended corridor over 100 miles wide, cutting off the German 6th army and part of the 4th Panzer army, and pressing the whole southern German front back to where it had stood in August. Paulus's immediate reaction had been to pull German forces rapidly back, out of the trap, but Hitler, failing entirely to see what the Soviet operation had achieved, and obsessed with the city, ordered him to stand fast and wait for rescue. Though Paulus's commanders urged him to break out, he stuck by the rulebook. An estimated 240,000 German and allied forces, with a mere hundred tanks and 1,800 guns, prepared for the desperate defence of a bleak area 35 miles long and 20 miles wide, the Stalingrad 'Cauldron'.[29]

*   *   *

The news from the Soviet front followed hard on the Allied landings in North Africa and the defeat of Rommel at El Alamein. Hitler, who had retired to his Bavarian redoubt, the Berghof, hurried back to his winter headquarters at Rastenburg in East Prussia. Here he took stock of the situation. At first, according to one witness, he was 'completely uncertain what to do'. But he was, he told Mussolini on 20 November, 'one of those men who in adversity simply becomes more determined'.[30] His decision to stand and defend Stalingrad was in part a statement of that determination, in part a recognition of what retreat or surrender would mean in terms of morale for Germany's war effort. To justify his order to Paulus he grasped at a slender straw, offered to him by the air force chief, Hermann Göring, who was at headquarters on 24 November to discuss the Stalingrad situation. Though Göring's staff doubted the ability of the Luftwaffe to do anything very much in the south after the losses sustained in the summer and autumn, their wayward chief assured Hitler that the besieged city could be supplied from the air, until such time as relief forces could be organised to break through. He promised to fly in 500 tons of supplies a day. Armed with this assurance Hitler reiterated to Paulus that day the need to stand fast.[31]

The airlift operation was a disaster. Instead of the 500 tons promised, the air force supplied fewer than 100 tons a day, and considerably below this by the end of December and during January. The Luftwaffe units in the east were down to only a quarter of their strength by the end of the year, flying in the teeth of a Russian winter, starved of fuel and technicians. German forces were compressed into a progressively smaller pocket of territory around the city, so the number of airfields available fell, until by mid-January German aircraft dropped supplies by parachute. Slow, poorly armed transport aircraft had to fly over 150 miles from crowded, ill-prepared airfields against a numerous enemy. Command in the air around Stalingrad, which German forces enjoyed until October, evaporated. By December there were fewer than 375 German fighter aircraft for the whole Soviet front, and many of those were unserviceable. Aircraft and crews were drafted in from training stations in Germany to augment the dwindling stock of pilots skilful enough to navigate the dangerous winter route. Soviet air forces set up a well-organised air blockade around the city, while ground forces fought to capture the remaining German airfields. In less than two months the Luftwaffe lost 488 transport aircraft and a thousand crew, while German forces in the Cauldron ran short of food, ammunition and medical supplies.[32]

While the Luftwaffe burned itself out in pursuit of a hopeless ambition, the German army planned a dramatic overland rescue. Its inspiration was Field Marshal Erich von Manstein, victor at Sebastopol. He was put in charge of the hastily formed Don Army Group, made up of forces scattered westwards by the Soviet offensive together with the remaining German reserves. A small group of Panzer divisions gathered south of the Soviet corridor at Kotelnikovo, under the command of General Hermann Hoth. Manstein planned to use this group to rush through the Soviet forces, link up with Paulus's army moving south-west and retreat back to a secure defensive line. The attempt began in heavy rain on 12 December, and made steady progress in difficult conditions. By 23 December Hoth's force advanced 40 miles, then ground to a halt in fierce tank fighting with Soviet reinforcements rapidly deployed into the area. The critical point of the whole Stalingrad campaign had been reached. After the initial shock of the German assault, Soviet commanders in some haste, and with poor intelligence on German intentions, succeeded in mounting a strong counter-blow. Though Manstein and Paulus later complained in their memoirs that a great opportunity to fight a successful retreat from Stalingrad was lost because of Hitler's insistence that Paulus stay put, the truth is more complicated.

When Zhukov and the Soviet General Staff first planned to encircle Paulus at Stalingrad, they realised that German forces would not stand idly by but would try to rescue him. The Uranus plan anticipated this. The corridor separating Paulus from the remaining German forces had to be wide enough to make rescue difficult, and it had to have a strong outer ring of defences and reserves to counter any German thrust. Over sixty Soviet divisions were poured into the breach, with almost a thousand tanks. Though Manstein's forces, with fewer tanks, made remarkable progress in driving blizzards and over difficult, toughly defended terrain, the Soviet high command was able to redeploy its reserves on the right flank and in front of the attacking force. By 24 December there was a very great danger that the whole of the force striking towards Stalingrad might itself be encircled. Manstein was forced to order withdrawal, the last gamble in ruins. All along the German southern front Soviet forces went over to the offensive, destroying Italian and Hungarian armies and weakened German divisions, driving towards Rostov-on-Don. The result was an operational triumph for Soviet commanders. Throughout 1941 and 1942 the German army had profited from the Soviet inability to coordinate their forces, to react to circumstances, to think through a plan and to execute it consistently. It was not just German mistakes that cost Hitler victory at Stalingrad. In the bitter attritional warfare of 1942 the Red Army was coming of age.[33]

For Paulus the end was only a matter of time. By late December his soldiers were beyond rescue. The German force, almost a quarter of a million men, was still an obstacle of sorts, but the conditions they experienced were destructive of strength and will. There were desperate shortages of food, the endless bombardments, the collapse of medical services for the wounded and sick, and the grim, unbearable realisation that there was no way out. Wilhelm Hoffmann, a German infantryman who arrived with his unit in Stalingrad in early September, wrote his last diary entry on 26 December: 'The horses have already been eaten. I would eat a cat, they say its meat is also tasty. The soldiers look like corpses or lunatics, looking for something to put in their mouths. They no longer take cover from Russian shells; they haven't the strength to walk, run away and hide . . .'[34] By January the temperatures were down to −30 degrees centigrade, and food supplies were reduced to 2 ounces of bread a day, half an ounce of sugar, and a little horsemeat. Each soldier was allowed one cigarette.[35] There was little fuel and ammunition. The German guns that had blasted the city flat fired only sporadically. The troops dug in to underground shelters and trenches and awaited the worst.

Stalin and Zhukov gave little thought to finishing off the Cauldron while the battles raged to seize back the Don steppe. Soviet intelligence informed the Supreme Command that only eighty thousand men were in the net, not all of whom were military combatants. It was assumed that 6th army would either surrender or be eliminated rapidly, at Soviet leisure. So the plan to destroy the German pocket, codenamed Operation 'Koltso', was not drawn up until 27 December. It was proposed that the armies of the Don front, to the north of Stalingrad, would attack from west to east, driving German forces back into the city, splitting them into smaller groups which would then be reduced by air and artillery bombardment, stage by stage. Difficulties in moving supplies and men into position postponed the opening of the offensive until 10 January, to Stalin's intense irritation. The force that gathered around the perimeter of the Cauldron was a hammer expected to crack a nut: 47 Soviet divisions, 5,600 heavy guns and mortars, 169 tanks and 300 aircraft.[36] Two days before the attack was unleashed Paulus was given an opportunity to surrender. A small delegation of soldiers approached German lines and was promptly fired upon. The following day, armed with a bugler and a red flag, they tried again. This time they were taken blindfold to German lines. Paulus would not meet them; the surrender terms were dismissed out of hand. The following day, 10 January, at eight o'clock in the morning, Soviet artillery began the heaviest barrage since the war began.[37]

When Soviet armies rolled forward they made slow progress against determined German resistance. Fierce fighting brought surprisingly heavy casualties from an enemy thought to be on his last legs. In terror at what would happen to them if they were caught, German troops fought with whatever desperate energy they could muster. It was now their turn to fight to the last man and the last bullet. After a few days Soviet commanders discovered why progress was so costly. From captured orders and the inter-rogation of prisoners it was discovered that 250,000 men had been encircled in November, not eighty thousand. This number exceeded the number of Soviet troops attacking the city. They were under orders not to surrender but to die as heroes. Most did die, from wounds, from frostbite, from starvation, hundreds of miles from Hitler's headquarters, where the Army Chief-of-Staff ostentatiously ordered his entourage to eat 'Stalingrad rations' while the battle raged.[38] Soviet forces pressed forward through the outer defence lines, until, under the weight of shelling and bombing, short of ammunition and heavy weapons, the German front collapsed. By 17 January the occupied pocket was less than half the size it had been at the start of *Koltso*. On 22 January the last push began into the city itself, where one-third of the original German force dug in. More and more German soldiers began to surrender in circumstances of utter hopelessness and panic. Those who stayed in their bunkers were incarcerated by the heavy tanks that passed overhead, or flushed out by flamethrowers and grenades. On 26 January the forward units of the attacking force finally met up with the 62nd army, which had continued to hold the west bank of the Volga, an anvil to the Soviet hammer. At 9.20 in the morning men of Rodimtsev's 13th Guards attacking the Red October Factory watched as panic broke out among the German units in front of them. Down the hill to the west of the factory settlement came a heavy tank column of the Soviet relief force. Soldiers from the two armies embraced in tears.[39]

Five days later the bulk of the remaining German forces surrendered. In his headquarters in the Univermag department store basement, with hundreds of frightened German soldiers huddled in the foetid air of overcrowded cellars, Paulus was found by a small detachment of Soviet troops led by Lieutenant Fyodr Yelchenko. Depressed, almost detached from his grim surroundings, Paulus refused to meet his captors face to face. His staff officers agreed to the surrender and asked for a car to take their chief away in order to prevent a lynching. Paulus and 23 German generals were taken into captivity, where Paulus publicly recanted his wicked ways. He ended up after the war living in Dresden, in communist East Germany, where he died in 1957.[40] Farther to the north of the city German forces refused to accept the surrender and battled on until 2 February until they had nothing left with which to fight. The Battle of Stalingrad was finally at an end.

There was no doubt that the Red Army had won a remarkable victory. In Moscow the journalist Alexander Werth found a pronounced psychological shift in the population: there was no longer grim, anxious desperation, but a renewed self-confidence. 'No-one doubted that this was *the* turning-point in World War II,' wrote Werth later.[41] The aura of German invincibility vanished – German losses were catastrophic. In the struggle since November, 32 divisions of German, Romanian, Hungarian and Italian forces had been annihilated and a further sixteen all but eliminated. Twelve thousand guns and mortars had been destroyed or had fallen into Soviet hands; according to Soviet sources, 3,500 tanks and three thousand aircraft had suffered a similar fate.[42] The talk in the Soviet press was all of Cannae, the ancient battle where

Hannibal's Carthaginians routed Rome. Early in February Werth was invited with other western journalists to visit the scene of Soviet triumph. On the approach to Stalingrad they were surrounded by a seemingly endless stream of men, lorries, horses, even camels, moving in an untidy mass westwards, to new battles. The temperature was -44 degrees centigrade. In Stalingrad itself the battleground was frozen into a gruesome still life of the conflict – dead men and horses stiff where they had fallen, burnt-out tanks and trucks, the litter of cruel combat. In the basement of the Red Army House, Werth was shown two hundred emaciated, disease-ridden Germans, their skin parched and yellowed, gnawing the frozen bones of one last horse while they waited in threadbare coats and rags for boots, to be taken off to prison camp. There seemed, Werth later reflected, a 'rough but divine justice in the yard of the Red Army House at Stalingrad'.[43]

The fall of the city reverberated worldwide. A year earlier a batch of maps sent from England to Moscow helpfully titled 'Follow the war with this map of the world on Mercator's projection' did not even include Stalingrad. Now the name was on everyone's lips. On 20 February British cities celebrated Stalingrad Day, the 'twenty-fifth anniversary' of the name (it was actually the seventeenth). The following day the Royal Albert Hall hosted a glittering array of the rich and the good in a salute to the valour of the Red Army.[44] Three weeks after the capture of Stalingrad, Stalin sent Churchill film of the battle which Churchill, laid up with pneumonia, watched with a private projector set up in his bedroom.[45] The effect on Hitler's allies, who shared in the collapse, was no less marked. The suffering inflicted on Romanian, Italian and Hungarian forces made them reluctant partners of the Reich. 'It is now certain', the Italian Foreign Minister told Mussolini on 8 February, 'that hard times will come.'[46]

Hitler, as usual, blamed everyone but himself for the catastrophe. When he heard the news of Paulus's surrender he could hardly contain himself with rage. The day before he had promoted Paulus to Field Marshal. Though he harboured to the end the illusion that 6th army could be rescued in the spring, he expected Paulus to fight to the very last if rescue proved impossible, leaving the final bullet for himself. He became obsessed with the view that the heroic sacrifice of so many German soldiers was besmirched by the failings of 'one single characterless weakling'.[47] For weeks he would not allow Göring's name to be mentioned for failing to keep the airlift going. His staff watched him age visibly over the period of crisis, crippled by gastric disorders which gave him pronounced bad breath, ill-tempered and depressed. His way of coping with the psychological blow was simply to pretend that Stalingrad had never happened. He refused to mention it again and busied himself in a flurry of insignificant staff work in preparation for further victory plans in 1943.[48] For the German public there was no disguise for the disaster. Too many were killed or captured to pretend that Stalingrad had not happened.

* * *

It has always been a temptation to signify Stalingrad as *the* turning-point of the Second World War. Field Marshal Keitel, Hitler's Chief-of-Staff, later confessed that this was the moment at which Germany 'played [her] last trump, and lost'.[49] But it was not a decisive victory on its own. It demonstrated a remarkable improvement in the operational skills and battle-worthiness of Soviet soldiers and weapons. The awesome scale of the carnage on both sides, fighting to the death for a city that no longer eixsted, indicates the special character of the savage contest between invader and victim. The victory had a moral and psychological impact well beyond the significance of the strategic triumph. It laid the foundations of Soviet self-belief for battles in 1943 that were really decisive.

Stalingrad was only one part of a much more ambitious Soviet campaign. The rapid success of Soviet armies encircling the 6th army in November encouraged the Soviet General Staff to exploit the advantages they enjoyed in winter fighting, and the obvious exhaustion of the enemy, by pushing the German front back along its whole length, from the Baltic to the Black Sea. With the enemy on the run in the south, Stalin seized the reins back from Zhukov and bullied and urged commanders to rush in hot pursuit before the enemy could regroup. In the north the siege of Leningrad was broken though not lifted; on the central front the German line was pushed farther away from Moscow. In the south the aptly named Operation 'Gallop' was launched at the end of January with the object of driving German forces from the industrial Donetz basin farther back into the Ukraine, and trapping and destroying the whole of the German southern front. The retreating German armies in the Caucasus region narrowly escaped the fate of Paulus, but so swift was Soviet pursuit that by the beginning of February large pockets of German forces were holed up on the wide peninsulas that form the gateway to the Sea of Azov.

There were fears among the German commanders that the whole of the southern front would collapse. One city after another was liberated. At the farthest point Soviet advance forces pushed the enemy almost to the river Dnepr. But Stalin had overplayed his hand. Soviet forces were in turn exhausted from the long gruelling battles in appalling winter conditions, against an enemy adept at rapid withdrawal and tough rearguard defence. It proved difficult to move supplies and reserves, many of which were stranded far in the rear of Stalingrad, so rapid had been the progress of Soviet armies. Stalin was furious at the delays, and ordered the NKVD to take over the running of the railways in the south, but their intervention, brutal but inexperienced, played havoc with already inadequate schedules.[50] The large advantage in number of tanks enjoyed by Soviet armoured divisions at the start of the offensive petered out on account of poor maintenance and heavy losses. Sensing growing Soviet weakness, von Manstein gathered together another strike force for a defensive counter-stroke. Soviet intelligence failed to detect the concentration of German forces south of Kharkov, preferring to believe that the armoured columns they had identified were part of a general retreat. Poor planning and constant pressure from Supreme Headquarters blinded local Soviet commanders to the danger. German armies, pushed back like a coiled spring towards the river Dnepr, were released on 20 February with devastating effect against an enfeebled enemy. By mid-March Kharkov was recaptured and the German line stabilised, not along the Dnepr as Stalin had hoped, but on the Donetz, 150 miles farther east.[51]

The German counter-attack was a timely reminder that the Red Army still faced a powerful and well-armed enemy. The spring

thaw at the beginning of April brought both sides to a standstill after
nine months of continuous warfare and enormous losses. The final
German counter-attack had left an untidy front-line. In the central
area opposite the city of Voronezh there existed a large Soviet
bulge far into the German lines, around the city of Kursk. This
salient was 120 miles wide from Belgorod in the south to within a
few miles of Orel in the north, and posed a threat to the German
forces on either side. It was here that the conflict was renewed again
in the summer.

After the large-scale offensives launched in 1941 and 1942,
German ambitions for 1943 were much more modest. The losses
suffered in the struggle for Stalingrad and the long retreat afterwards
had been debilitating. In January 1943 only 495 of the tanks along
the whole length of the eastern front were serviceable. Even with
extensive repairs and reinforcements, by May the German army
could raise a force of only 2,500 tanks, the lowest figure for two
years.[52] Manpower was also hit. By the summer of 1943 the field
army totalled only 4.4 million men, against the 62 million available
for the campaign in 1942. German army leaders were almost all
agreed that the best that could be hoped for was to hold the line in
the east, and to frustrate the Soviet offensive by undertaking limited
but powerful counter-blows. For once Hitler accepted advice.
After Stalingrad his interest in the eastern front perceptibly
declined. The initiative was left to von Manstein, who proposed
pinching out the Kursk salient before moving further south to
regain some of the lost territory. Hitler accepted the plan, which
was given the codename 'Citadel'; but he insisted that there should
be no more defeats after the humiliation at Stalingrad. Although
Manstein wanted to attack in April or early May, before the enemy
had time to regroup and dig in around Kursk, Hitler postponed the
offensive until mid-June, and then again until July, in order to build
up German tank forces to a level that he felt would reduce risk of
failure.[53]

During March and April Stalin and the Soviet General Staff tried
to assess what their enemy would do next. After two years of
disastrous miscalculation, Stalin was more receptive to advice. He
recognised that he had been the main barrier to interpreting
German intentions correctly. This time opinion was sought from

staff officers and front commanders as well as his immediate circle. There emerged a consensus among them all that the Kursk salient was the only place on the front where German forces were in a position to launch an attack with any prospect of success. This view was based for the first time on solid intelligence. From aerial reconnaissance, partisan activity and radio interception Soviet commanders pieced together a much clearer picture of German dispositions. From the concentrations of Panzer forces and infantry divisions around Orel and Kharkov it was clear from where the main thrust would come. From two years' experience of German operational planning, Soviet commanders predicted with remarkable accuracy how German forces would begin the attack. It was assumed that two heavy armoured thrusts, north and south of the neck of the salient, would converge to encircle Soviet forces in the bulge. But after that German intentions could only be guessed at.[54]

The next step was to decide how to respond. Stalin instinctively sought an offensive solution: an attack launched pre-emptively against German positions, followed by hot pursuit. Zhukov and the General Staff rejected this, and it says much for Stalin's judgement that he bowed again to his deputy. On 8 April Zhukov suggested the plan which a few weeks later was adopted as Soviet strategy for 1943. First, Soviet forces would absorb the German punch at Kursk on deep defensive lines, to wear down enemy tank numbers, and then respond with a counter-punch of annihilating power that would leave the enemy sprawled on the canvas. It was a bold plan, for the Red Army had previously failed to defeat German forces in summer time, when ground and climate suited the invaders more. But it was a plan based on solid experience. The lessons learned in the long Stalingrad campaign had been absorbed. Zhukov insisted that the Kursk battle should be placed under the control of the Supreme Headquarters, with special attention to the central planning and coordinating of all aspects of the campaign. There was an operational and logistical depth to the preparations for Kursk that previous Soviet plans had lacked. After a string of glaring intelligence errors in the past, the Soviet High Command also insisted on the topmost priority for intelligence gathering of all kinds. Some was provided from Britain, but the most detailed information was culled from the Ultra decrypts and forwarded by the Soviet spy John Cairncross (later revealed as the 'Fifth Man' of the Cambridge spy ring), who had found himself a job at the Bletchley Park intelligence centre in 1942. But most of the intelligence on German troop concentrations came as a result of great improvements in Soviet intelligence gathering, particularly in reading low-level German radio traffic, and in using small storm detachments to capture Germans from the front-lines for interrogation.[55] Though Soviet commanders never guessed the exact striking point for the German attack, they had a much clearer idea of what they were up against than in previous campaigns.

The whole crux of the Soviet plan was the carefully prepared battlefield. Troops began digging in even before the plan for Kursk had been finalised. Inside the salient seven armies were concentrated on the Central and Voronezh fronts, which defended the north and south of the salient. On either side of the bulge additional strong forces were placed, in the Bryansk front to the north and the South-western front to the south, ready to launch the counterpunch. Many miles to the rear of the salient was Zhukov's trump card, a reserve force with two infantry armies, a tank army and the 5th air army, organised into the Steppe front. This massive reserve was to be thrown forward once the German attacks had been blunted on the defence lines. Inside the Kursk bulge the population stayed put, a decision taken partly to prevent demoralisation at the sight of yet another eflux of refugees, and partly because civilians of both sexes and all ages were needed to help the troops prepare the defensive ramparts.

The defensive system was one of considerable tactical sophistication. No fewer than six defensive zones were established in the salient, up to a depth of 50 miles, with two further defensive lines in front of the reserve Steppe front. The first three zones were the most crucial, for it was here that the Soviet armies' main strength was concentrated. Each zone had continuous trenches and anti-tank obstacles, connected with a system of communication passages. On the Central front alone troops dug some 3,000 miles of trenches, and laid 400,000 mines. Each line was protected by barbed wire, some of it electrified. The main object was to hold up and destroy the enemy's armour. Every kind of device was used – ditches fitted with 'dragon's teeth' (wooden stakes slotted together like a giant set of jacks), small dams to flood the ground in front of advancing tanks, and abatises (trees piled one upon the other with branches pointing towards the enemy) in every wood. Great attention was paid to the field of fire for artillery and anti-tank weapons. A criss-cross system was devised of artillery strongpoints, from which all lines of advance could be covered, providing what one front commander called 'an impenetrable curtain of fire'.[56] And last, but not least, an intensive programme of training was undertaken for Soviet anti-tank and artillery forces.

Within a matter of weeks the Kursk area was turned into a vast fortress. Inside its walls massed more than a million Soviet troops. They were better armed than at any time since the beginning of the conflict. The infantry had improved automatic weapons and more effective communications equipment. The standard Soviet tank, the T-34, had undergone small but very significant improvements. Despite its strong armour and high speed the T-34 performed poorly in combat up to 1942. It carried a crew of only two (three was standard on German tanks), which meant that the tank commander also had to fire the gun. The turret was cramped and afforded poor visibility, and when the tank commander stuck his head out of the top hatch all he could see forwards was the hatch cover. Very few tanks carried radios, so that they were on their own once engaged with the enemy. During 1943 Soviet engineers slowly rectified these blunders. To allow another crew member, a larger turret was added, with a modified hatch allowing 360-degree visibility. A great many more tanks carried radios, and a battlefield communications system was set up allowing commanders to direct larger and more complex tank operations, and to call up aircraft assistance against stubborn pockets of resistance.[57] Soviet tank forces knew that they faced a new enemy in 1943, the next generation of German heavy tanks, the 56-ton 'Tiger' and the smaller, 45-ton 'Panther'. Both had large-calibre guns and good armour, more than a match for the T-34. To cope with the threat a new tactic was devised. At close-range the large German guns could not be deployed so easily. Soviet tankmen were trained to drive in close and to fire point-blank at the vulnerable parts of the enemy tank, in the sides and rear. The anti-tank artillery were taught the same. In a head-on attack the Soviet 45-millimetre anti-tank gun was ineffective; but ambushing the Tigers or Panthers from the side or rear, as they rolled past, could destroy or disable them with a lucky strike.[58]

If mines, guns and ditches failed to stop enemy tanks, the Soviet high command added the famous Ilyushin Il-2 Sturmovik dive-bomber. The 1943 version was faster and better armed, with a 37-millimetre tank-busting cannon and the new 'PTAB' anti-tank bomb. There were almost a thousand of them assembled for the Battle of Kursk, along with over a thousand fighters and almost a thousand bombers. To support this array of largely modern equipment stood a huge reserve air force of another 2,750 aircraft, waiting, like the Steppe front, to deliver the knock-out blow. The number of radio communications stations, a facility which at last permitted air commanders to coordinate attacks, rose from 180 to 420, and they were equipped with effective radar for the first time. Over 150 airfields were built for Kursk, and fifty dummy fields to mask Soviet strengths and intentions.[59] In the air and on the ground Soviet forces were both tactically and technically a more formidable barrier than German commanders realised.

The small details mattered a great deal in the final outcome of the confrontation, for there was not much to choose in size between the two opposing forces. Each had watched what the other was doing over the late spring. More and more soldiers and equipment were drafted in, until what Manstein had thought of as a local counter-offensive assumed anything but local proportions. Kursk became, almost by default, the largest pitched battle of the war. On the German side were massed fifty divisions of 900,000 men, 2,700 tanks and 10,000 guns, and 2,000 aircraft. On the two main defensive fronts, Central and Voronezh, there were 1,336,000 men, 3,444 tanks, 19,000 guns and 2,900 aircraft.[60] The Soviet commanders allocated 40 per cent of Red Army manpower, and three-quarters of its armoured force, to the contest. Victory or defeat for either side became critical.

One piece of intelligence the Soviet high command was desperate to learn: when would German armies attack? This question was impossible to answer firmly, since Hitler, prey to anxieties of his own about the unpreparedness of German forces, kept postponing the date, first 3 May, then 12 June, finally early

July. Soviet forces were kept in a state of semi-alertness for much of this period; regular intelligence scares brought them up to full alert more than once. The longer Stalin waited, the more impatient he became for action. But Zhukov knew that Soviet forces must be kept on a leash, or the whole carefully orchestrated operation would be spoiled. He hammered the point home again and again until Stalin treated the plan as his own. During June all Soviet intelligence sources confirmed that German forces were in place and ready to attack, but the waiting game went on. The Soviet side became restless, suspecting some unanticipated ruse. The mystery deepened when, on 23 June, a Soviet spy ring in Switzerland reported, through agent 'Lucy', that Hitler had changed his mind, and the attack was off. In a week of high tension Zhukov kept calm. All the indicators from the front-line showed that German forces were moving to battle stations and that an attack could be expected at any time between 3 July and 6 July. Troops were on full alert for every day from 2 July. Then, abruptly, on 4 July all activity on the German side ceased. A strange silence descended on the front-line, disconcerting and menacing.[61]

Soviet leaders knew it could only mean one thing. A prisoner seized on the Voronezh front told his captors that the German soldiers had all been issued with battle rations and a portion of schnapps. At ten o'clock in the evening another infantryman was seized by a Soviet scouting party. His interrogation revealed that the German assault would start with an artillery barrage in four hours' time.[62] Stalin and Zhukov waited anxiously at headquarters, unable to sleep. At 2 a.m. the order went out for Soviet guns to open fire with a spoiling barrage against German positions. The thud of bombs and shells, the whoosh of rocket salvos, the rumbling of aircraft engines, all merged into one, 'like the strains', Zhukov later recalled, 'of a symphony from hell.'[63] The waiting Germans were taken completely by surprise. For some time German commanders thought that by the wildest of coincidences the Red Army had begun an offensive at exactly the same time as them. Order was slowly restored, and the German offensive began at 4.30 on the morning of 5 July to the sound of ten thousand guns and the roar of two thousand aircraft.

★   ★   ★

The German assault was an awesome demonstration of modern military power. From the assembly points around Belgorod in the south and Orel in the north a heavily armed Panzer fist, strongly supported by aircraft, was hurled against the Soviet rampart. To the north Field Marshal Model's 9th Panzer army led the charge. In his way stood the 13th army of General Rokossovsky's Central front. Zhukov's patient preparations paid off almost at once. The German forces, spearheaded by battle groups of Tiger tanks and the powerful new 'Ferdinand' self-propelled guns, and backed by smaller tanks and motorised infantry, found themselves subjected to a fearsome wall of fire. Anti-tank gunners followed instructions by attacking the new heavy tanks at suicidally close range. Mobile shock units, armed with petrol bombs and portable anti-tank obstacles hurried about the battlefield, immobilising enemy vehicles. At the end of the first day Model's forces had moved forward only 4 miles and were pinned down under heavy and accurate fire. Each German thrust brought a rapid response from the enemy, who moved whole divisions to block German progress. The next day 2nd and 9th Panzer divisions mounted an attack with three thousand guns and a thousand tanks on a 6-mile front but made slow progress against the main defensive zone. The following day Model brought the full weight of his armies to bear on two small towns, Ponyri and Olkhovatka. Four Panzer divisions ground slowly forward against stiffening resistance, the area behind them littered with burning tanks and guns. Caught in the chief zone of defence, the tanks were bombarded by Soviet fighter-bomber aircraft, and picked off by carefully camouflaged anti-tank nests. After five days of bitter fighting the German drive into the north of the salient petered out. By 12 July the roles were reversed. Rokossovsky still had forces in reserve and now Soviet armies slowly prised open the tenuous grip of the depleted Panzer divisions, forcing them back across the grisly field of war.[64]

In the south, where more German armour was concentrated, the attack made greater headway. The defences on Vatutin's Voronezh front were more thinly spread, for it had been difficult to guess exactly where the German blow would fall. At five o'clock in the morning, 4th Panzer army, led by the same General Hoth who had failed to rescue Paulus in December, pushed forward on a narrow

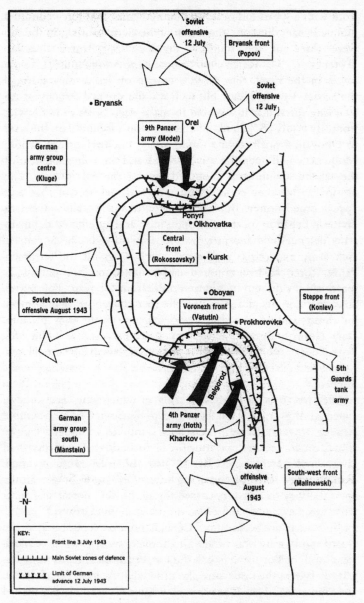

9 Battle of Kursk, July–August 1943

front with a spearhead of seven hundred tanks. His force consisted of nine Panzer divisions, the cream of the German army. In the van were three of the toughest divisions, SS Panzer divisions 'Totenkopf', 'Das Reich' and 'Leibstandarte Adolf Hitler'. Like the forces in the north they faced a curtain of fire as they stormed the Soviet defensive lines, but such was the size and ferocity of the attacking force that it was able to make deep holes in the Soviet front. By 7 July SS Totenkopf division won a foothold on the road to Oboyan, a small town 20 miles inside the Soviet front, whose conquest would open the way to Kursk and the vulnerable rear of the salient. Confidence mounted among the SS soldiers. The possibility that they could repeat the great successes of 1941 and 1942 loomed nearer. But the following day the leading German divisions began to meet stiffer opposition as they entered the main army defence field and progress suddenly slowed. The Soviet 1st tank army and 6th Guards army, pitched against the terrifying Panzer force of five hundred tanks massed on a 4-mile-wide spearhead, held them up a dozen miles from Oboyan. The Soviet Supreme Command ordered up the first reserves to help combat the threat, and for two days a grim slugging match took place to deny German forces the vital road northwards. By 9 July SS Totenkopf crossed the Psel river, south of Oboyan, and established a small bridgehead. This was the farthest point the attacking force reached.[65]

After five days of furious fighting in which little quarter was expected or given, both sides paused to regroup. The high cost of pushing towards Oboyan persuaded Hoth to swing his Panzer armies on to another axis, north-east towards the small town of Prokhorovka. Hoth hoped that the classic pincer movement would destroy and encircle the remaining Soviet forces in front of him and open the way to Kursk once and for all. Soviet commanders saw what was happening, and ordered in the reserves they had created for just such an eventuality. General Nikolai Rotmistrov's 5th Guards tank army was ordered to move rapidly from its staging areas south of Voronezh to halt the Panzer advance. Units from the vast reserve Steppe front were also ordered up, against the advice of its commander, Ivan Konev, who wanted to commit the whole front later, as a single blow. Rotmistrov's Guards faced a gruelling march. As they pressed forward on the dry steppe they were enveloped in clouds of hot dust, which settled in a thick layer on the tanks, lorries and soldiers. 'It was intolerably hot,' recalled Rotmistrov. 'Soldiers were tortured by thirst and their shirts, wet with sweat, stuck to their bodies . . .'[66] In 48 hours they marched over 150 miles. Maintenance crews rode with the tanks, keeping them on the move. Like the proverbial cavalry, 5th Guards arrived in the nick of time.

While Soviet reserves poured in to repair the breach made by 4th Panzer army, German forces prepared for their final push. On the morning of 11 July the three SS Panzer divisions marched towards Prokhorovka. The dry dusty heat gave way all at once to swirling winds and driving rain. German tanks plunged forward, to the west and south-east of the town, but could go no farther against determined anti-tank defence and in the face of heavy losses. Zhukov and the Chief of the General Staff, Vasilevsky, assumed control of the battle which approached its climax on 12 July. Zhukov organised ten regiments of artillery as tank-busting fists around the town, while 5th Guards tank army, now supported by Zhadov's 5th Guards infantry, prepared to counter-attack the SS forces with a mass of 850 tanks and self-propelled guns, the largest tank force assembled for a single engagement since the beginning of the war.

On the morning of 12 July Rotmistrov set up his headquarters on a hill overlooking the main site of battle. 'From a solid dug-out in an apple orchard, with trees half-burnt and half cut down', Rotmistrov viewed the scene below on the undulating plain, broken with small groves and tree-lined ravines, an area in peacetime of rich farmland. On the fringes of the plain, hidden in woods and high grass, lay his tank force. At 8.30 in the morning, as the two rival air forces battled in the cloud above, he ordered the repeated codeword 'steel, steel' sent out to the tanks.[67] It was the signal to move. Below him through the grass and crops he saw the T-34s race from their hiding-places. By chance, at exactly the same time the German tanks moved on the far side of the plain. Two vast armoured forces rushed into a head-on collision. Nothing could be done to avoid it. It proved impossible to control the ensuing battle. The tanks swirled around in a sea of exploding shells, crashing willy-nilly into their attackers. The T-34 crews took advantage of the mêlée to attack at point-blank range, blasting the Tigers and Panthers at the side and rear, where the ammunition was stored. When Soviet tankmen ran out of shells they rammed German vehicles, or attacked them on foot with grenades. From on high dive-bombers from both sides tried to sort out friend from foe. Amidst the fury of battle fierce thunderstorms could barely be heard. The plain was littered with the burning hulks of tanks from both sides. The 3 square miles below Rotmistrov turned black from green. Everything was on fire: woods, fields, villages. Even up on the hill the air stank of soot and smoke.

While the two tank forces stayed locked in combat, barrel to barrel for eight hours, Soviet forces south of Prokhorovka blocked an attempt by other Panzer divisions, hastily brought up from reserve, to push through to the east of the town. Still more Soviet divisions were sent west of the town to prevent 11th Panzer corps from outflanking the great tank battle and attacking 5th Guards tank army from the rear. Rapid Soviet redeployment frustrated each German assault. Though minor gains were made against Soviet positions here and there, Soviet tank forces began to push back the Panzer divisions at the critical points of the battlefield. Illuminated by fire, the battle continued on through the evening. With the coming of deep night the firing gradually ceased. Both sides withdrew to lick their wounds. The following night Zhukov came to visit Rotmistrov. The two walked out on to the plain, through the corpses and the wrecked machinery of war, the tanks burning fitfully in the summer rain. Zhukov was visibly moved. He removed his cap, and stood, for some moments, in thought.[68]

The German failure to seize Prokhorovka marked the end of Citadel. German forces continued to probe Soviet defences for several more days, but they had been decimated in the gruelling armoured confrontations. Over three hundred German tanks were destroyed on the 12th alone. The crack divisions were exhausted. SS Totenkopf was so severely battered that it had to be withdrawn from combat to recuperate. Over half its tanks and vehicles lay lifeless on the steppe. Other divisions suffered even more. The 3rd Panzer division had thirty tanks left out of three hundred; 19th Panzer was left after the battle with just seventeen.[69] The struggle for Kursk tore the heart out of the German army. Heinz Guderian, Inspector of Armoured Forces in 1943, described the failure of Citadel as a 'decisive defeat'. The losses at Kursk could not quickly be restored, and the balance of armoured vehicles had tipped decisively to the Soviet side. On the whole of the eastern front in August the German army could muster only 2,500 tanks and self-propelled guns against 8,200 on the Soviet side.[70]

Soviet success at Kursk, with so much at stake, was the most important single victory of the war. It ranks with the great set-piece battles of the past – Sedan in 1870, and Borodino, Leipzig and Waterloo from the age of Napoleon. It was the point at which the initiative passed to the Soviet side. German forces were certainly capable of a sustained and effective defence as they retreated westward; but they were now too weakened and overstretched to inflict a decisive defeat on their enemy. The Red Army at Kursk demonstrated that it was a formidable modern fighting force.

The day after Prokhorovka Hitler cancelled Citadel. Three days before, an Anglo-American force landed in Italy. A real threat now existed of a frontal assault on German-held Europe across the English Channel. Faced with dangers in east and west, even Hitler could see that timely retreat in the face of defeat at Kursk was a necessary course. But no one on the German side could have predicted what followed. In the months before the Kursk battle the Soviet Supreme Command carefully planned a massive counter-offensive to be launched once the defensive stage of the campaign was over. The scale and geographical extent of the plan was concealed from German intelligence by careful use of camouflage and misinformation, at which Soviet forces had become much more adept. Even before the battle for the south of the Kursk salient was over, the neighbouring Soviet fronts in the north, the Western and Bryansk fronts, tore into the rear of the German forces grouped around Orel. German defensive lines, constructed in depth like those in the Kursk bulge, gave way before the onslaught, overwhelmed by numbers. German Panzer reserves were thrown in to stem the tide, but by 5 August Orel was again in Soviet hands and the whole German force was in retreat.

On the south of the salient von Manstein began to pull back German forces as more intelligence reached him about huge Soviet reserves moving westward. Stalin, anxious to speed up the timetable again, was all for pursuing the beaten enemy at once. 'With the greatest reluctance', Zhukov recalled, Stalin had to accept the need for careful preparation.[71] Reserves, fuel, ammunition and the whole array of rear services were brought up behind the armies. The knock-out punch that was held back throughout the bitter Kursk battles was to be released only when Zhukov was ready. As German forces tried to recuperate and dig in around Belgorod and Kharkov, the Soviet steamroller was fired up. On 3 August the southern offensive, Operation 'Rumyantsev', was ready. The Voronezh front constituted the Soviet battering ram, mirroring the combined tank and air offensive employed by their enemy. Within hours a hole 30 miles deep was gouged out of German lines. By 5 August Belgorod was captured. A little over two weeks later Konev's Steppe front pushed through to Kharkov, where under cover of darkness Soviet soldiers rushed in to liberate the city. In recognition of the liberation of Orel and Belgorod, Stalin ordered military salutes to be fired in Moscow as each city was freed, the first of many. It was these battles, fought against the Panzer armies still smarting from the combat at Kursk, that completed the operations Zhukov had first suggested in April. The next stage was to capitalise on the destruction of German forces by driving westward into the Ukraine and towards Byelorussia.[72]

The German front in the east was forced, after Kursk and the seizure of Kharkov, to fall back along its entire length. Between August and December the Red Army gave the retreating enemy no rest. Stalin urged the Soviet armies forward across the vast battle-field stretching from the Baltic to the Black Sea. German forces simply lacked the manpower and equipment to hold fast against an enemy with larger resources and an increasingly confident grasp of large-scale mobile operations. By October they were pressed back to the Dnepr river. Stalin promised to award the coveted title Hero of the Soviet Union to the first soldiers to cross it. In September the broad river barrier, which Hitler had insisted would be the final rampart against Bolshevism, was breached in several places. Early in November an attack in strength was made across the river which routed the German force concentrated around the Ukrainian capital of Kiev and the city was seized on the 6th. Though Red

Army commanders still made mistakes, and Stalin drove his armies on to the point of exhaustion, the traffic was almost all one way, against a demoralised and under-armed enemy. There was a long way to go to take the fight into Europe and to reach Berlin, but Kursk had unhinged the German front irreversibly. Von Manstein ordered the best that could be hoped for: an active defence to wear down the Soviet attackers, a strategy that was maintained through the long, wounding trek back to the Reich.

At the height of this pursuit Stalin agreed for the first time to meet both his Alliance partners, Churchill and Roosevelt, at Teheran to discuss face to face the course of the war and Allied strategy. On 24 November his train left Moscow for the south. The following day it rolled into Stalingrad, a city of ghosts, draped in snow. No attempt was made to survey the ruins; Stalin ate dinner in his carriage and thirty minutes later the train pulled on. From Baku the party flew to the Persian capital, where on 28 November the conference assembled in the Soviet Embassy. At each sentry post stood a Soviet, an American and a British soldier, side by side.[73] The Soviet delegates came in a mood of evident self-assurance. They were the victors of Stalingrad and Kursk, and had defeated the overwhelming bulk of German armed power. Stalin was in a position to give advice, and to make demands. Roosevelt and Churchill agreed to launch a western attack on Europe in May of the following year. Among the formalities of the conference was the presentation to Stalin of a gift from the British King, George VI, of a gleaming Sword of Honour for the victory at Stalingrad. In front of a room full of officials and soldiers, and in the presence of the American President, Churchill handed the sword to Stalin, who raised it to his lips and kissed it. Roosevelt was visibly moved as the sword was solemnly escorted from the room by a Soviet guard of honour.[74] A few days later Stalin returned to Moscow. He told Zhukov that Roosevelt had given his word that the Allies would attack German-held France in the spring. 'I believe he will keep his word,' Stalin continued, 'but even if he does not, our own forces are sufficient to complete the rout of Nazi Germany.'[75]

*   *   *

Soviet victory in the campaigns at Stalingrad and Kursk effectively determined the outcome of the war. There is little dispute on either the German or Soviet side that this was the major turning-point. But there is a great deal of argument about the reasons for it. The conventional view has been to blame Hitler for gross strategic mismanagement, or to ascribe Soviet success to crude weight of numbers. Manstein in his memoirs blamed defeat in 1943 on 'the extraordinary numerical superiority of the enemy'. General von Mellenthin, veteran of the tank battles he later described, wrote that defeats 'were not due to superior Russian tactical leadership', but to 'grave strategic errors' and 'the gigantic Russian superiority in men and material'.[76] The underlying assumption is not that Soviet forces won the contest in 1943, but simply that the German side lost it.

From everything now known about the eastern front from the Soviet side such an explanation is no longer tenable. Hitler may well have appeared a liability to German commanders, but until the failure at Stalingrad, as he told his staff the day that Paulus surrendered, 'We were always superior . . .'.[77] Moreover for the Citadel operation he left the planning and execution largely to the professionals, Manstein and Zeitzler. The crude weight of Soviet numbers cannot be the answer either. Soviet forces on paper greatly outnumbered the German attackers in 1941, but were cut to pieces; at Stalingrad and Kursk, though the margin in equipment slightly favoured the Soviet side, the gap was too small to blame German defeat on Soviet 'masses'. It is true that German forces found themselves taking loss rates that would have seemed inconceivable in 1941. The rest of Europe was conquered for less than the losses they suffered in December 1942 and January 1943. But Soviet losses were also exceptional, totalling in the end almost nine million dead and twenty million wounded. During the critical battles of 1942–1943, long before Britain and the United States had set their own armed forces and populations fully on the scales, the balance of population between the Axis states and the Soviet Union was approximately 130 millions each, while Germany could also exploit the labour-power of the Soviet areas it had conquered, with more than 60 million people. Soviet divisions were seldom more than a fraction of their formal establishment, and the quality of much Soviet manpower (over-age or recuperating soldiers) declined as the war went on. The idea that the German armed forces were simply swamped by human numbers is a myth.[78]

The reasons for Soviet victory on such a scale in 1943 are active Soviet reasons, the result of a remarkable resurgence in Soviet fighting power and organisation after a year and a half of shattering defeats. When Marshal Zhukov wrote his reminiscences of the campaign he could point to solid achievements: better central planning of operations, and their careful supervision by the General Staff; very great improvements in Soviet technology and the tactics for using them, exemplified nowhere more fully than on the prepared defensive ground around Kursk; and the ability at last to deploy millions of men and thousands of tanks and aircraft, with all their supplies and rearward services, in lengthy and complex operations, without losing control of them.[79] To this Zhukov might have added the argument that Soviet planning and central direction, generally viewed unfavourably today, were the final factors that turned a demoralised population and its shattered economy into a great armed camp, providing the weapons and food and labour to sustain 'deep war'. No other society in the Second World War was mobilised so extensively, or shared such sacrifices. The success in 1943 was earned not just by the tankmen and gunners at the front, but also by the engineers and transport workers in the rear, the old men and the women who kept farms going without tractors or horses, and the Siberian workforce struggling in bitter conditions to turn out a swelling stream of simply constructed guns, tanks and aircraft.

Doubtless some of this energy was conscripted at the point of a gun and through fear of the Gulag, but this cannot explain on its own the remarkable effort of will expressed by ordinary Soviet citizens and soldiers in the face of the German threat. That effort was fuelled by the very visible consequences of invasion. When Ilya Ehrenburg, the Soviet writer, visited the Kursk salient after the battle he was horrified by what he saw: 'Villages destroyed by fire, shattered towns, stumps of trees, cars bogged down in green slime, field hospitals, hastily dug graves – it all merges into one, into deep war.'[80] The sight of so much wanton destruction in the areas once occupied by the invader provoked Ehrenburg to write in *Red Star*: 'Now the word "German" has become the most terrible swearword. Let us not speak. Let us not be indignant. Let us kill . . . if you have killed one German, kill another!'[81] When General Chuikov crossed the Volga in September 1942 to join the 62nd army he had been moved to tears by the sight of Soviet refugees huddled on the jetties waiting to escape the bombing, small children without parents, their eyes filled with a blank despair.[82] For Soviet soldiers war was not something happening in another country; it was happening in front of them, to their villages and cities, and to their families.

The drive to succeed in the battles in 1943 stemmed from violent emotions and a directed hatred. The stubbornness of Soviet resistance astonished German commanders; the ferocity of the confrontation led to barbarisms on both sides. The contest came to assume the character of that very struggle of nature which Hitler believed lay at the root of all human life, the survival of the fittest. The Soviet will to win, which emerged painfully from the wreckage of Soviet fortunes before Stalingrad, was not a mere abstraction but a spur to efforts that both sides, Soviet and German, would have thought impossible a year before. The Soviet people were the instrument of their own redemption from the depths of war.

4

THE MEANS TO VICTORY
Bombers and Bombing

'. . . our supreme effort must be to gain overwhelming mastery in the air. The Fighters are our salvation, but the Bombers alone provide the means to victory.'

Winston Churchill, 3 September 1940

O N 6 OCTOBER 1942, at the height of the dangerous struggle on the Don steppe in front of Stalingrad, *Pravda* published a cartoon by Yefimov ridiculing the failure of Russia's British ally to help the desperate Soviet forces by attacking the German rear. The picture showed three Blimpish British officers, General What-if-they-lick-us, General What's-the-hurry and General Why-take-risks, confronted by two go-getting young Americans, General Guts and General Decision. The Soviet public needed no caption. They understood that Britain, perhaps from timidity, certainly in bad faith, had failed to help her ally at her hour of greatest need.[1]

It is easy to understand the disappointment felt by Soviet leaders and the Soviet people. Germany had almost won the contest in 1941; she was poised for what might be the decisive victory in 1942. The feeling in Moscow was that anything the British or Americans could do to tie down or divert German forces away from the Russian front should at least be attempted. Ever since the initial German assault in June 1941, Stalin had pleaded with the western powers to do something. In June 1942 a half-hearted hint was given from London and Washington that a 'Second Front' might be started in Europe in the course of that year. But actions spoke louder than words and only a month earlier Britain had suspended supplies to the Soviet Union on the difficult Arctic route to Archangel because of German submarine and air attack from Norway. And on 21 June, the anniversary of the German invasion, Rommel succeeded in capturing Tobruk with only limited resistance. In Moscow the 'gutless' surrender of Tobruk was contrasted with the heroic defence of Sebastopol. Stalin, not one to mince words, accused the British of cowardice.[2]

In truth Britain and America lacked the trained manpower and shipping, particularly landing craft, to do anything very effective in Europe; British forces in North Africa were stretched to the limit containing a mere four German divisions. British and American military planning was based on the assumption that a large-scale landing in Europe could not be contemplated before the summer of 1943 at the earliest.[3] But Churchill was bitterly provoked by the suggestion of western cowardice. On 30 July he proposed a face-to-face meeting with Stalin at a place of his choosing. The following day Stalin extended a formal invitation to come to Moscow. Churchill accepted. On 12 August he flew from the Middle East in a converted B-24 Liberator bomber which was so noisy that the passengers could only communicate by passing handwritten notes. The plane landed at Moscow in the early evening, and Churchill was whisked away in a limousine with windows of 2-inch-thick glass to State Villa No. 7 on the outskirts of the Russian capital. Churchill found everything prepared with 'totalitarian lavishness' – caviar, vodka, veteran servants with 'beaming smiles', and the novelty of mixer taps, which he later installed at Chartwell. At seven o'clock he was driven to the Kremlin to meet Stalin.[4]

Churchill's task was an unhappy one. At one and the same time he was to repudiate Stalin's accusations of bad faith and announce that there could be no Second Front. The meeting went badly from the start. Churchill outlined in detail the reasons for abandoning any prospect of helping Russia in 1942, and promised an offensive at some point in 1943. Stalin argued with everything, bluntly, 'almost to the point of insult': 'You can't win wars if you aren't willing to take risks'; 'You must not be so afraid of the Germans.'[5] Sullen and restless, Stalin rejected all Churchill's arguments, but accepted that he could not actually force the west to come to combat. It was at this point, with the discussion deadlocked, that Churchill revealed what the west could actually offer: the heavy bombing of Germany and Operation Torch, an Anglo-American landing in North Africa late in 1942.

The mood of the meeting lightened. Stalin liked Torch. He saw straight away that it would secure the defeat of Rommel, and speed up the withdrawal of Italy from the war. But what he liked most was the bombing. The American envoy, Averell Harriman, who was present throughout the meeting, wired Roosevelt the following day that the mention of bombing elicited 'the first agreement between the two men'.[6] Stalin came to life for the first time in the conference. He told Churchill to bomb homes as well as factories; he suggested the best urban targets. 'Between the two of them', Harriman continued, 'they soon had destroyed most of the important industrial cities of Germany.' The tension eased. Stalin accepted that the British could, as Churchill put it, only 'pay our way by bombing Germany'. His visitor promised a 'ruthless' bombardment to shatter the morale of the German people.[7] After four hours the meeting broke up with more cordiality than it had started with.

There were three more days of meetings. Great pressure was exerted to get Churchill to change his mind about a Second Front. Stalin returned to the theme of Britain's yellow streak, her manifest 'reluctance to fight on the ground', remarks which drew from Churchill such a heated and dignified rebuttal that Stalin was provoked to his famous response: 'Your words are of no importance. What is important is your spirit!'[8] But by the end of the visit relations between the two leaders were better than Churchill could have hoped. On the evening before Churchill's departure, 15 August, Stalin took the unprecedented step of inviting his guest to visit his private apartment. A long dinner, liberally accompanied by every kind of alcohol, created a more relaxed atmosphere, though Stalin still flung taunts at British courage ('Has the British Navy no sense of glory?'). At one o'clock in the morning Stalin began to eat heavily from a sucking pig, at what was his usual dinner hour. At 2.30 with a splitting headache, Churchill made his farewells, and flew off exhausted from Moscow airport three hours later. When he arrived in Cairo he told officers of the 8th army of his renewed commitment to bombing: 'Germany has asked for this bombing warfare . . . her country will be laid to ruins.' On his return to London he ordered RAF statistics to be sent to Stalin, and called for the large-scale reinforcement of Britain's heavy bomber forces.[9] For the moment bombing constituted the Second Front.

*　　*　　*

Churchill had always been an enthusiast for bombing. In World War I, as First Lord of the Admiralty, he authorised the early bombing attacks by Britain's fledgling air forces against the German Zeppelin bases in Germany. As Minister for War in 1919 he defended the newly formed Royal Air Force from the predatory claims of the army and navy, who resented the parvenu service, and hoped to stifle its claims to independence.[10] During the 1930s it was Churchill who led the campaign to increase British air rearmament to match German strengths. Where Chamberlain as Prime Minister refused to unleash the air weapon, Churchill had no such scruples. Only five days after he became Prime Minister in 1940, on 15 May, he ordered the restrictions on British bombing lifted, and Britain embarked on a five-year campaign of air bombardment against Germany. It was Churchill's view, expressed in the bleak summer of 1940, that the only thing that would defeat Hitler was 'an absolutely devastating, exterminating attack by very heavy bombers upon the Nazi homeland'.[11]

For all Churchill's uncritical enthusiasm for air power, the whole strategy of the bombing offensive was shot through with paradoxes. At the time when Churchill promised Stalin his 'ruthless' campaign the whole issue of bombing was in the balance. Little had been achieved by the offensive between the summer of 1940 and the spring of 1942. Churchill told the Chief of Air Staff in March that bombing was probably 'better than doing nothing', but after all the extravagant expectations, which he had done much to fuel, it would clearly not be 'decisive'.[12] There was strong pressure from the other services in Britain and the United States to disperse the bombing forces to support the land and naval campaigns in the Pacific and the Mediterranean. In May 1942 a committee of inquiry into bombing accuracy headed by Judge John Singleton revealed that fewer than a quarter of the bombs dropped fell within 5 miles of the designated target, and only 30 per cent hit built-up areas.[13] Powerful voices began to urge a diversion of manpower and industrial capacity away from the unproductive bombing effort and towards building a large army which could provide, in the end, an effective Second Front. When Churchill made his promise to Stalin, as a means to sweeten the bitter pills he had prescribed, the effect was to lend bombing political support at a critical moment in its evolution.

The paradox of British bombing ran deeper than this. With the important exception of the United States, none of the other warring states saw much merit in long-range 'strategic' bombing. On Stalin's own initiative the heavy-bomber development programme in the Soviet Union was wound up in 1937, and the hapless designers, headed by Alexander Tupolev, shunted into a labour camp, where they continued their work from behind barbed wire. Experience in the Spanish Civil War, where Soviet pilots fought for the Republican side against Franco in 1936, had convinced Stalin that aircraft were best employed at the front-line, helping the army. German and French observers arrived at more or less the same conclusions. During the Spanish Civil War the German air force tried out in combat their new dive-bomber, the Junkers Ju-87, with its fearful siren that heralded the coming bomb. The aircraft was the brainchild of Colonel Ernst Udet, a larger-than-life aviator who thrilled cinema audiences in the 1920s with his flying stunts, and was a noted cartoonist. His support for Hitler's cause brought him the job of directing production for the German air force. His infatuation with aerial acrobatics led to the neglect of long-range bombardment in favour of smaller bombers that could dive down and destroy even small targets on the battlefield.[14] When the head of the German air force, Hermann Göring, insisted that Germany develop a long-range four-engined bomber as well, Udet told the German designers that it, too, had to be able to dive, an instruction that held up its development for three years.[15] When war broke out Germany gave priority to air force cooperation with the army, a strategy that persisted there and in the Soviet Union right to the war's end.

Why, then, did Britain and the United States fly in the face of conventional military wisdom and persist with bombing? At best the answer is a patchwork. Public opinion in both states was unusually susceptible to the science-fiction view of air power, first popularised by writers such as H.G. Wells, whose *War in the Air*, published in London as long ago as 1908, painted a lurid picture of 'German air fleets' destroying 'the whole fabric of civilisation'.[16] Wells was father to a whole generation of scaremongers, who traded on popular anxiety that bombing was somehow a uniquely unendurable experience. Writing in the American journal *Liberty* in

December 1931, the retired head of the Irish Free State air corps, Colonel James Fitzmaurice, offered 'An Expert's Prophetic Vision of the Super-Armageddon That May Destroy Civilization':

A Hideous shower of death and destruction falls screeching and screaming through space and atmosphere to the helpless, thickly populated earth below.

The shock of the hit is appalling. Great buildings totter and tumble in the dust like a mean and frail set of ninepins . . . The survivors, now merely demoralized masses of demented humanity, scatter caution to the winds. They are seized by a demoniacal frenzy of terror. They tear off their gas masks, soon absorb the poisonous fumes, and expire in horrible agony, cursing the fate that did not destroy them hurriedly and without warning in the first awful explosions.[17]

This was hardly a basis for military strategy. Yet British airmen and politicians stuck by the view in the inter-war years that indiscriminate bombing, to achieve a rapid 'knock-out blow', was likely to be the central feature of the next war. To counter the threat of bombing you needed bombers of your own. The roots of postwar deterrence reach back to the British decision in the 1930s to build a powerful striking force of bombers to frighten potential enemies into restraint.[18]

There were, to be sure, more solid reasons for pursuing bombing than fear of being its victim. No one wanted to repeat the awful bloodletting of the Great War. A bombing war, for all its manifold horrors, promised a quicker, cleaner conflict. Colonel Fitzmaurice comforted his readers with the reassurance that his 'Super-Armageddon' would end wars of attrition once and for all, ensuring they were 'buried securely in the mud, slush, and war cemeteries of Flanders'. In comparison with the wastage of young lives in the trench stalemate, bombing might bring a week of horror followed by surrender. Here was war on the cheap, saving not only lives, but money: an economical strategy to appeal to the democratic taxpayer and the parsimonious treasury alike.

All of this begged a very large question: what should the bombers bomb? In Germany and France, with politically powerful armies, the issue was easily resolved: aircraft attacked enemy aircraft, and when they had finished doing that they bombed enemy armies. This met the Clausewitzian test of war, the concentration of effort against the military forces of the enemy. But British and American airmen, much freer from the smothering constraints of old-established armies, wanted a strategy that gave them real independence, one that complemented the novelty and modernity of the weapon itself. They chose what became known as 'strategic' bombing, to distinguish it from mere 'tactical' bombing in support of armies and navies. The target of the strategic bomber was the very heart of the enemy state, its home population and economy. The Great War had opened the way to a new kind of conflict, total war, in which the distinction between civilian and soldier was said to have been eliminated. The bomber was the instrument of total war *par excellence*, capable of pulverising the industries of the enemy and terrorising the enemy population into surrender. Air strategists were fond of using the analogy of the human body. Bombing did not just weaken the military limbs of the victim, but destroyed the nerve centres, the heart, the brain, the arteries (lungs remained a curiously neglected metaphor) – the very ability of a nation to make war at all.[19]

Much of this was fantasy. In the 1920s the RAF, and in particular its overbearing, truculent commander, Sir Hugh Trenchard, clung to the quite unproven assertion that the moral effect of bombing was twenty times greater than the effect of material damage. It was an assumption that lingered on past Trenchard's retirement in 1930. Though air leaders in Britain, and later in the United States, hesitated to urge attacks on civilian populations for the sake of terror, the whole logic of independent, strategic bombing was to crush popular willingness and capacity to wage war. If there was not quite a straight line between Trenchard's dogmatic views on morale and the ghastly apotheosis of bombing at Hiroshima, the curve was barely noticeable. For those airmen with qualms about attacking civilians, there developed in the 1930s a second assumption, that bombers could precisely attack selected economic targets, whose destruction would dry up the supply of arms and drain away the material lifeblood of the enemy's armed forces. Whatever the justification, the object of attack was the home front and not the military shield in front of it. For all of the 1930s, and even in the early years of war, this strategy was technically unfeasible and operationally amateur. Nevertheless in December 1937 British Bomber Command was ordered to plan for the destruction of the German economy through bombing, and from that date the groundwork was laid for what became the Combined Bomber Offensive launched in 1943.

When British airmen came to think seriously about targets in Germany, they were faced with a bewildering choice. The Air Ministry drew up a broad list of objectives, the so-called Western Air Plans, which included in Plan W5 instructions for 'Attacking Enemy's Manufacturing Resources in the Ruhr, Rhineland and Saar'. The question still had to be resolved about which of these resources might prove the most strategically profitable. At first Bomber Command favoured the German electric power system; then during 1938 oil, chemicals, the metals industry, the engineering and armaments industry and transport were added to the list. Priority was given to electricity supply and oil in the area of the Ruhr, which was home to the greatest concentration of German heavy industry, and was the easiest to reach from air bases in southern England or France.[20]

This was whistling in the dark, and Bomber Command knew it. For all the exaggerated claims attached to bombing strategy, woefully little was done in the 1930s to prepare the bombing campaign. When war broke out in September 1939 Bomber Command had a mere 488 light-bombers, whose range was generally too poor to reach even the Ruhr from British bases, and whose bomb-loads were too small to do more than insignificant damage. There were no effective bomb-sights; there were few bombs bigger than 250 pounds; only a handful of bases in Britain could handle the larger aircraft; and there was even a severe shortage of maps for navigating in north-west Europe. Bombing trials betrayed a wide margin of inaccuracy even when bombing in bright sunlight from a few thousand feet with no enemy interference. Though there was better equipment in the pipeline, it was still years away. At the onset of war Bomber Command recognised that bombing attacks over Germany would be virtually suicidal, and reluctantly accepted the order to use bombers as the French and Germans did, in support of the battle on land.[21]

Even had Bomber Command been more prepared to attack Germany's industrial heartland, the government would have prevented it. In London, and in Paris, the politicians were terrified lest war should unleash Germany's 'knock-out blow'. It was what they had expected throughout the last years of peace, that Germany would launch a merciless assault with gas, germ bombs and explosives to paralyse western capitals and bring the war to a hasty, despairing conclusion. At the very moment that Chamberlain publicly declared war on the morning of 3 September, the air raid sirens sounded in London. People dived into ditches and trenches, scrambled to put on their gas masks, cowered in underground sanctuaries. It was a false alarm, triggered by the approach of two friendly aircraft (there were similar panics in Berlin and Paris), but it was symptomatic of the deeper fear that a foe as apparently ruthless and evil as Hitler would not hesitate to bomb civilians into surrender when it suited him. Chamberlain insisted that Bomber Command do nothing against Germany that might provoke retaliation.

Bomber Command might well have remained in limbo had it not been for the catastrophes that overtook the British war effort in the spring of 1940. The British defeat in Norway in April brought the fall of Chamberlain. In his place came a man much less scrupulous about the use of force. When the Germans broke through into northern France and threatened to inflict a disastrous defeat on Allied armies, Churchill searched desperately for anything that might halt the onslaught. Bomber Command, chafing at the bit, offered the prospect of hitting the German homeland, even perhaps of drawing the German air forces away from the land battle to defend Germany against the bombing threat. Churchill grasped at the straw. Following a devastating attack by the Luftwaffe on the Dutch port of Rotterdam on 15 May, Churchill ordered Bomber Command to begin a long-range offensive against German industrial and military targets in the Ruhr. On the night of 15-16 May a force of 96 twin-engined bombers was dispatched to destroy oil and power installations. The Ruhr was shrouded in its usual industrial haze and was thickly defended with anti-aircraft guns. Only 24 crews later claimed even to have found the target area, and six aircraft failed to return. It was an inauspicious baptism of fire.[22]

The early bombing of Germany demonstrated all the limitations of Bomber Command. Daylight raids were soon abandoned once it was found that some attacks suffered the loss of almost the entire force. Bombing by night, on the other hand, though safer for the crews, could only be attempted when it was clear and moonlit. Without the moon it was difficult to locate the area of the target, let alone the target itself, and without the clear night sky navigation by the stars was impossible. There were no radio navigation aids, no radar, not even effective bomb-sights. So inaccurate was British bombing that German intelligence had considerable difficulty in grasping exactly what strategy the British were pursuing. The offensive was continued only because, following the defeat of French and British forces in June 1940, there was no other way left for Britain to demonstrate to the world her willingness to continue fighting: 'We are hitting that man hard,' Churchill wired Roosevelt at the end of July.[23]

In fact Bomber Command was doing anything but. Moreover the pinprick attacks did exactly what Chamberlain had always feared: they provoked fierce German retaliation. Between 25 August and 4 September Bomber Command launched five attacks against the German capital following German bomber attacks on the outlying suburbs of London. Hitler pretended to be outraged, although the German air force had already been bombing military and economic targets in other British towns and cities for two months. On 4 September Hitler used the occasion of a speech in Berlin inaugurating wartime relief work to pledge revenge: '. . . we shall obliterate their cities! We shall put a stop to these night pirates, so help us God.'[24] Over the following six months the Luftwaffe subjected British cities to the first real independent bomber offensive, killing 40,000 people and destroying wide areas of London, Bristol, Coventry, Liverpool, Glasgow, and a dozen other urban centres. If there remained any moral scruples or strategic second thoughts on the British side about whether they should continue bombing, they were instantly dispelled by the Blitz. The bombing of Germany not only promised the one slim prospect of eventual victory in the absence of powerful allies, but it was also justified in the eyes of the British public as legitimate retaliation for German attacks.

With the coming of the Blitz the British got the war they had expected all along. The German attacks were costly and terrifying, but the Blitz was not the Super-Armageddon. When Gallup polled Londoners early in 1941 about what made them most depressed that winter, the weather came out ahead of bombing.[25] Bombing did little to dent British production, and if anything morale was stiffened rather than reduced. Hitler, who had started the Blitz in a rage at British 'terrorism', soon became disillusioned with city bombing. What the Blitz *did* achieve was the firm commitment of both British and American leaders to the idea that strategic bombing as a way of conducting war was here to stay. The bombing of London shocked world opinion in ways that the bombing of Berlin evidently could not. In thousands of American cinemas the newsreels showed the famous image of St Paul's cathedral, majestically, incredibly rising above the tide of flame that licked all round it. The American newsman Ed Murrow's poignant reports of ordinary British courage and tenacity kept alive western optimism at a dark moment of the war. The public perception of the Blitz in the United States perhaps did more than anything to switch American opinion to the view that bombing mattered, and that the United States should do it too.

During the course of 1941 Roosevelt and his military advisers began seriously to make ready for war. Roosevelt himself was the inspiration behind the planning of a bombing offensive as a central part of American preparation for war. On 4 May 1941 he ordered the production of five hundred heavy-bombers a month – British production was just 498 for the whole of 1941 – in order to achieve what he called 'command of the air by the democracies'. Money now flowed from the US Treasury to fund the development and production of a vast heavy-bomber force. It is still not entirely clear why Roosevelt, a man of peace and good-neighbourliness, who had long campaigned to get aerial bombing outlawed by international agreement, should have become so enthusiastically committed not just to air power but to its unlimited use against civilians. But there seems little doubt that this is what happened. His confidant Harry Hopkins reported in August 1941 that the President was 'a believer in bombing as the only means of gaining a victory'. Roosevelt told his Treasury Secretary, Henry Morgenthau, that 'the only way to break the German morale' was to bomb every small town, to bring war home to the ordinary German.[26]

Closer examination shows that Roosevelt had been influenced for some time by the scaremonger's view of bombing. At the time of Munich, and to the horror of his Cabinet colleagues, he cursed Chamberlain for not circling Germany with bombers and threatening to smash her cities if Hitler did not see sense. Roosevelt had a strong personal dislike of Germans, whom he regarded as arrogant bullies. And he deeply feared their scientific genius and lack of scruple. At another cabinet meeting, in October 1939, he scared his colleagues again with rumours that the Germans had invented a stratospheric bomber that could stay aloft for three days, strewing bombs over American cities, and a concussion bomb that could kill every living thing in Manhattan.[27] During 1940 and 1941 American army intelligence fed the President with greatly inflated figures of German aircraft output, which might well explain his enthusiasm for massive American striking power. Germany was believed to be producing 42,500 aircraft in 1941, including a force of twelve thousand long-range bombers, (the true figure was a mere 11,776 aircraft in total, and no truly long-range bomber).[28] With a frightening scenario before him of massed air attacks against America's eastern seaboard, and with a ready willingness to believe that Hitler would not hesitate to bomb America if he could, Roosevelt opted for massive retaliation.

Without the support of Roosevelt and other senior American politicians for a policy of bombing it is unlikely that it would ever have materialised as one of the key elements in the western war effort. Without Churchill's championship of 'a ceaseless and ever growing air bombardment', whose strategic necessity he regularly endorsed in his correspondence with the American President, Bomber Command would almost certainly have remained auxiliary to the activities of the army and navy.[29] Even with such powerful allies, the bombing offensive found the senior services snapping at its heels throughout the period of trial and error. Admirals and generals pleaded with Roosevelt and Churchill to release the bombers to help them overcome the armed power of the Axis at the battlefront. Without the urgent pressure of Soviet demands for a Second Front in 1942, which bombing was used to placate, even Churchill and Roosevelt might have given way. Strategic bombing emerged as a major commitment not from proven operational success but from political necessity. It was chosen by civilians to be used against civilians, in the teeth of strong military opposition.

*　　*　　*

In its early stages the bombing offensive faced a daunting learning curve. Poor technical resources, the long periods of training, the diversion of bomber crews to other pressing objectives, forced a slow build-up of the force. Though Bomber Command was anxious to paint its operations every shade of rose, it proved impossible to disguise the disquieting reality. In the summer of 1941 a British civil servant, D.M. Butt, produced a report for Churchill's scientific adviser, Lord Cherwell, on the accuracy of Bomber Command attacks on Germany. The resulting report was a devastating indictment of the Command. By examining photographs taken in night attacks over Germany, and extrapolating from the statistics of each operation, Butt concluded that only one aircraft in three got within 5 miles of its target, and in the Ruhr, whose smog-bound industries formed the Command's main target, the figure was a mere one in ten. On hazy nights, or when the moon was waning, the proportion fell to one in fifteen. Taking conditions as a whole the average proportion of aircraft that managed to make an attack within the 75 square miles surrounding a target was a mere 20 per cent.[30] The figures spoke for themselves. Even Churchill was deeply affected. Early in October 1941 he told the Chief of the Air Staff that he deprecated 'placing unbounded confidence in this means of attack', though he would stick by the strategy regardless.[31] Six months later the Singleton report confirmed these critical conclusions.

If Bomber Command was handled roughly at home, it met yet stiffer resistance over the skies of Germany. When the first night attacks were made against the Ruhr in May 1940 the German authorities were caught by surprise. Very little specific preparation had been made to meet such attacks, though Germany bristled with anti-aircraft guns. In June 1940 air force General Josef Kammhuber was given the job of constructing a system of air defences against British attacks which became popularly known as the Kammhuber Line. He began modestly enough with a squadron of heavy

Messerschmitt fighters, Me-110s, converted to a night-fighter role. They were dependent on whatever the searchlights could pick up in clear weather. German defences were transformed by the addition of radar in the autumn of that year. The large *Würzburg* radar units were set up, one in each of a series of grid 'boxes' drawn on a map, roughly 20 miles across. Within each box a night-fighter could be directed to intercept any bomber picked up on the screen. A central control room kept each fighter within its sector, circling round, waiting for its prey.[32]

The system worked so well that Kammhuber extended it to cover the whole area from Paris to the Danish coast. It was ideally suited to exploit the tactics of Bomber Command attacks. British aircraft approached the target singly, at intervals, giving the night fighters plenty of time to track and attack each one. Bomber Command experienced rising losses – 492 bombers in 1940, 1,034 in 1941. Not all of these losses came from fighter attack. German anti-aircraft guns accounted for a third of them. The manifold dangers of night-flying, often in cloud and with poor navigation equipment, took a depressing toll, even when aircraft had survived the German defences. On 14 July 1941 six aircraft set out for Hannover from Oakington airfield near Cambridge. They all succeeded in returning but found East Anglia shrouded in cloud. Only one made it back to base; another ran out of petrol near the coast, one crashed in the town centre of Northampton, and two more were damaged landing on other airfields.[33] By the end of the year losses were running at such a rate that Bomber Command was facing difficulty in replacing aircraft. It seldom had more than a hundred operational at any one time. By the last months of 1941 the bombing offensive was petering out.

The fortunes of the campaign were revived dramatically in the spring of 1942. It is usual to attribute this renaissance to the extraordinary impact of the man appointed as Commander-in-Chief of Bomber Command on 23 February that year, Air Marshal Arthur 'Bert' Harris. Though there are other explanations of a technical and tactical nature, there can be little doubt about the striking effect that Harris's arrival had on the force. He was fifty years old when he assumed command, the son of a civil servant in the India Service. He began his military career, improbably, as bugler in the 1st Rhodesian Regiment, established in 1914 with the onset of the Great War. He later joined the Royal Flying Corps, survived, and continued as a career airman between the wars. He was an enthusiast for bombing in the 1930s, helping to steer the plans for what became the first generation of heavy-bombers, the Wellington, Stirling and Halifax. At the outbreak of war he held the conviction that 'the surest way to win a war is to destroy the enemy's war potential', and that this could only be done with the heavy-bomber, concentrated in great numbers.[34] He had a clear grasp of the limitations of bombing, particularly as it had been conducted since 1940, combined with a resolute belief in its ultimate vindication. It was this quiet confidence that perhaps explains his effect on the Command, for he was not a flamboyant man. Very few flyers ever got to see him; he was reserved, even dour; he suffered fools not at all. From the moment he assumed office he doggedly defended bombing strategy from its detractors, while he patiently set about creating a force of sufficient size, operational effectiveness and technical sophistication to carry out the task for which he felt Bomber Command was best equipped, the destruction of Germany's industrial areas.

Much has been written about Harris's responsibility for the so-called 'area bombing', the practice of attacking whole cities rather than specific targets. After the war he carried the blame for launching a campaign of indiscriminate terror-bombing against Germany's urban areas in the pursuit of a strategic chimera, the critical fracturing of morale. Neither of these accusations carries conviction. The strategy of area bombing was already in place months before Harris took command. It was the product of operational reality. The inaccuracy of attacks on single factories or rail centres forced Bomber Command to adopt a policy designed to cause general disruption and the demoralisation of the factory workers. In May 1941 Lord Trenchard, Chief of Air Staff until 1930, sent Churchill a memorandum on how best to conduct air warfare, which reiterated his hoary dictum that morale was what mattered most, spiced up with the view that Germans were particularly prone to 'hysteria and panic'. The report was read approvingly by the service chiefs, and in July Bomber Command was directed to attack 'German Transportation and Morale'.[35] In

February 1942, a week before Harris's appointment, Bomber Command was formally directed to concentrate all its efforts 'on the morale of the enemy civil population'.[36] Harris, then, inherited a commitment to area bombing; he did not originate it. Neither did Harris have the slightest doubt that 'morale' was a hopelessly ill-thought-out objective, a 'counsel of despair', he later recalled. He had no confidence that German morale was as brittle as his colleagues hoped, and even if it was he doubted that any strategic good would come of its collapse, 'with the concentration camp round the corner'.[37] Harris stuck to his view that Germany's material ability to wage war was what counted, and this could be undermined only by heavy and persistent bombing of industrial centres – factories, transport facilities, supporting services and workers' housing. Demoralisation was, in Harris's view, a by-product of the war of attrition against the German economy.

This was the campaign that Harris waged. It involved civilian deaths, though few people, Harris included, would have found this as unacceptable in 1942 as it properly appears to the liberal conscience today. Indeed Harris's choice of targets and the new technology which at last became available in 1942 made a high level of civilian casualties unavoidable. Harris took up his appointment in Bomber Command at just the time that the new heavy-bombers, at whose conception he had played a key part, came of age. In February 1942 there were just 69; by the end of the year almost two thousand had been produced, including 178 of the new four-engined Lancaster, which became the mainstay of Bomber Command for the rest of the war. The 'heavies' made it possible to carry very much larger bomb-loads very much farther, and to concentrate a great weight of both high explosive and incendiary bombs against a major industrial centre. It was concentration of force that Harris wanted to achieve. Just before he assumed command a new radio navigation device, known by the codeword 'Gee', was introduced. It was one of a number of target-finding devices hastily developed by British scientists in the early years of war. Its range was short – barely as far as the Ruhr – and on the first raid when it was used, against Essen on the night of 8 March, the target was missed altogether. Gee did not improve bombing accuracy very greatly, but it did allow Harris to concentrate his force. Bombers no longer had to run the gauntlet of the Kammhuber Line in vulnerable isolation – with the help of Gee they swept across the German coast in a great stream, swamping German defences, and arriving, in most cases, all together over the target area.

Harris himself made a number of operational improvements. He insisted that bombers should carry a specially trained bomb-aimer to relieve the harassed navigator of this extra burden. He began to organise a group of elite crews – 'Pathfinders' – to lead the attacking force and to illuminate the target for the bombers that followed. But he realised that Bomber Command in 1942 was still in its infancy. He saw 1942 as an experimental year, until the force could be provided with large numbers of heavy aircraft and the more effective navigation aids which he knew were in the pipeline. Without this equipment Bomber Command had little chance of redeeming its tarnished reputation.

One great advantage Harris enjoyed over his predecessors was that in 1942 Bomber Command was no longer fighting alone. During the course of that year the United States began to organise its own air forces in Britain with the object of hitting hard at Germany as soon as the force was trained and assembled. This was easier said than done. It took months after Pearl Harbor to agree a strategy for Anglo-American forces. The initial aim of setting up a huge force of heavy-bombers had to be modified on account of the more urgent demands for equipment in the Pacific and North Africa. Not until June did General Carl Spaatz arrive in London to set up the 8th air force, and it was by then assumed that all the bombers would do would be to soften up Germany before ground forces stormed across the English Channel in the following spring. Spaatz was followed only slowly by men and machines. For much of 1942 and 1943 the 8th air force relied on British station equipment, radar and communications. The main bomber force had to be flown from the United States across the inhospitable Arctic, first to Goose Bay in northern Canada, then in small groups to Greenland, then Iceland and finally to northern Scotland. One-fifteenth of the force was lost in the icy waters and glacial landscape before it even reached Britain. By the end of August 1942 only 119 bombers had reached their destination.

The 8th air force, for all the painful delays in its establishment, was a blood transfusion for the bombing campaign. It was not just that American airmen, brash, confident, eager for combat, brought a breath of fresh air to a stale campaign. It was also that they brought with them the enthusiastic conviction that bombing, whatever its limitations hitherto, was the answer to destroying German power. Unlike the British, American airmen believed that the destruction of Germany's vital industries, her transport system and her fuel supply could be precisely calculated in terms of aircraft despatched and bomb tonnages dropped. The source of that conviction was the technology available. American bombing forces were composed largely of the Boeing B-17 'Flying Fortress' which had been developed as early as 1935. It was robust, had a high ceiling and long range and was so heavily armed that it was assumed that it could fly in daylight and defend itself effectively on the wing. Flying by day meant a better view of the target; the use of the Norden bomb-sight, which the American air force refused to pass on to the RAF, increased accuracy even more. Bombing tests in the United States demonstrated not quite the ability to hit a 'pickle-barrel' that some American airmen claimed, but were claimed to be more than satisfactory.[38] Armed with equipment for accurate, daylight bombardment, the 8th air force deliberately distinguished their practice of 'precision bombing' of individual targets from the more random efforts of the Royal Air Force. The choice of daylight tactics would prevent the two forces from treading on each other's toes: the RAF bombed industrial areas by night, the 8th air force would pick off factories and railway stations and oil depots by day.

In practice the American air forces seldom did anything of the sort. Though their learning curve was shorter than Bomber Command's, it had to be endured. American bombers launched their first attack in Europe on 17 August 1942, against the giant railway marshalling yard at Rouen, in northern France. Twelve aircraft took off, with the commanding general of 8th Air Force Bomber Command, Ira Eaker, along for the ride. They were heavily escorted by four squadrons of RAF Spitfire fighters. All the bombers reached the target, and approximately half the bombs fell within the target area. All twelve aircraft returned to base. The only casualties were the bombardier and navigator of one aircraft who were showered with glass when the nose of their plane was struck by a pigeon.[39]

For the rest of the year the 8th air force confined itself to attacks against French targets which were within range of American fighters. This was however a poor preparation for more distant attacks over Germany, where there could be no fighter cover, and where anti-aircraft artillery was much more densely concentrated. The issue of how well the B-17 could defend itself was hardly proven from the French attacks. On one occasion, the attack on Lille on 9 October 1942, the B-17 crews claimed to have destroyed up to 102 enemy fighters. Since this was 15 per cent of German air strength in western Europe, sceptical eyebrows were raised. A later report reduced the claims to 25 definitely destroyed. The number was in fact two. The spurious claims were used to vindicate the concept of the 'fighting' bomber, just as the occasional hit on the centre of the designated target was used to justify the choice of precision bombing. After four months of hard training, the 8th air force was keen to take the fight to Germany.

By the end of 1942 both bomber forces were poised to mount a much more dangerous and effective campaign than anything they had carried out hitherto. Yet even at this late stage, with large resources mortgaged to the endeavour, it was far from clear what strategy bombing was supposed to support. For much of the autumn and winter of 1942–3 the bombers were directed to attack targets to help in the Battle of the Atlantic, which was reaching its desperate finale. Wave after wave of heavy-bombers all but obliterated the French ports of Lorient and Brest where the German submarine pens were based, without once inflicting serious damage on the reinforced concrete structures that protected the U-boats.[40] And while the arguments continued about opening a Second Front in Europe, bombing was generally regarded by the military chiefs on both sides of the Atlantic as a mere adjunct to that greater enterprise, a means of weakening the enemy before invasion. The decision to embark on Torch rather than invade mainland Europe in the foreseeable future, left the bombing offensive in limbo once again, with no clearly defined objective.

The turning-point for the bombing effort came in January 1943 when Roosevelt and Churchill at last agreed to give priority to the bombing campaign. The decision was taken when the two men met at Casablanca, in Morocco, in January 1943. The meeting was called to decide the basic pattern of western strategy. The location was Roosevelt's doing. Churchill had travelled all over the world in his months of premiership. Against the advice of his staff the President wanted to match him, by making a trip of his own. Not only that, but he insisted on going by plane, the first time he had flown since becoming President in 1933. His personal adviser, Harry Hopkins, recalled that his boss 'was sick of people telling him it was dangerous to ride in airplanes'. He left Washington by train on 9 January, and flew from Miami in the famous Pan-Am Clipper flying boat, first to Trinidad, then to Brazil. He was so excited by the experience that 'he acted like a sixteen-year-old'. From Brazil there followed an exhausting eighteen-hour flight across the south Atlantic, arriving in Bathurst (now Banjul), capital of the Gambia. A crudely converted C-24 army transport took the presidential party the last seven hours of the trip, flying, on Roosevelt's insistence, high over the Atlas mountains. Once in Casablanca he was driven in a limousine, with windows blacked-out with mud for extra security, to a comfortable villa, 50 yards from Churchill's.[41]

At Casablanca the two leaders agreed to postpone the Second Front, almost certainly until 1944, and to continue with the Mediterranean strategy of attacking what Churchill called the 'soft underbelly' of the Axis 'crocodile'. To bridge the now widening time-lag between Torch and the invasion of northern Europe the two leaders agreed to intensify the air campaign with the 'heaviest possible bomber offensive against the German war effort'. Though neither leader expected bombing to bring Germany to defeat on its own, the task it was allocated at Casablanca went well beyond anything yet asked of it. On 21 January the Combined Chiefs-of-Staff issued a directive to the air forces to bring about 'the progressive destruction and dislocation of the German military, industrial and economic system, and the undermining of the morale of the German people to a point where their capacity for armed resistance is fatally weakened'.[42] The go-ahead was given to see what bombing could do; a high priority was given to the production of the tools for the job. The two leaders touched at Casablanca on the issue of atomic weapons, though no decision was taken, and precious little information was exchanged about the progress of what was now largely an American project. But Roosevelt had in prospect a weapon that made his call for the 'unconditional surrender' of the Axis powers, first publicly announced at the final press conference at Casablanca, appear more than mere rhetoric.

*   *   *

In postwar arguments about the morality of bombing a great deal has always been made of the unequal nature of the contest between bombers and defenceless civilians, and there is no disguising the fact that the collapse of German civilian morale was one of the objectives agreed when the Combined Bomber Offensive was launched at Casablanca. But the actual combat was not between bombers and ordinary people, but between bombers and the defensive forces of the enemy, the fighters and anti-aircraft fire. When the bombers streamed into Europe, their crews knew that they had to fight their way to the target and fight their way back. They sat in uncomfortable, surprisingly noisy, vulnerable aircraft, cold and cramped. They had to avoid heavy anti-aircraft fire from the massive concentration of fifty thousand anti-aircraft guns in the German defences; with the constant refinement of the Kammhuber Line and German fighter tactics they faced periods of brief but fierce engagement with an enemy who held all the advantages of timing and manoeuvrability. On the return flight poor weather, fuel shortages, or damage from anti-aircraft fire added to what hazards the enemy could still supply. If it now seems disingenuous of surviving crewmen to claim, as they often do, that they were too absorbed with battle against the elements and the enemy to think of those they were bombing, each bombing mission *was* a military confrontation with a statistically increased risk of death the more missions each crewman flew.

When the Combined Bombing Offensive began in 1943 the German forces had the upper hand. They had compelled RAF Bomber Command to fly at night; they exacted a high toll from the attacking forces; and after a few months even the advantages of Gee were lost when German defenders took counter-measures to jam the radio signals on which the system relied. The heavier Allied attacks became, the greater was the concentration of fighters on the German home front, and the more sophisticated was their equipment. By the spring of 1943 70 per cent of German fighters were in the western theatre of operation, leaving a much smaller force facing the Red Army. Night-fighters were equipped by 1943 with airborne radar known as '*Lichtenstein*', which protruded from the nose of the aircraft like stunted television aerials. The new equipment allowed them to hunt the bomber streams in large packs rather than in ones or twos. When British scientists found a way of jamming this radar, a new mechanism – 'SN2' – was introduced with even greater success.[43] Over the target area itself the German air force employed single-engined fighters at night, using a tactic known as '*Wilde Sau*' ('wild sow'). Attacking bombers were illuminated by searchlights and flares, and in the artificial pyrotechnic glare the fighters gunned down the silhouettes.

In practice German successes reflected the weaknesses of the attacking force. German defences provided at best a thin and brittle shield. Hitler was hostile as ever to the idea of defence for its own sake. He refused to release additional aircraft for the campaign against the bombers in the summer of 1943. He urged the air force Commander-in-Chief, Hermann Göring, to meet like with like, terror with terror. In the autumn of that year he ordered a fresh bombing assault on Britain, Operation 'Steinbock', in the belief, expressed to one of his generals in July 1943, that 'the English will stop only if their cities are knocked out, and for no other reason'.[44] The growth of a fighter defence force in the Reich, and the diversion of large quantities of anti-aircraft artillery to protect German industry, was achieved piecemeal, in defiance of the Führer's preferences. If the German air force had been able to prepare the defence of Germany more thoroughly and extensively the Combined Bomber Offensive might well have been stopped in its tracks.

As it was, the resumption of mass attacks on the Ruhr by Bomber Command in the spring of 1943 had only mixed results. The accuracy of the RAF's attack was improved greatly by the introduction of two new navigational aids, 'Oboe' and 'H$_2$S', during the course of 1943. The first was a target-finding device, with two fixed radio beams along which the special Pathfinder aircraft would fly to the target, where they dropped flares to guide in the bombers. Its range was limited to the nearest German targets because of the curvature of the earth. The second device was altogether more sophisticated. It was an airborne radar device which freed the bomber from reliance on ground control stations. By using radar pulses reflected from the ground below, H_2S permitted blind-flying in cloud and industrial haze. It was fitted in a small number of Pathfinder aircraft which led the bomber streams to the target. Both devices were employed from the beginning of 1943, but they took time to perfect. Not until 1944 did the radio and radar war swing firmly the Allies' way. What success Bomber Command enjoyed in its area attacks in 1943 came from the simplest of tactical innovations, the introduction of a device that jammed German radar almost completely, known by the nickname 'Window'.

Window consisted of huge quantities of small strips of aluminium foil dropped from bomb bays to smother German radar with false information. A few hundred strips simulated the radar image of a Lancaster bomber. Thousands of strips rendered radar unusable. The device was first dreamt up by a young government scientist R.V. Jones, but its employment was bitterly resisted on the grounds that its use not only compromised the whole radar research programme, but also would allow the Germans to retaliate in kind. The argument was only resolved at a meeting with Churchill in June 1943, in which it was finally agreed that the saving of Bomber Command lives outweighed the risk to British radar. A month later, Window was employed for the first time, with devastating effect, against the North Sea port of Hamburg.[45]

The attack on Hamburg on the night of 24–5 July, with the unfortunate codename Operation 'Gomorrah', was launched as German forces began the slow retreat from the carnage at Kursk. In the first attack 791 bombers were sent with instructions to jettison packets of Window at one-minute intervals. The defences were thrown into confusion. Only twelve aircraft were destroyed. Bomber Command returned with increasingly deadly loads on the 27th and the 29th, and their offensive was supported by attacks from the 8th air force by day. The overall loss rate was reduced to only 2.8 per cent, well below the losses sustained in other area attacks that year. But on the ground a terrible toll was exacted. The combination of high-explosive bombs and large numbers of incendiaries overwhelmed the emergency services in the city. The fires raged unchecked for two days. Gradually they merged together, creating a heat of greater and greater intensity. The fierce heat sucked in the surrounding air, working like giant bellows on the cinders of the city. The fiery storm devoured everything in its path. When the heat died away almost three-quarters of the city had been destroyed and forty thousand of its inhabitants consumed in the inferno. One million were rendered homeless. The glow from the fires could be seen 120 miles away. 'We all entertained the idea of an apocalypse,' wrote one eye-witness a few weeks later. 'The events of our time suggested it.'[46]

As the stream of wretched, incredulous refugees spread out into the neighbouring provinces, a widespread panic gripped the German public. Though the destruction of Hamburg was not the turning-point of the air war, not decisive in any general sense, it shocked the German leadership. 'It put the fear of God in me,' Albert Speer, Hitler's Armaments Minister, confessed in his memoirs. Three days after the attack he told Hitler that the destruction of another six cities on the same scale would bring armaments production 'to a total halt'.[47] The Luftwaffe Chief-of-Staff, Hans Jeschonnek, was deeply affected by a tragedy beside which 'Stalingrad was trifling'. Two weeks later, following a destructive attack by Bomber Command on the German rocket research station at Peenemünde, Jeschonnek shot himself.[48]

Speer's fears proved to be unfounded. There were few other major cities within range of the Oboe apparatus that had not already been attacked heavily that year. Moreover, despite the temporary shock produced by Window, which took the German night-fighters back to where they had been before radar, German forces quickly adapted to the new situation and once again began to inflict higher and higher losses on Allied bombers both by day and by night. When Bomber Command turned its attentions to the German capital from September onwards, its loss rates began to reach unacceptably high levels. In the raids in the early autumn, 14 per cent of the attacking force were destroyed or damaged. The slow attrition of Bomber Command strength was an expensive drain on production and trained crews. It took its toll on the morale of the Command. The long and dangerous circuit to Berlin and back gave the defending forces a much greater opportunity to find and intercept the bombers. The new H$_2$S navigation aid proved much less effective as a means to accurate bombing than had been expected, because the radar gave only a very general picture of the ground below. Out of 1,719 bombers sent to Berlin in the first wave of attacks only 27 dropped their bombs within 3 miles of the target centre.[49]

The fortunes of Bomber Command reached their lowest point in March of 1944. The last attack on Berlin on the night of 24 March cost the bomber force a loss of almost 10 per cent, and extensive anti-aircraft gun damage to the returning planes. The German defenders had adopted a new and lethal tactic. One or two controllers kept in contact with the entire night-fighter force which could then be directed as a whole towards the bombers once their route and target were visually confirmed. On some nights British radio jammers succeeded in talking over the German controllers and directing the fighters to the wrong town; on some occasions they played recordings of the Führer's speeches to apoplectic pilots, who were forced to circle round in an aerial no-man's-land. But such a ruse was very hit-and-miss. On the night of 30 March it failed entirely. Some 795 bombers were sent on an unusually direct route to bomb Nuremberg. Instead of the expected haze there was a clear moonlit night; the bombers left a conspicuous condensation trail. The German controller was able to call in night-fighters before the bomber stream had even reached the German frontier. On the long run towards Fulda, north-east of Frankfurt, the night-fighters fought a spectacular running battle with the bombers. In all 95 bombers, 11 per cent of the force, failed to return. The German authorities claimed their 'greatest success'. Harris gloomily contemplated what amounted to the defeat of his Command. In April he warned the Air Ministry that such casualty rates could not be sustained for long. He urged 'remedial action'.[50]

The fortunes of 8th air force in 1943 were no less chequered. For much of the year the force was still in the process of building up its organisation, manpower and bomber strength. Where Bomber

Command could routinely organise attacks with five or six hundred aircraft, most American attacks involved fewer than half this number. From June some of these attacks were at last directed against targets in Germany. The 8th air force stuck to its guns over bombing precision targets by day. It was planned to give priority to German air force industries and their chief suppliers, but in practice most attacks were directed at a wider range of targets in northern Europe where the force could enjoy fighter cover. When American bombers began to penetrate deep into German air-space they came up against the same problems afflicting their night-bombing partners.

The first such attack, against the Focke-Wulf aircraft plant at Oschersleben, south of Berlin, saw the loss of fifteen out of 39 bombers. On 17 August American bombers made the trip right across Germany to attack the Messerschmitt aircraft works at Regensburg and the ball-bearing plants at Schweinfurt, which produced half Germany's requirements of bearings, a component vital in aircraft manufacture. The force of 376 B-17s set off from England with a heavy escort of fighters which took them as far as the German border. The German fighter defences were alerted. A swarm of three hundred hovered over Frankfurt along the path of the bombers. There followed a ferocious air battle, the greatest yet seen over German soil, which left sixty B-17s destroyed (19 per cent of the attacking force) and over a hundred damaged, for the loss of 25 German fighters.[51] For the first time bombers were brought down by German air-to-air rocket fire.

Loss rates on such a scale were insupportable. Yet two months later, in the hope that repeated attacks would render the revival of ball-bearing production impossible, the daylight bombers returned. The Schweinfurt battle of 14 October 1943 was a victory for the Luftwaffe. Almost three hundred B-17s left Britain in two separate streams, 30 miles apart. The escorting fighters left them at Aachen, and almost immediately the German fighter forces pounced. Rocket fire, air bombing from above the bomber stream, and heavy cannon fire broke up the bomber formations. Then the fighters hunted in packs, waiting for the injured or straggling prey before moving in to the kill. They had perfected their tactics. Although 220 bombers managed to reach Schweinfurt, and to inflict heavy damage, the force lost sixty bombers and suffered damage to a further 138. These were catastrophic losses. The 8th air force suspended attacks into Germany. The Combined Bomber Offensive reached a critical impasse.

★   ★   ★

In the winter of 1943–4 the two bomber forces were compelled to rethink the campaign. Schweinfurt for the Americans, Berlin and Nuremberg for the British, showed the limitations of bombing, even though the defence had been stretched to the limit. It gradually dawned on the western forces that one thing, and one thing only, would secure the skies over Germany: the defeat of the German air force. They had ignored Clausewitz at their peril. Concentration of force against the armed forces of the enemy was not mere dogma; it was proved again and again in the harsh crucible of combat.

To be fair, American service chiefs saw this sooner than the British. When bombing plans were first drawn up in August 1941, American planners pointed to the Luftwaffe as an intermediate priority, whose destruction would permit uninterrupted assault on the economy. But British airmen had always dismissed attacks against an enemy air force as a waste of time. The targets were regarded as too dispersed and too small, and, as they had ably demonstrated in the Battle of Britain, recuperation from attack was relatively easy. When the 8th air force began operations in the summer of 1943, the Combined Bomber Offensive directive (codenamed 'Pointblank') placed the undermining of German air power as a top priority. American attacks that year did have some success in limiting the expansion of German fighter production; the attack on Schweinfurt on 14 October reduced ball-bearing output by 67 per cent.[52] But the defeats suffered in these attacks were a reflection of the complete failure to confront the direct sources of German air power, the fighters, pilots and airfields of the Luftwaffe.

The defeat of the German air force became a matter of urgency. Without that defeat the projected invasion of Europe in the late spring of 1944, Operation 'Overlord', would be placed in jeopardy. Without the bombing offensive, which was supposed to weaken

German resistance to invasion and re-conquest, the western Allies faced a long and bloody contest. There was no alternative but to defeat the Luftwaffe if western strategy was not to be thrown into disarray. On New Year's Day 1944 the Commander-in-Chief of the American air forces, General 'Hap' Arnold, sent the following to his commanders in Europe: 'My personal message to you – this is a MUST – is to, "*Destroy the Enemy Air Force wherever you find them in the air, on the ground and in the factories.*" '[53]

Here, crudely expressed, was the central issue of this zone of conflict. It was resolved not by some revolution in air warfare, but through a tactic of remarkable simplicity: the introduction of additional, disposable, fuel tanks on Allied fighters. It might well be asked with hindsight why it took the Allies so long to recognise that they needed fighters to fight German aircraft, and that their range could only be increased by giving them more fuel. The straight answer is that no one thought it was possible without sacrificing speed and manoeuvrability. It was a technical possibility quite early in the war, but neither the Allies nor the Germans saw any real gains in pursuing it. For too long it was assumed that heavily armed bombers could defend themselves. In fact the bombers were like unescorted merchant ships: to deliver their bomb cargoes they needed to be convoyed.

By good fortune more than by sound planning the convoy vessels were to hand. The standard American fighter aircraft, the P-38 Lightning and the P-47 Thunderbolt, had been fitted with extra fuel tanks under the wings when they were ferried long distances. By repeating this simple expedient the escort range of both fighters was increased from 500 miles to 2,000 miles. But the attention of the bombing commanders became focused on a hitherto neglected fighter, the P-51 Mustang. Built originally for the RAF in American factories, the marriage of an American airframe and the British Rolls-Royce 'Merlin' engine produced a fighter of exceptional endurance and capability, even when loaded with heavy supplies of extra petrol. The first group of adapted Mustangs joined the 8th air force in November 1943, and flew a mission to Kiel and back in December, the farthest any fighter had yet flown and fought in Europe. A crash production programme in American factories turned out enormous quantities in a matter of months. By

March 1944 the P-51, with a maximum armed range of 1,800 miles, flew with the bombers all the way to Berlin and back.[54]

The long-range escort fighter transformed the air war overnight. When Thunderbolt fighters were shot down over Aachen in September 1943 Göring refused to believe they could have flown that far, insisting instead that strong westerly winds had somehow blown the damaged aircraft to the east. But high and rising fighter losses in the last months of 1943 showed that the new threat was real enough. In November the Luftwaffe lost 21 per cent of the fighter force, in December 23 per cent. Production of fighters declined too, from a peak of 873 in July to 663 in December. German forces were faced with a vicious cycle of attrition from which there was no escape. American production escalated beyond anything the Germans had expected. Eighth air force fighter strength quadrupled in eight months. The Lightnings, Mustangs and Thunderbolts roamed at will over northern and central Germany, forcing the German fighters to unequal combat. By the spring the German fighter force was decimated. Half the fighters and a quarter of the pilots were lost each month.[55] Despite frantic efforts to reorganise fighter production the initiative lay firmly with the Allies. With a heavy heart the General of Fighters, Adolf Galland, reported to his superiors the irreversible crisis of German air power: 'The ratio in which we fight today is about 1 to 7. The standard of the Americans is extraordinarily high. The day fighters have lost more than 1000 aircraft during the last four months, among them our best officers. These gaps cannot be filled. Things have gone so far that the danger of a collapse of our arm exists.'[56]

The German air force never regained the initiative. The defeat of German air power in the skies over the Reich starved other theatres of desperately needed aircraft. The invasion in the west could be met with only three hundred aircraft against over twelve thousand Allied planes.[57] By April there were only five hundred single engine fighters left on the eastern front facing over thirteen thousand Soviet aircraft. If this were not enough, the bombing offensive was now renewed with fresh vigour. The target for the daylight bombers was the German aircraft industry. In a concentrated attack, later nicknamed 'Big Week', escorted bombers launched a devastating attack on the leading aircraft and aero-engine plants. The attacks did not halt aircraft production altogether, but they forced a desperate and improvised dispersal of the plants which put to an end any thought of producing five thousand fighters a month, as the German Air Ministry wanted. Most new aircraft were shot out of the skies in a matter of days, and nine thousand were destroyed on the ground by marauding US fighters. A temporary respite, while the bombers were withdrawn for the destruction of transport targets to support the invasion of Europe, ended when the bombers returned in May to hammer the German oil industry. In June the production of aviation fuel was reduced to a trickle, and the Luftwaffe from then on relied on accumulated stocks. The defeat of the German air force was now an accomplished fact.

During the last year of the war the bombing campaign came of age. With the thin veil of German air defence in tatters, the economy and population was at last exposed to the full fury of the bombers. Most of the bomb tonnage dropped by the two bomber forces was disbursed in the last year of the war – 1.18 million tons out of a total for the whole war of 1.42 million. These attacks did not go entirely unresisted. There were over fifty thousand heavy and light anti-aircraft guns at the peak, organised into heavy batteries, concentrated around the most important industrial targets. There remained an exiguous fighter force by day and night (including a few of the new jet fighters), whose pilots continued with almost reckless bravery to pit their tiny strength against the air armadas. The force now available to the Allies was one of over-whelming power. 'It is a truly awe-inspiring spectacle which confronts us,' wrote one German fighter pilot in his diary early in 1944. 'There are approximately 1,000 of the heavy bombers flying eastwards along a wide frontage with a strong fighter escort . . . Against them we are forty aircraft.'[58] A few months later he led his depleted squadron of five aircraft in a final do-or-die battle with sixty Mustangs and Thunderbolts and was shot down in minutes.

Over the last year of war the bombing of Germany was relentless. Both Bomber Command and the American air force finally had the aircraft they needed. Eighth air force was supplemented by 15th air force from the Mediterranean theatre, which could attack over the

Reich at long range. At the end of the war, in March 1945, the Americans had over seven thousand fighters and bombers on hand for the offensive. The RAF had over 1,500 heavy-bombers, which could carry bombs of up to 20,000 pounds in weight. This massive firepower was thrown against the industrial fabric and urban inhabitants of Germany. Oil supplies for Germany's war effort were critically reduced. Chemical production was emasculated, reducing the output of explosives by half by the end of the year. From the autumn, attacks were concentrated against transport targets, which could now be hit accurately by fighters and fighter-bombers flying unmolested in German air space. The railway system was fatally debilitated. By December 1944 the number of freight-car journeys was half that of the previous year, and only half the quantity of coal needed by German industry could be moved by rail.[59] Bombing gradually dismembered the economic body. By the winter of 1944–5 Germany was carved up into isolated economic regions, living off accumulated stocks, while frantic efforts were made to divert essential military production into caves and saltmines and vast, artificial, concrete caverns built, like the pyramids, by an army of wretched slaves. But by January 1945 the end was near. Albert Speer sent Hitler a memorandum on 30 January, the twelfth anniversary of the Nazi seizure of power, in which he declared that 'The war is over in the area of heavy industry and armaments . . . from now on the material preponderance of the enemy can no longer be compensated for by the bravery of our soldiers.'[60] By the time the Allies mounted their last bombing attacks late in March 1945, Bomber Command was back to flying by day, for the first time since 1940.

With so much power to inflict massive physical destruction it was difficult to resist the temptation to turn the bombing weapon against Japan. For long geography had been Japan's salvation. But by the spring of 1945 the American forces in the Pacific had finally secured island bases from which it was possible to hit Japanese cities. The air forces, now equipped with a new heavy-bomber, the B-29 Super-Fortress, were ordered to do what they had done in Europe, to soften up mainland Japan to ease the American conquest of the home islands. But against a flimsy air defence quite unprepared for the aerial onslaught the bombers were able to destroy almost at will.

In the end air power alone was able to bring Japan to the point of surrender.

In the six months between April and August 1945 21st Bomber Command, under the direction of General Curtis LeMay, devastated most of Japan's major cities. One by one the urban areas were ticked off as vast quantities of incendiary bombs were poured on to houses of wood, bamboo and paper. So poor were the defences that the B-29s could fly in at just 7,000 feet to release their bombs. Against the firestorms they caused there was little the emergency services could do. Using only a fraction of the bomb tonnage disgorged on Europe the Command destroyed 40 per cent of the built-up area of 66 cities. In desperate efforts to halt the spread of the conflagrations the Japanese authorities knocked down over half a million houses to form fire breaks. The terrified populations fled into the hills and the countryside. Over eight million refugees clogged the villages; in the factories absenteeism rose to 50 per cent. A combination of sea blockade and bombing reduced most Japanese industries to a mere fraction of their wartime peak – by July aluminium production was reduced to 9 per cent, oil refining and steel production to 15 per cent.[61] All except the most diehard militarists could see that Japan was defeated.

In Tokyo the politicians and generals squabbled over the terms of a possible surrender while American bombers remorselessly ate away at the very fabric of Japanese life. The equivocation cost Japan dear. At long last the Americans had not only a 'super-bomber' but also a 'super-bomb'. Science fiction became science fact. On 16 July in the heart of the New Mexico desert the first atomic device was successfully exploded. A second was shipped in a lead casket across the Pacific. Months before, LeMay had been told to reserve some Japanese cities for special treatment, and the fire-bombers left alone Kyoto, Hiroshima and Niigata. On 24 July the 20th air force was ordered to prepare for atomic attacks on two Japanese cities. Hiroshima was selected for the first, Niigata was rejected as too far north. For the second LeMay proposed Nagasaki, farther south down the coast, which he had not yet reached on his city hit-list.[62] America's new President, Harry Truman, gave his approval to destroy what he insisted on calling a 'military base', when everyone knew Hiroshima was an ordinary city, like so many of those already torched.[63]

For the inhabitants of Hiroshima their neglect by LeMay's bombers was welcome but unnerving. Superstitions quickly spread. It was widely believed that rubbing a pickled onion over the scalp, to symbolise a bombing, rendered immunity. The rumour spread through the city that President Truman's mother was Japanese and lived in seclusion in Hiroshima.[64] The reality was nightmarishly, almost absurdly different. On the morning of 6 August a single atomic bomb was loaded on to a B-29, christened the *Enola Gay* a few days before. As it approached Hiroshima the air-raid alarms began, but on the sight of only one aircraft the all-clear was sounded. A few minutes later a single bomb destroyed in seconds 50 per cent of the city and killed forty thousand people. Here was Super-Armageddon with a vengeance. Everything around the bomb was shrivelled to ash. The thick black clay tiles which covered most Japanese roofs boiled and bubbled over a mile from the explosion. Windows shattered 5 miles away. Demented survivors staggered into the suburbs, ghoulish, doomed. A young student observed them: 'Their faces were all burned and the meat on their faces was hanging down, the lymph dripping all over their bodies.' As for Hiroshima, 'everything – as far as the eye could reach – is a waste of ashes and ruin.'[65]

The Japanese authorities could scarcely believe what had happened. Frantic efforts were made to find a surrender formula, but too slowly to avoid a second, equally devastating attack on Nagasaki on 9 August. By the end of 1945 seventy thousand people, half the city's population, were dead. Even Truman was horrified by the results and called a halt to atomic attacks. Yet five days later, as the Emperor, Hirohito, prepared to broadcast the surrender of Japan, a thousand bombers bearing incendiaries inflicted a final, retributive, flourish. The bombers might have failed to bring Germany to defeat unaided, but the ruthless destruction of Japanese cities from the air made direct invasion here redundant. The war was over.

* * *

Did bombing help the Allies to win the war? The arguments began even before the war was over, when American and British technical intelligence teams scoured the bomb sites trying to decide what effect bombing had had on the enemy war effort. The air force commanders wanted the civilian investigators to confirm that if bombing had not quite won the war, it had at least made a major contribution to victory. The civilians, drawn in the main from academic or business backgrounds, were at best sceptical of air power claims, at worst hostile to bombing. Their concluding reports damned with faint praise: bombing had certainly contributed to undermining resistance in Germany in the last months of war, but until then it had done nothing to reverse the sharp upward trajectory of German production, and it had clearly not dented morale sufficiently to reduce production or produce revolution.[66] It was estimated that Germany lost only 17 per cent of its production in 1944, which could hardly be regarded as critical. The view has persisted ever since that bombing was a strategic liability, a wasteful diversion of resources that might more fruitfully have been used building tanks or laying down ships.

Bombing has also occasioned a chorus of moral disapproval. The Western allies killed over 800,000 civilians from the air, including 70,000 Frenchmen and 60,000 Italians in lesser-known campaigns.[67] Yet to understand what bombing achieved it is necessary, though agonisingly difficult, to lay aside the moral issue. Its strategic achievements are distinct from its ethical implications, however closely entwined they have become in the contemporary debate. Even on the strategic questions there is a great deal of confusion. There persists the myth that bombing was supposed to win the war on its own, and was thus a failure in its own terms. This was never the expectation of Allied leaders. Bombers were asked to contribute in a great many ways to what was always a combined endeavour of aircraft, ships and armies. The popular view of bombing has concentrated far too much on the narrow question of how much German production was affected by bomb destruction, at the expense of looking at bombing's other functions, and the other theatres of combat in which it was used. To give a more balanced answer to the question of how much bombing contributed to victory we must be clear about the nature of its achievement.

To take Allied strategy first. Bombing met some significant objectives of western engagement in war. Though the costs were far from negligible – over six years the death of 120,000 American and British airmen, and the loss of 21,000 bombers – bombing did reduce the overall level of western casualties in Europe and in the Far East, by weakening German resistance and by knocking Japan out of the war before invasion. Western losses were far lower than those of the other fighting powers. Bombing also permitted Britain and the United States to bring their considerable economic and scientific power to bear on the contest. The campaign was capital intensive, where the great struggle on the eastern front was based on military labour. This suited the preferences of the west, which did not want to place a much higher physical strain on their populations. For all the criticism directed at the waste of resources on bombing, the whole campaign absorbed, according to a postwar British survey, only 7 per cent of Britain's war effort.[68]

The effect of bombing on the Axis states was uniformly negative. About Japan there is little argument. Bombing of Japanese cities clinched victory and almost certainly shortened the war in the Far East. Of course the early naval victories were essential to turn the tide on Japan's embattled perimeter. But as the island-hopping became costlier for the American forces and Chinese resistance in Asia weakened in 1944, bombing recovered the initiative for the Allies and struck the awesome *coup de grâce*. In Italy, where heavy bombing began in the summer of 1943, there was almost no effective resistance, or adequate civil defence. Production in the northern industrial regions was cut by an estimated 60 per cent, largely due to the urban workforce sensibly decamping to the countryside. There is little dispute that Mussolini's fall from grace in July 1943, and the eventual surrender of Italy in October that year, owed a good deal to the effects of bomb attack.[69] Neither of Germany's junior partners had the means to obstruct bombing, and both paid the price. With Germany the case was different, and it was here that bombing mattered most.

The bombing offensive was for most of its course a fighting contest between the two western bomber forces and the German defences. From the middle of 1943 the defeat of the German air force became a central objective. Until that date German air power, deployed in the main as a tactical offensive arm, was a critical factor in German success on land and sea. The bombing offensive caused German military leaders to drain much needed air strength away from the main fighting fronts to protect the Reich, weakening German resistance in the Soviet Union and the Mediterranean. Though Stalin remained sceptical of Churchill's claim that bombing somehow constituted a Second Front, the facts show that German air power declined steadily on the eastern front during 1943 and 1944, when over two-thirds of German fighters were sucked into the contest with the bombers. By the end of 1943 there were 55,000 anti-aircraft guns to combat the air offensive – including 75 per cent of the famous 88-millimetre gun, which had doubled with such success as an anti-tank weapon on the eastern front. As the bombing war developed, the whole structure of the Luftwaffe was distorted. On the eastern front it was the bombers that had caused the damage to Soviet forces in 1941 and 1942. The shift to producing fighters reduced the German bombing threat over the battlefield. In 1942 over half the German combat aircraft produced were bombers; in 1944 the proportion was only 18 per cent. The German air threat at the battle of Kursk and in the long retreat that followed visibly melted. By compelling Germany to divide its air forces there were reductions in effectiveness on all fronts, which could not be reversed even by the most strenuous production effort.[70]

Once the Allies had the long-range fighter pouring out in numbers from America's industrial cornucopia, German air power could be blunted once and for all. The result was not a single, spectacular victory, but a slow and lethal erosion of fighting capability. For the Allies this was an essential outcome, if their re-entry to the Continent was not to face hazardous risk. For all Stalin's impatience, the western Allies knew the stunning effect on their war effort that a German victory on the invasion beaches would generate. Without the defeat or neutralisation of the German air force the Allies might well have hesitated to take the risk. Without the successful diversion of the heavy-bomber force to the job of pulverising roads, railways and bridges, in order to stifle German efforts to reinforce the anti-invasion front, D-Day might have failed at the first attempt. All of these factors, the defeat of the

German air force, the diversion of effort from the eastern front at a critical point in that struggle, the successful preliminaries to D-Day, belie the view that bombing was a strategy of squandered effort. It is difficult to think of anything else the Allies might have done with their manpower and resources that could have achieved this much at such comparatively low cost, which is why they accepted the unscrupulous arguments for pursuing a strategy at odds with their pre-war views on restricting air action.

Beyond these military gains lay the German economy and German morale. No one would argue with the view that for much of the war the bombing forces exaggerated the degree of direct physical destruction they were inflicting on German industry. Neither can it be denied that between 1941 and 1944, in step with the escalating bombardment, German military output trebled. The effect of bombing on the German economy was not to prevent a sustained increase in output, but to place a strict ceiling on that expansion. By the middle of the war, with the whole of continental Europe at her disposal, Germany was fast becoming an economic super-power. The harvest of destruction and disruption reaped by bomb attack, random and poorly planned as it often was, was sufficient to blunt German economic ambitions.

The effects of bombing on the economy were both direct and indirect. The direct physically reduced the quantity of weapons and equipment flowing from German factories; the indirect forced the diversion of resources to cope with bombing, resources which German industry could have turned into tanks, planes and guns. Direct effects were felt from both area bombing and precision bombing, as the city attacks hit water, gas and electricity supplies, cut railway lines, blocked roads, or destroyed smaller factories producing components. Much of this, it is true, could be made good within weeks, sometimes within days. But for the German manager in the last two years of war there were two battles to fight: a battle to increase production, and a battle against the endless inconveniences produced by bombing, the interruptions to work, the loss of supplies and raw materials, the fears of the workforce. Where the businessman in America or Britain could work away at the task of maximising output, German managers were forced to enter an uncomfortable battlefield in which they and their workers were unwitting targets. The stifling of industrial potential caused by bombing is inherently difficult to quantify, but it was well beyond the 10 per cent suggested by the postwar bombing survey, particularly in the cluster of war industries specifically under attack. At the end of January 1945 Albert Speer and his ministerial colleagues met in Berlin to sum up what bombing had done to production schedules for 1944. They found that Germany had produced 35 per cent fewer tanks than planned, 31 per cent fewer aircraft and 42 per cent fewer lorries as a result of bombing. The denial of these huge resources to German forces in 1944 fatally weakened their response to bombing and invasion, and eased the path of Allied armies.[71]

The indirect effects were more important still, for the bombing offensive forced the German economy to switch very large resources away from equipment for the fighting fronts, using them instead to combat the bombing threat. By 1944 one-third of all German artillery production consisted of anti-aircraft guns; the anti-aircraft effort absorbed 20 per cent of all ammunition produced, one-third of the output of the optical industry, and between half and two-thirds of the production of radar and signals equipment. As a result of this diversion, the German army and navy were desperately short of essential radar and communications equipment for other tasks. The bombing also ate into Germany's scarce manpower: by 1944 an estimated two million Germans were engaged in anti-aircraft defence, in repairing shattered factories and in generally cleaning up the destruction.[72] From the spring of that year frantic efforts were made to burrow underground, away from the bombing. Fantastic schemes were promoted which absorbed almost half of all industrial construction and close to half a million workers.[73] Of course, if German efforts to combat the bombing had succeeded the effort would not have been wasted. As it was the defences and repair teams did enough to keep production going until the autumn of 1944, but not enough to prevent the rapid erosion of German economic power thereafter, and not enough to prevent the massive redirection of economic effort from 1943. Bombing forced Germany to divide the economy between too many competing claims, none of which could, in the end, be satisfied. In the air over Germany, or on the fronts in Russia and

France, German forces lacked the weapons to finish the job. The combined effects of direct destruction and the diversion of resources denied German forces approximately half their battle-front weapons and equipment in 1944. It is difficult not to regard this margin as decisive.

The impact of bombing on morale is a different question altogether. The naive expectation that bombing would somehow produce a tidal wave of panic and disillusionment which would wash away popular support for war, and topple governments built on sand, was exposed as wishful thinking. Neither in Germany nor Japan did bombing provoke any serious backlash against the regime from those who suffered. But there can surely be little doubt that bombing was a uniquely demoralising experience. No one enjoyed being bombed. The recollections of its victims are unanimous in expressing feelings of panic, of fear, of dumb resignation. The chief ambition of ordinary Germans in the last years of war was survival, *das Überleben,* the desperate struggle to secure food and shelter, to cope with regular and prolonged cuts in gas and light, to keep awake by day after nights of huddling in cramped shelters.[74] The last thing on the minds of those living under the hail of bombs was political resistance. Not even the prospect of vengeance against the bombers could sustain morale.[75] Bombed populations developed an outlook both apathetic and self-centred; each night they hoped that if there had to be bombing, it would be on someone else.

If the real thing was not bad enough, the survivors were subjected to a second bombardment when the war was over, this time of surveys and questionnaires. Ordinary Germans and Japanese were selected for close interrogation by an army of American officials, anxious to learn at first hand what being bombed felt like. The Irish-American writer James Stern was among their number. He found the fatuous questions and the endless stream of tired, unhappy, defeated Germans almost more than he could bear. 'What do you do and say with all that Galluping nonsense on the table to be answered?' he wrote two years after the war, 'and across the table the forlorn life with nothing to live for . . .' Nevertheless Stern, and a host of others, went on to extract the answers.[76] They were in the main predictable. Many Japanese respondents placed bombing at the top of the list of factors that had made them doubt the possibility of victory (34 per cent of those polled). In a second poll on 'Reasons for Certainty that Japan Could Not Win' 47 per cent put bombing.[77] In Germany, 36 per cent of interviewees explained the decline of morale by the impact of bombing, and one out of three claimed that their personal morale was affected by bombing more than by any other single factor. When they were asked the more specific question 'What was the hardest thing for civilians during the war?', 91 per cent said bombing.[78]

The impact of bombing was profound. People became tired, highly strung and disinclined to take risks. Industrial efficiency was undermined by bombing workers and their housing. In Japan absenteeism from work rose to 50 per cent in the summer of 1945; in the Ford plant in Cologne, in the Ruhr, absenteeism rose to 25 per cent of the workforce for the whole of 1944. At the more distant BMW works in Munich the rate rose to one-fifth of the workforce by the summer of 1944. A loss of work-hours on this scale played havoc with production schedules. Even those who turned up for work were listless and anxious. 'One can't get used to the raids,' complained one respondent. 'I wished for an end. We all got nerves. We did not get enough sleep and were very tense. People fainted when they heard the first bomb drop.'[79] For the bombed cities the end of the war spelt relief from a routine of debilitating terror and arbitrary loss. No one could doubt who walked through the ghost towns of Germany and Japan, past the piles of rubble and twisted concrete, the rusting machines and torn up rails, miles and miles of burned-out houses, their few frightened inhabitants eking out a half-life in the cellars and ruined corners, no one could doubt but that bombing shattered civilian lives.

*   *   *

There has always seemed something fundamentally implausible about the contention of bombing's critics that dropping almost 2.5 million tons of bombs on tautly-stretched industrial systems and war-weary urban populations would not seriously weaken them. Germany and Japan had no special immunity. Japan's military economy was devoured in the flames; her population desperately longed for escape from bombing. German forces lost half of the weapons needed at the front, millions of workers absented themselves from work, and the economy gradually creaked almost to a halt. Bombing turned the whole of Germany, in Speer's words, into a 'gigantic front'. It was a front the Allies were determined to win; it absorbed huge resources on both sides. It was a battlefield in which only the infantry were missing. The final victory of the bombers in 1944 was, Speer concluded, 'the greatest lost battle on the German side . . .'[80] Though there should be necessary arguments over the morality or operational effectiveness of the bombing campaigns, the air offensive appears in fact as one of the decisive elements in explaining Allied victory.

5

ALONG A GOOD ROAD . . .
The Invasion of France

'We are going along a good road . . . The history
of war never witnessed such a grandiose operation.
Napoleon himself never attempted it. Hitler envisaged
it but was a fool for never having attempted it.'

Stalin to Averell Harriman, 10 June 1944

LATE IN MARCH 1942 a small convoy of ships steamed south from the Clyde estuary in Scotland destined for the invasion of French territory. Convoy WS17 carried two thousand Royal Marines and a wide assortment of naval and military supplies. The ships were a motley collection, small armed escort vessels swaying side by side with smart passenger liners crudely converted to the dull costume of war. Mercifully unattended by submarines, the convoy ploughed on, past the continent of Europe, past the Azores and on into the South Atlantic. On 19 April the convoy arrived in Cape Town where it met up with the rest of the invasion fleet, the aircraft carriers *Indomitable* and *Illustrious,* an ageing battleship and two cruisers. The 34 ships left for Durban, on the east coast of South Africa. In the last week of April they sailed in two separate groups to take part in Operation 'Ironclad', the invasion of Madagascar.

Their destination was the northernmost tip of the island, Cap d'Ambre, which was almost separated from the rest by a deep inlet that formed the natural harbour of Diego Suarez. This large sheltered anchorage was used by the French as a naval base. It was overlooked by the small port of Antsirane, where the French garrison and a handful of aircraft were stationed. This colonial backwater might well have remained untouched by the war save for the threat from Japan. Following the rapid Japanese conquest of south-east Asia and the East Indies it was feared that Japanese forces would fan out into the Indian Ocean, seizing Ceylon or Madagascar in order to cut the vital shipping lines that sustained Britain's fragile war effort in the Middle East and India. Madagascar suddenly became the key to British survival, and Churchill signalled his strong approval of its occupation.

Ironclad was a tricky operation. The island was protected in the north by natural fortifications of shoal and reef. The harbour itself was dominated by large naval guns set in coastal fortresses, and its long winding entrance was easy to defend. Armed with the element of surprise, the task force was detailed to land on the undefended western coast and attack the port from the rear. D-Day was fixed for 5 May – every operation had a D-Day and H-Hour to signal its beginning – and the flotilla arrived punctually off the coast at two in the morning. Minesweepers marked a channel through the treacherous waters and the small transport vessels gingerly steered past the buoys to reach the undefended beaches of Courrier and Ambararata bays. Three mines exploded in the approach but no one on the shore noticed. The landings were carried out unopposed, for the French regarded the western shore as unnavigable. Not until the marines had advanced 3 miles towards the port did they suddenly meet stiff resistance. Any hope that the defenders would come over to the Allied cause evaporated. For most of the following day British Commonwealth forces were pinned down with heavy casualties. The operation was rescued from disaster only by an act of desperation. The destroyer *Anthony* was sent with fifty marines aboard to run the gauntlet of the harbour guns and seize the port under the noses of the French forces. In darkness and in swirling seas *Anthony* rose to the occasion; the marines were disembarked on the jetty and seized the naval depot and the commanding general's house. Attacked from the rear the startled garrison began to crumble. By 3 a.m. on 7 May resistance was almost over. The port was surrendered. A brief naval bombardment the following morning silenced the harbour guns.[1]

The seizure of Diego Suarez effectively forestalled the Japanese. It was the first successful amphibious assault of the war for the western Allies, and the first genuinely combined operation, using aircraft, ships and soldiers working together. It came at a dark time in the war for the Allied cause and was, Churchill later recalled, the only bright spot in Britain's war effort 'for long months'.[2] But it was small comfort. The whole operation had come close to disaster. The ships supporting it had almost run out of fuel and water by 7 May; casualties were surprisingly heavy, 107 killed and 280 wounded, some 20 per cent of the attacking force; and contrary to expectations the French governor of the island not only refused to surrender but continued hostilities. The doughty Monsieur Masset retreated south with his forces leaving behind a trail of blown bridges and booby-trapped roads. He survived the fall of his capital in September. South African forces, depleted by illness and plagued by clouds of dry red dust, finally cornered the remnants of the French army in the very south of the island. Here the governor solemnly surrendered on 5 November, exactly six months and one minute after the onset of hostilities in May. Under French law the island's defenders were now entitled to higher pay and awards for enduring more than half a year of combat.[3]

It would be unkind to argue that Ironclad was the best the British could do in the summer of 1942, but it was not far short. For all those critics of British policy, then and since, the invasion of Madagascar is a salutary reminder of just how slender were British resources in 1942, and how inexperienced were its forces for a major amphibious assault of the kind that Stalin urged against German-held Europe. The British Chiefs-of-Staff were even hostile to an operation as modest as Ironclad because of the disruption to shipping. As it was, the Royal Naval task force at Gibraltar had to be severely depleted to support the Madagascar invasion. If the Japanese navy had chosen to intervene, Britain could have done little to obstruct it, and the whole operation would have produced a strategic nightmare.[4] How much greater were the risks and costs of a cross-Channel assault against a strongly defended coastline with limited resources. When later in the year a substantial raid was mounted on Dieppe by Canadian forces stationed in Britain the outcome was disastrous. Until the landings in North Africa later in the year, the invasion of Madagascar remained the one solid victory for the Allied cause, and it was fought not against hardened soldiers of the Axis but against an assortment of French colonial troops who had started the war as Britain's allies and had little stomach for the conflict.

The western Allies knew that at some point they would have to invade Europe and face their most dangerous enemy. But in the summer of 1942 they were not even sure they could save themselves against the onrush of Japan in the Pacific and Axis forces in North Africa. The choice of Madagascar was an admission of weakness, not strength. Ironclad disrupted the war effort elsewhere but was ultimately successful. The invasion of France in 1942 was operationally impossible. It took another two years before the secretive, hazardous assault on the beaches of Cap d'Ambre was writ large in Normandy.

* * *

From the outset the outlook for frontal assault on Hitler's Europe seemed singularly unpromising. Before Pearl Harbor the British had discounted it as beyond their means, but even with American participation from December 1941 delays were unavoidable. The training and equipping of an Anglo-American force large enough to be confident of remaining ashore in northern Europe was a time-consuming process. The air and naval back-up needed to carry the invaders to Europe and shield their lodgement there did not yet exist in 1942, and no amount of hustling improvisation could supply them. In 1943, with subsidiary demands in the Pacific and North Africa, existing forces were spread dangerously thin. Even with large trained armies invasion held all kinds of hazard. The English Channel was an unpredictable sea. Though only 20 miles wide at its narrowest point, it had defied all attempts at invasion from Europe since 1066, a fact so well known that its significance is sometimes overlooked. Beyond the sea lay an enemy with powerful coastal defences, good communications and a large army in waiting, seasoned with men battle-hardened from the fearful contest in Russia. Tactical advantage would lie with the defender, established in prepared positions to drive back an enemy confined to narrow beach fronts and with the sea behind him.

Worse still from the Allied point of view in 1942 were the unpredictable elements in their strategy, each of which might well have rendered invasion out of the question. It was essential that the front in the Soviet Union held. If it did not Germany would be able to swing large forces from the east to the west and make the northern French coast virtually impregnable. It was necessary for the western Allies to defeat the submarine and recapture the Atlantic sea-lanes for the shipping of troops and supplies from

North America. Finally it was vital to do something about the German air force whose close support of German ground forces had been, since 1939, a critical element in German military success. In the end the Allies were able to turn the tide in all three of these campaigns, but such an outcome could not have been taken for granted in 1942 or for much of 1943. Until the picture was clearer and more favourable, preparations for invasion had more the character of contingency plans than a firm commitment. The final decisions for invasion rested on the achievement of victory in the Battle of the Atlantic, in the Red Army campaigns of the summer and autumn of 1943, and on the attrition of German air forces over the Reich in the months before June 1944. Without them invasion was a gambler's throw.

It is only in the light of these many problems that sense can be made of the prolonged and often bitter arguments between the two western Allies about invasion strategy. Though united in a common cause, they were anything but united on how best to prosecute it. At times relations between the two partners were strained almost to breaking point. Their first battles were fought across the conference table. There was no question but that both sides wanted to defeat Germany. The passion with which advocates from each camp defended their strategic views was testament to their conviction that they held the surer key to German collapse, and that only the obduracy and *amour propre* of their colleagues was preventing general approval. The differences of opinion were sharply, but honestly, held.

From the start American military leaders and politicians were more enthusiastic about the invasion of northern Europe than the British. The American rearmament programme authorised by Roosevelt in June 1941, portentously named the Victory Programme, was based on the assumption that at some time western forces would need vast army and air equipment for re-entry to the European continent. Once the United States was at war Roosevelt's Chief-of-Staff, General George C. Marshall, predicated American war planning on a European invasion as the only sure means of bringing German forces to defeat. His head of War Plans, Colonel Dwight D. Eisenhower, who had spent much of his military career in the Pacific, was an early convert. 'We've got to go to Europe and fight . . . a land attack as soon as *possible* [*sic*],' he wrote in his diary in January 1942.[5] Marshall was all haste for an attack in July 1942 and Roosevelt reacted with sympathy. When it became clear that the material resources and trained men could not be procured in time, the American Chiefs-of-Staff agreed on a build-up of forces in Britain ('Bolero'), with a view to mounting a cross-Channel assault in the spring of 1943 ('Roundup'), or an emergency landing late in 1942 if the Soviet Front cracked ('Sledgehammer'). To this the British half-heartedly agreed.[6]

British leaders were always more cautious about invading across the Channel. They rejected the prospect of any serious attack in 1942. In July they succeeded in persuading Roosevelt that the invasion of North Africa would at least see American troops in action that year, and nothing the dismayed American Chiefs-of-Staff could do would dislodge their President's commitment. The British agreed to begin preliminary invasion planning, but not until April 1943 was a staff established with a clear brief to explore the prospect for a frontal assault on Hitler's Europe. A British general, Frederick Morgan, was put in charge of planning on behalf of a Supreme Allied Commander who had yet to be appointed. Morgan found the Americans 'all in favour', the British 'cautious'; he thought the risk of invasion 'appalling'.[7] British leaders favoured a more indirect route to the defeat of Hitler, through the Mediterranean and the bombing of the German homeland. They stressed flexibility, and regarded American fixation with invasion as a mortgage on the future development of the war.

During the course of 1943 the gap between the two sides grew wider. At the Casablanca conference in January 1943 no clear commitment to invasion was given. It remained one option among a number: Churchill was keen to exploit the Allied presence in North Africa by pressing into Italy and the Balkans; Roosevelt neither rejected nor confirmed this approach. Their respective staffs were left with no very clear impression of future plans. An invasion at some point in 1943 was not ruled out entirely, but both sides saw that with the current situation of forces heavily committed in North Africa and the Pacific, and with the submarine war at its most dangerous, invasion was a last resort. It was impossible to ignore the difficulties. The build-up in Britain of forces convoyed from America was much slower than anticipated. In January 1943 there were only 390,000 American servicemen in Britain, but well over half a million in the Pacific. By September that year the American army had 361,000 men in Britain, but 610,000 in the Mediterranean and over 700,000 fighting the Japanese.[8] The balance of forces was dictated by the circumstances of war, but it made invasion in 1943 unlikely.

The shortage of men was compounded with a serious lack of the means to transport them across the Channel. This could only be done in purpose-built landing craft: large landing ships to transport stores and troops to the invasion zones, and smaller landing craft to carry tanks, guns and men on to the invasion beaches. Neither ally had attached much significance to the production of landing craft, but during the course of 1943 the issue came to dominate the whole invasion plan. In the spring of 1943 most landing craft were in the Pacific supporting the island-hopping campaign. In the Atlantic theatre, the American navy had only eight converted merchant ships suitable for the task and the British just eighteen. The remaining craft were supporting the Allied forces in the Mediterranean. When the deficiency finally dawned on the Allies a crash programme of production was ordered from American shipyards. By April 1943 8,719 had been built, from 4,000-ton ships to carry eighteen tanks or 33 lorries, to small beach-craft of 7 tons that could move one lorry or 36 men on to the beaches.[9] Over the following year another 21,500 were procured from American shipyards, most of them smaller vessels. It was an extraordinary production achievement, but in the event it only just sufficed. All three areas of Anglo-American combat involved amphibious assault. Most landing craft went to the Pacific. When Morgan's planners did the shipping sums there was little left for the cross-Channel assault, but each month the attack was delayed more landing craft left the shipyards for the training waters along the St Lawrence river.[10]

The American military chiefs argued that more could have been done without the diversion of men and shipping to the Mediterranean, and this is no doubt the case. They deplored what one American general called a strategy of 'scatterization'. They recognised sooner than Roosevelt that the Mediterranean campaign once begun would be difficult to scale down. Having reluctantly agreed the first step it became progressively more difficult for them to argue against the proposal to attack first Sicily, then the Italian peninsula. But the effect of each step was to make the invasion of northern Europe a more distant prospect. The most to which the British would agree, when the two sides met at Washington in May 1943 for another round of strategic argument, was to develop a plan for possible cross-Channel invasion in May 1944. It was not a firm commitment, and American chiefs continued to doubt British good faith, but it did allow them to set in motion at last large-scale programmes of training and production with an invasion in mind. Morgan was now given something to do with his planning staff in London.

Over the summer months they laboured away, exploring the operational options. This generated heat as well as light. The two invasion sites chosen as the most likely to yield solid results were the Pas de Calais, the closest area to England, with the wide flat plain behind it pointed toward Germany's heartland, and the area from the mouth of the Seine to the Cotentin peninsula in Normandy. The second of these had a number of advantages: it provided some natural shelter from the Channel weather and tides; it was less well defended than the Calais region; and it contained the valuable port of Cherbourg, which was larger than the northern Channel ports. To Morgan's surprise his staff divided in fierce defence of the two options. He himself was not sure that the operation was feasible at all, whichever option was chosen. The argument broke down along national lines again. American planners favoured Normandy; British planners favoured Calais. The threatened stalemate was broken only by the intervention of Lord Louis Mountbatten, Chief of British Combined Operations. He invited Morgan's staff to his headquarters at Largs in Scotland. There, in two days of bitter wrangling, the invasion strategy took shape. The outcome was victory for the American view. When Morgan prepared the final version of the invasion plan to present to the next Anglo-American summit in Quebec in August it was based on a narrow assault on Normandy with three divisions in May of 1944.[11] Beyond this nothing was yet decided. America might have won a battle with its ally, but it had not yet won the war.

It had taken more than a year for invasion even to reach the planning stage. To Soviet leaders western hesitancy was viewed with deep suspicion. When Stalin was told in June 1943 that no invasion was possible that year he warned Roosevelt of the 'negative impression' the statement would make on Soviet opinion and he refused to entertain excuses.[12] What Stalin could not see was that the failure to invade came not from cowardice or bad faith, as he believed, but was the offspring of a sharp difference in strategic outlook between his two Allies. The contrast was in part a product of differing experience. For two years British forces in Europe had faced one defeat after another. They had been forcibly expelled from Norway, from France and then from Greece. These experiences bred an understandable caution. British leaders wanted to run no risks of further defeat. If an invasion were bloodily repulsed the effect on opinion at home and abroad might be disastrous; defeat would certainly jeopardise prospects for any second attempt. The circumstance of America's entry into the war raised quite different expectations. There was no humiliating retreat to redeem. American opinion stood six-guns blazing, eager for combat. American military leaders reacted with an instinctive belligerency. The contrast, according to Roosevelt's Secretary for War, Henry Stimson, was between a 'fatigued and defeatist government' on the one hand and 'a young and vigorous nation' on the other.[13]

The differences ran deeper than this, At stake were two ways of warfare that were difficult to reconcile. British preference was for war on the periphery, using naval power to develop a flexible and opportunistic strategy, as they had done in the wars against Napoleon. With a relatively weak army and limited economic resources, short winnable campaigns against light enemy forces made greater strategic sense. Britain hoped that Germany would be worn down by losses sustained elsewhere, so that British forces would never have to face the full weight of the German army on land. The watchwords were attrition and dispersion, 'scatterization'.[14] The American traditions were those of the Civil War, of Ulysses Grant and the massive rolling front. American soldiers looked for the main enemy force in order to concentrate all efforts upon its comprehensive destruction. This could be achieved, they believed, only by ceaseless and vigorous action by powerfully armed formations driving the enemy to a standstill: the bludgeon rather than the rapier. They found little merit in dividing forces unnecessarily. Where the British saw calculated attrition, American soldiers saw action that was piecemeal and indecisive. British arguments for a Mediterranean strategy reflected British priorities; cross-Channel invasion was the American way, and America had the resources to invade not once but, if necessary, twice.[15]

By the autumn of 1943 the whole issue of invasion still hung uncertainly in the air. The two allies had agreed to meet at Quebec in the middle of August to thrash out once again the direction of Allied strategy. The American delegation travelled to Canada determined to force a showdown. A few days before, Henry Stimson called on Roosevelt to present a letter in which he laid out the issue. The British could not be trusted to direct the invasion, he argued, for 'their hearts are not in it'. Instead America should assume strategic leadership; the Supreme Commander should be an American; and the British should be told that 'pinprick warfare' did not work and be made to accept the big battle.[16] Roosevelt agreed with it all. At Quebec the British delegates found their opponent more determined and better informed. The two sides sat down for their discussion almost literally face to face across an unusually narrow table. As the Americans expected, their ally tried to water down commitments to invasion. After three days of argument the British accepted the principle of a cross-Channel invasion, Operation Overlord, based on Morgan's final plan. But they refused to allow the words 'overriding priority' to be applied to the allocation of men and material to the campaign, preferring to use the more modest term 'main object'. They also insisted that invasion should only go ahead under certain important conditions: the German fighter force should be much reduced in strength first, and there should be no more than twelve German divisions in reserve to oppose the attacking force, Since they had won the main commitment, the American negotiators gave way on the rest. The British accepted Overlord with deep reservations, but they saw the conditions they had imposed as a final defensive line against excessive risk.[17]

No sooner had the ink dried on the Quebec agreements than the

British negotiators began to chip away at the idea of Overlord. In private the British deplored the enterprise. Churchill cursed 'this bloody second front!'; 'all this "Overlord" folly must be thrown "Overboard",' wrote Alexander Cadogan, head of the British Foreign Ministry, in his diary when bad news came from Italy in October. Field Marshal Sir Alan Brooke, Chief of the Imperial General Staff, told the British Chiefs-of-Staff in November that they should not regard Overlord as 'the pivot of our whole strategy'. Not for nothing did Stimson suspect that the British even after Quebec wanted nothing more than 'to stick a knife into the back of Overlord'.[18] Though British leaders could not openly flout publicly agreed strategy, they did everything they could to suggest other courses of action that could only have the effect of inhibiting plans for invasion. By October Churchill was once more all for planning adventures elsewhere, in the Balkans, in Yugoslavia, in Italy. By November, with the Allies bogged down on the Italian front, and little sign of German weakening elsewhere, the British began to argue that the putative conditions laid down at Quebec in August might not be met after all.

Whether or not the British would in the end have baulked at Overlord remains an open question. By late 1943 a great deal of planning and force preparation had already been carried out, and they risked a serious breach with a watchful ally, growing more confident of its power month by month. But in the end the decision was taken out of their hands. At the end of November the three Allied leaders agreed to meet at Teheran. Rather than argue any more with the British, American leaders planned to out-manoeuvre them. The two western Allies met first at Cairo to discuss issues from the Far East and, so the British expected, the Mediterranean. Relations between the two military staffs were poorer than ever. Brooke became uncharacteristically intemperate; Admiral King, commander of the American navy, came close on one occasion to striking him. But on issues to do with Overlord and the Mediterranean the Americans remained silent, leaving the floor to their ally. When pressed they replied that the issues would be discussed when they met with Stalin.

On 26 November the Presidential party flew to Teheran, followed a day later by Churchill's. Teheran was taken over by

Soviet troops. On every corner stood NKVD men in plain clothes, or in their distinctive dark blue and khaki uniforms. Security was tight. In the ride from the airport Roosevelt's place in the limousine was taken by an American security agent wearing the President's familiar hat and cape. So anxious were American agents for Roosevelt's safety that when the Soviet authorities announced the discovery of an assassination plot the President's party agreed to move to the Soviet Embassy compound. The British were housed next door in their Legation buildings. The close proximity of Soviet and American staffs gave some cause for concern because it was assumed that Roosevelt's rooms might be bugged. It is now known that Soviet security agents concealed the devices so carefully they were never detected; Stalin received a report every day, astonished at his ally's carelessness. 'It's bizarre,' he is reported to have said, 'They say everything in fullest detail!'[19]

More worrying for Brooke and Churchill was the failure to discuss in advance with the Americans any of the issues that were to be laid before Stalin. For once, a heavy cold had rendered Churchill almost speechless. In bad health and worse temper he attended the opening session of the three leaders on 28 November with foreboding. The three sat at a baize-covered table. There was no agenda, at Roosevelt's insistence. Before Churchill could say anything Roosevelt outlined Overlord. Stalin's reply was everything the Americans could have hoped for. He said that Overlord was essential to bring the German army to defeat: 'Make Overlord the basic operation for 1944.' When Churchill finally growled out an Aegean alternative it was too late. Stalin dismissed the whole idea as a wasteful diversion. The following day the Soviet–American alliance held the field. When Churchill tried to raise the Mediterranean again, Stalin stared directly across the table at him: 'Do the British really believe in Overlord?' Churchill was undone. He 'glowered, chomped on his cigar' and finally spat out that it was indeed 'the stern duty' of his country to invade.[20]

In a few hours of negotiation Stalin achieved what the Americans had failed to get in eighteen months of frustrating argument. Stalin liked Overlord; so too did Roosevelt. There was little Churchill could now do to keep other strategic options open. The following day a magnificent dinner was put on at the British Embassy for

Churchill's 69th birthday. This time it was Stalin's turn to feel discomfited. He was so ill-at-ease with the array of cutlery on the table before him that he asked the British interpreter to instruct him when each item should be used. Churchill was good-humoured at last, briefly the centre of attention.[21] Exaggerated toasts were drunk to the good health of each party. But for Churchill respite was short-lived. The following day Stalin insisted on a firm date for Overlord. The first of May was agreed. Then he demanded the immediate appointment of a commander for the operation as an indication of western good faith. In return he promised a Soviet offensive to coincide with the invasion. When the three leaders left Teheran they did so with a common strategy for the first time in the whole course of the conflict. Overlord was approved not on its strategic merits alone, but also to seal the alliance.

★　　★　　★

After two years of messy argument and uncertainty the final six months before invasion brimmed over with a bustling sense of purpose. The first step was to appoint commanders. The British had agreed at Quebec that when the time came the Supreme Commander should be an American. It was expected that Roosevelt would recommend the army Chief-of-Staff, General Marshall, who was anxious to prove that he was more than a desktop soldier. But the President realised at the last moment how much he relied on his military manager, and chose instead the man who had assumed the overall command of Allied efforts in the Mediterranean, General Eisenhower. He was a natural choice as the senior American general in Europe. After a year in the field he had much more experience than Marshall. He had a reputation as a good manager of men, a good chair for a committee. A tall, balding figure, Eisenhower ('Ike' to almost everyone) looked at 53 like a school headmaster in uniform – even more so when he donned his round-rimmed spectacles to read. Born in Denison, Texas, in 1890, the son of a failed storekeeper, his rise to supreme commander had much of the American Dream about it. With no money and a modest mid-West education behind him, he stumbled into an army career in which he quickly showed himself to be an energetic organiser. The First World War ended before he got to Europe. He swore to himself that he would 'make up for this', but he spent a fruitless twenty years stuck at the rank of major. There was nowhere to fight and little to fight with.[22] On the outbreak of war he was posted to the War Department to take over as Deputy for War Plans, but not until August 1942 did he get a field command, Supreme Commander for the Torch landings in North Africa. When he arrived in Africa in November to take up his command he had never seen armed combat. His talents were managerial. His inexperience was self-evident; Brooke complained in his diary that Eisenhower had 'absolutely no strategical outlook'. His strength was his ability to achieve 'good cooperation' from subordinates and allies alike. When Brooke later annotated his wartime journal he admitted that such a talent had been at a premium in preparing Overlord.[23]

It was a talent that was stretched to the limit in Eisenhower's relationship with the commander chosen by the British to storm Fortress Europe, General Bernard Montgomery. He was once again a natural choice. 'Monty' was the victor of the Alamein campaign which turned the tide in North Africa; he was enormously popular with the troops under his command and with the British public. Three years older than Eisenhower, he had had a fuller military career. The son of a clergyman, he followed a conventional path from public school to the British army academy at Sandhurst. In 1914 he was a lieutenant in the Royal Warwickshire Regiment. He saw fierce fighting on the Western Front, was severely wounded, returned to the front and ended the war as a divisional chief-of-staff with the rank of major; two years later he saw combat again, against Sinn Fein in the struggle for Irish independence. Between the wars he was a successful staff officer; when war broke out again he was a major-general. As with Eisenhower, real responsibility came only in 1942 when Churchill chose him to take over the 8th army in Egypt and turn back the Axis armies advancing on Suez.[24] He was a good organiser and a careful strategist. His bloody baptism of fire in 1914 taught him not to gamble with the lives of his men. He suffered fools not at all, and had little respect for rank and distinction. He believed that officers should get close to their men, but with fellow commanders he could be prickly and arrogant. He possessed a strong self-belief which he communicated to those below him, but it was a quality that made him intolerant of allies and colleagues where Eisenhower was a model of appeasement. The eventual success of their awkward partnership owed more to Eisenhower's self-restraint than it did to any diffidence on the part of Montgomery.

On one thing both men were agreed: the invasion plan drawn up in the summer of 1943 was not adequate for the task. Montgomery first saw the plan on New Year's Eve in Marrakesh where Churchill, whose cold had deteriorated into pneumonia, was convalescing. He spent the night reading it through and reported to the Prime Minister the following morning that the plan as it stood was impracticable. Instead of a narrow beachhead and three assault divisions Montgomery recommended a much broader front with a larger attacking force of five divisions, and a concentration of air power sufficient to win air supremacy over the lodgement area.[25] Eisenhower arrived at the same conclusion when he had leisure to study the plan in January. While the two men settled into new London headquarters – Montgomery at his old public school, St Paul's in West Kensington, Eisenhower at the Supreme Headquarters (SHAEF) on the outskirts of London at Bushy Park – they got down to the task of drawing up a firm plan. On 21 January the command team met and agreed to widen the front of attack from the mouth of the Seine to the east coast of the Cotentin peninsula, and to put ashore five divisions in the first assault phase instead of three. The aim was to capture the port of Cherbourg as soon as possible, and to seize the town of Caen in order to provide suitable bases from which to build up Allied air power. Once ashore it was planned to build up a secure lodgement using 37 divisions already stationed in Britain, until sufficient strength was ashore to break out from the bridgehead and push the German armies back across France.[26]

The decision to expand the invasion force made unavoidable a further postponement. The extra divisions and the wider front meant more landing craft and larger supplies. So narrow was the margin in the supply of vessels for amphibious assault that the necessary craft could only be secured by waiting for an extra month for the next consignment from American dockyards, and by abandoning the idea of a subsidiary assault on southern France which Stalin had warmly supported at Teheran. Instead of a pincer movement, the Allies had to settle for a frontal assault. The southern attack, codenamed 'Anvil', was put on ice, though not terminated. On 1 February the two staffs agreed on a date of 31 May for the invasion, subject to the state of the tides and the moon. These were small but critical considerations. There were only a small number of days in the early summer when adequate moonlight for the crossing coincided with low-tide at dawn to permit the landing craft to negotiate the many obstacles thrown across the beaches by the enemy. The ideal dates were easily calculated. The nearest to 31 May were the 5th, 6th and 7th of June, but the first of these dates provided optimum conditions. When in May Eisenhower had to make a final decision on the date he chose as D-Day 5 June, with H-Hour, the very moment of attack, at 5.58 in the morning.[27]

It was a plan of attack fraught with difficulty. An army of officials and officers, more than 350,000 men and women, laboured behind the scenes on routine issues of supply and recruitment. The organisation of training and troop deployment across 3,000 miles of ocean constituted a major feat of logistics. Dull though it seems on paper, the work of the long tail of non-combatants behind the Allied fist was vital to the success of the operation. Between January and June 1944 almost 9 million tons of supplies were shipped across the Atlantic, and some 800,000 troops, while almost four and a half million soldiers waited at bases in the United States.[28] But the critical issues remained those of operations rather than organisation. Allied leaders recognised that the margin between success and failure was slim indeed, and would rest in the end on three factors: the ability to maintain the momentum of supply to the Allied beachhead before capture of a major port on the French coast; the ability to restrict the build-up of German reserves for a powerful counter-thrust; and finally the necessity, forlorn though the prospect seemed at first, of keeping hidden from the enemy the direction and timing of the invasion.

The first of these was essentially a naval matter. Although the primary object of Overlord was to move a large army to fight in France, it relied for success on the movement of a vast armada of shipping, and the regular supply of seaborne troops and equipment for weeks after the initial landing. Indeed the whole campaign could only be contemplated by major naval powers. For all the attention lavished by historians on the land battle in Normandy, Overlord was a classic example of Admiral Mahan's famous dictum that the sea rules the land. The naval operation was given a different codename, 'Neptune', and was placed under the command of a British officer, Admiral Sir Bertram Ramsay. He had no illusions about what lay ahead: it was to be 'the largest and most complicated operation ever undertaken'. His task was to marshal and load almost seven thousand vessels with men and supplies, move them from around the British coastline to pre-arranged assembly points in the Channel, and then to shepherd them through marine pathways cleared of mines towards an enemy shoreline in unpredictable seas.[29] Thanks to the experience of four years of convoy planning, the movement of ships could be arranged almost like clockwork. This was a task for which British and American seamanship was well equipped. The difficult part of Ramsay's brief was to keep those supply lanes open day and night with nowhere for ships to sit at anchor secure from the elements. During the invasion itself eight convoys a day sailed for France. Even a captured port – Cherbourg, or Le Havre – would need time to be returned to working order following almost certain German demolition.

The solution arrived at seemed so fantastic that the German authorities never guessed it. The Allies brought their harbours with them across the Channel. The idea was the brainchild of a British naval officer, Commodore Hughes-Hallett, who in 1942 began to work on plans for the construction of artificial harbours. The scheme was not finally approved until the Quebec conference, when Allied staffs agreed to establish two harbours, codenamed 'Mulberries', off the French coast. Each one was constructed from long concrete sections 200 feet in length, which had to be towed by tug across the Channel and secured in place. These formed the harbour walls. Inside each were large floating steel structures designed to provide additional shelter, from which ran long piers and roadways to the shore. It was subsequently decided to set up a second breakwater to reduce the tidal strain on the Mulberries. These were known as 'Gooseberries', and were to be made *in situ* from the hulks of old ships. The 55 merchant and naval vessels chosen for sacrifice were gathered together in Scottish ports and sailed to France on D-Day, where they were scuttled in five different places. In all some four hundred separate pieces made up the artificial anchorage. Together they weighed 1.5 million tons; ten thousand men were employed on D-Day towing them to France and building the harbours.[30] Though in the event only one worked fully, enough supplies rolled across the 12 miles of artificial pier to keep Allied armies provisioned in the first critical weeks.

The second task, limiting the German build-up after D-Day, was more difficult. Great hope was placed upon strategic deception, which was orchestrated in ways to keep alive Allied threats to the Balkans and Norway, and to disguise the exact landing point in France. The success of these efforts, as we shall see, kept German forces dispersed around Fortress Europe well after invasion had begun. But such deceits could not stop the immediate reinforcement of the invasion area. For this the Allies turned to air power. Eisenhower was convinced that the only sure way to retard the German build-up was to damage the French communications system so severely that troop movements that should take a matter of hours would take days instead. In January 1944 his deputy commander, the British Air Marshal Tedder, and the British scientist Solly Zuckerman, drew up a plan to attack the railway system in France for ninety days before invasion, hitting over a hundred marshalling and repair yards before turning to the coastal areas and the destruction of German air defences. Straightforward enough on paper, the plan turned out to be anything but straightforward in practice.[31]

Eisenhower had reckoned without the self-interest of all those parties with a stake in the bombing effort. Spaatz and Harris, who commanded the American and British heavy-bombers, were united in their hostility to the transport plan. Some of the attacks on rail targets were the responsibility of the fighter-bombers and light-bombers of the tactical air forces assigned to Overlord, but Eisenhower had no doubt that the heavy-bombers were essential to the success of the operation. The problem was not one of feasibility. Spaatz thought the Overlord attacks were 'child's play' compared with strategic bombing; Harris had recently introduced tactical changes in bomb attack which greatly increased the accuracy of night-time bombing of limited targets.[32] Precision was less of an issue though thousands of French people died in the attacks. The problem was one of strategic priorities. The bomber commanders wished to continue attacks on the German aircraft industry and oil supply in the belief that this would weaken German resistance to invasion more surely than attacks on railway lines. They regarded the Overlord campaign as an unnecessary diversion of effort when bombing was on the point of achieving decisive, possibly war-winning results. They won powerful allies, including Churchill and Brooke. Eisenhower's calm imperturbability was, for once, thoroughly punctured. By March he was all for resigning rather than battle any longer with the 'prima donnas': 'I will tell the Prime Minister to get someone else to run this damn war. I'll quit!'[33]

But there was worse to come. Not only would Harris and Spaatz not accept the strategy for Overlord, they also refused point-blank to relinquish command of the heavy-bombers to Eisenhower's air commander, Air Marshal Trafford Leigh-Mallory. Again they had Churchill's backing, a spoiling action that Eisenhower deeply resented. Early in March he squabbled so violently with the Prime Minister over command of the heavy-bombers that he finally threatened 'to go home'.[34] The threat was enough. On the issue of command a compromise was reached. Leigh-Mallory was excluded, but for operations in support of Overlord Eisenhower and Tedder had temporary command of the heavy-bombers. This still left unresolved the more serious issue of what to bomb. Eisenhower's headquarters were bombarded with objections to the transport plan throughout March. Spaatz told his staff that Overlord was doomed. On 25 March Eisenhower called all his critics together for a final decision, almost at the end of his tether. The result was another compromise. The transportation plan was accepted, while the bomber forces were allowed to continue attacks against oil installations and the aircraft industry when the opportunity arose. Honour appeared satisfied.[35]

Eisenhower had reckoned without Churchill. No sooner was the strategic issue settled than the Prime Minister refused to sanction a campaign that might kill '100,000 Frenchmen' and 'smear the good name of the Royal Air Force around the world'. There was deadlock. By early May, only five weeks before invasion, no final decision had been taken; Churchill wanted attacks restricted to targets where fewer than a hundred Frenchmen would be killed. Only under pressure from Roosevelt did he concede on 11 May, with scarcely time to implement the air plan that Eisenhower regarded as critical to the whole success of Overlord. Fortunately the bombers had been at work since March attacking railway targets in the Ruhr and north-eastern France, killing Frenchmen without formal approval When the Free French General Koenig was asked for his view he replied laconically: *'C'est la guerre.'* During May attacks were stepped up. By early June rail traffic was down to just over one-third of the amount in January.[36] French casualties numbered at least ten thousand.

The results of the pre-invasion bombing campaign were mixed. Civilian rail traffic declined sharply, but German mobile repair teams were able to maintain the operation of rail transport for military priorities. Real success came in the final campaign of what was called interdiction, the destruction of bridges and tunnels connecting the invasion area with the east. There was even argument about this until, on 7 May, eight American P-47 fighter-bombers dropped two 1,000-pound bombs each on the railway bridge across the Seine at Vernon, proving beyond doubt that these targets could be destroyed with pin-point accuracy and few bombs. Over the next three weeks 74 bridges and tunnels were destroyed, effectively isolating the whole of north-western France, and fatally curtailing the prospects of rapid German reinforcement.[37] There was in this a profound irony, for the most successful part of the whole transport plan was executed not by the heavy-bombers, which had plastered 72,000 tons of bombs on the hapless French railways, but by the small battlefield bombers under Leigh-Mallory's command, whose 4,400 tons of bombs achieved all of what Eisenhower wanted, in a matter of days. The most important contribution of the heavy-bombers, as Spaatz maintained all along, was the progressive destruction of the German air force over the Reich, which gave the Allies overwhelming air supremacy throughout the invasion.[38]

None of these efforts would have availed a great deal if the

German Supreme Command had known well in advance where and when the Allies would strike, and disposed their forces accordingly. The element of surprise was critical. And yet an operation designed to move 4,000 ships, 2 million men and 12,000 aircraft to France, from a base only a few minutes' flying time from German airfields, appeared an impossible secret to keep for six long months. The fact that German leaders failed to guess the main invasion target and the date of invasion owed something to good fortune; but it also owed a good deal to an elaborate web of misinformation and deceit spun by Allied intelligence.

Deception was the instrument of surprise. The British had begun to use it extensively in the Middle East, with mounting success, and it was their growing experience in the techniques and tactics of deception that persuaded them, and eventually their sceptical American allies, that it could be used on a grand scale for Overlord. There was no chance that preparations for invasion could be disguised. Even the most casual intelligence would indicate a massive build-up of men and material in southern England during the first half of 1944. The intelligence staffs instead chose to mislead the Germans about the destination of the invasion forces, and to conceal the date of attack. By this stage of the war the initiative lay very much with the Allies. All German spies in Britain had been rounded up, and many of them acted as double agents, purveying to their credulous German controllers a diet of innocuous, but true, information, spiced with plausible falsehoods, all of it supplied by the British secret service. The effectiveness of the Double-Cross System, as it became known, relied on British access to the German Enigma cryptographic communications (Ultra), which confirmed the success or failure of the campaign of misinformation. By 1944 German secret service codes were routinely broken. By contrast, German intelligence on Britain was poor. The high level codes for Allied communication proved impenetrable; and air reconnaissance remained scrappy and unsystematic, thanks to a wall of radar and fighter defence.[39]

The deception plan, codenamed 'Bodyguard', was drawn up in December 1943 and finally approved by the Allied staffs late in January 1944. The core of the plot was to persuade the German leadership that the Allies intended to attack across the narrow neck of the English Channel, between Dover and Calais (codenamed 'Fortitude South'), with a diversionary attack against Scandinavia ('Fortitude North'). The object was to tie down German forces in northern Europe, and to keep the bulk of German troops in France and Belgium guarding the Pas de Calais and the Belgian and Dutch coastline. The Scandinavian deception had limited success. Hitler had been obsessed for years with the idea of an Allied landing in Norway. The mass of false information about a British 'Fourth Army' in Scotland played on this fear but did not succeed in enlarging it. Twelve divisions remained in Norway, but they were not supplemented by reinforcements taken from France. German radio operators were tuned in to the Soviet front, so that they missed the phantom radio traffic beamed from Scotland to sustain the illusion of a northern invasion.[40] The Channel deception was far more successful. It hinged on persuading German intelligence that large Allied formations existed in south-east England for an invasion towards Calais, while concealing the size of the forces and shipping concentrated in the south-west. Credentials for an entire force, one million strong, christened First US Army Group (FUSAG) were fed, piece by piece, into the German intelligence system, where they lodged, until well past the real invasion, as actual fact. Nationally stationed across south-east England and East Anglia, FUSAG was from first to last a figment of the deceivers' imagination.[41]

The creation of FUSAG was a work of artistic mendacity. Across the whole of south-eastern England sprang up dummy camps and supply depots, false headquarters, tanks made of rubber (fashioned by set designers from the Shepperton film studios), landing craft of wood and fabric, and airfields complete with concealed lighting to give an air of authenticity to the ruse. Lights blazed and camp stoves smoked, visible to any intruding enemy aircraft. But west of a line drawn from the south-coast city of Portsmouth every effort was made at concealment: tents were darkened, food was dispensed from smokeless stoves, army issue white towels were replaced with khaki.[42] To confirm the visible falsehoods, the double agents radioed details of unit insignia and FUSAG organisation to their German contacts. Eisenhower chose one of his most successful commanders, the profanely outspoken General George S. Patton, to command FUSAG. Patton fumed at the inactivity this imposed upon him, but the appointment of a commander of this quality added inestimable weight to the whole FUSAG deception. Finally, every effort was made to get German intelligence to inflate their estimates of Allied forces in order to include the false divisions. In the United States the FBI rounded up an elderly Dutch spy recruited by the German secret service, who obligingly handed over the German codes and instructions, and was operated as a double agent. Under the codename 'Albert van Loop' he relayed regular information about units sailing for England, including those contributing to FUSAG. Such a ploy was made possible by the American system of numbering army divisions. The regular army was listed from 1 to 25; National Guard units absorbed into the army from 26 to 75; reserve units from 76 upwards. By 1943 only numbers 1 to 45 and 76 to 106 had been used, leaving ample room to feed false division numbers into the Allied order of battle without arousing suspicion.[43] By January 1944 German military intelligence in the west believed there to be 55 divisions in Britain, when there were still only 37; by May 1944 they had identified 79, when in fact there were only 47.[44] The German 15th army, facing the Calais coast, stayed put to fight thirty invisible divisions.

The success of the FUSAG deception, as its perpetrators always realised, relied on the extent to which it reinforced preconceptions in the mind of the enemy. Hitler was convinced that the Allies would land somewhere on the northern French coast; he was at one with his generals in assuming that the Allies would come by the shortest route, to provide better air cover, to make early detection of the invasion fleet more difficult, and to bring Allied forces quickly within striking distance of the Ruhr and the heart of German resistance. Hitler thought like an army commander; the possibilities open to a naval power were foreign to his outlook. The FUSAG deception fed a conviction already forming that the Allied plan was to invade the Pas de Calais. Nevertheless by the spring of 1944 there was unmistakable evidence of large troop concentrations and exercises in south-west England. Hitler came round to the view that a subsidiary or diversionary landing was planned for Normandy, and he instigated moves to strengthen the coastal defences there. The idea of Normandy as a secondary target actually strengthened the deception, by fixing German eyes on the Pas de Calais as the main invasion site long after the Normandy beaches had been stormed. The campaign of misinformation did just enough to prevent German sources ever discovering the focal point of invasion or the precise timing. Had they been able to do so the concentration of forces to oppose Allied landings would have made Overlord too dangerous to attempt.

By May of 1944 most of Eisenhower's conditions had been met: German forces were dispersed along the whole northern and north-western coastline; reinforcement of Normandy by German armies was likely to be a lengthy and cumbersome process; behind the Mulberries and Gooseberries, the Allied rate of supply could more than match that of the enemy. He still regarded Overlord as hazardous, but now possible.[45]

★　★　★

Hitler had no illusions about what was at stake in 1944. Defeat of the Allied invasion would amount to 'a turning-point of the war', a psychological shock from which British and American opinion would not recover. Victory would free German forces for a renewed offensive on the eastern front.[46] By the end of 1943 he was sure 'the attack will come' and that it would 'decide the war'. On 3 November he published the last of his numbered war directives, No. 51, on the war in the west. His strategic solution was simple: once invasion started all German forces were to concentrate in one major counter-blow to throw the enemy 'back into the sea'.[47]

This was a campaign long awaited. Work had begun in 1942 on strengthening the coastal defences of northern France, the so-called Atlantic Wall. Hitler himself designed the pill-boxes and concrete casements, which were faithfully reproduced by the Corps of Engineers. An army of conscript labour used up 17 million cubic yards of concrete and 1.5 million tons of iron building it. Along the coast Hitler planned fifteen thousand strongpoints, bristling with machine guns and flamethrowers; every port was to be defended by huge naval guns. But the project was too large for German resources. Most of the defensive wall was concentrated between the river Seine and the river Schelde, where the invasion was generally expected. The rest of the coastline took what was left over. The same went for the men defending it. In late 1943 the Commander-in-Chief West, Field Marshal Gerd von Rundstedt, had in all some 46 divisions under his command. They were spread thinly along the French coast: the area of north-eastern France had one division for each 50 miles, but the Normandy front had one for every 120 miles, and the rest of the coast one for every 217 miles. Many of these divisions were largely made up of older soldiers – a quarter of the German army was over 34 by 1944 – and some of the units manning the defences, composed of wounded from the eastern front and men of poorer physical condition, had low combat effectiveness. One whole division on the Normandy sector was made up of men with stomach complaints; other units were methodically constructed from men with lung or ear conditions. Most of the combat divisions had been reduced to eleven thousand men from the conventional 17,200 of the early campaigns.

Once it was clear that invasion could be expected at some time in the spring or summer of 1944, great efforts were made to make good these deficiencies. To keep up numbers large contingents of volunteer battalions from the occupied part of the Soviet Union – Tatars, Turcomen, Cossacks, Georgians – were drafted in. By the time of the invasion the German 7th army deployed in Normandy had a sixth of its numbers recruited from such units.[48] On 15 January Hitler appointed Field Marshal Erwin Rommel, a flamboyant tank commander who grew into a popular legend with the success of the Afrika Corps in 1941–2, as commander of the forces facing the invasion threat, designated Army Group B. A man of great energy, impetuous, brave to the point of foolhardiness, Rommel was a particular favourite with Hitler. His choice as commander of the invasion front had a political edge to it. Hitler wanted to be able to exercise a close watch on the new campaign preparations, and saw his relationship with Rommel as a way of by-passing the Commander-in-Chief in Paris. Rommel shared his master's view that the decisive moments in the invasion would be the first hours on the beach. Unusually for a man who had made his reputation as the master of mobile warfare, he favoured a water-line battle from static defences. This meant a considerable strengthening of the Atlantic Wall. He ordered the construction of armed concrete bunkers, or 'resistance nests', along the coast; he demanded fifty million mines along the shore, and heavy mining of the coastal seas; he accelerated the erection of obstacles of all kinds on the beaches to obstruct a landing at high tide. Some of this was certainly achieved, but mostly in the area already more strongly defended, in north-east France. The three new minefields laid at sea were all along this stretch of coast. The resistance nests were neglected farther west; only one out of a planned 42 was built on the Cotentin coast. And instead of fifty million mines, he got only one-tenth this number.[49]

To man these defences Rommel had two armies: in the Normandy and Brittany area 7th army under General Dollmann; from Le Havre to the Dutch border 15th army under General Hans von Salmuth. Between January and June 1944 the number of divisions in the whole of France and the Low Countries was raised from 46 to 58. A number of new divisions with younger manpower and enhanced weaponry were assembled to add striking power to the defences. The elite 3rd parachute division was sent to the western army, generously equipped and with a full complement of seventeen thousand men, but the bulk of new divisions, including the Panzer units, were well under strength in both troops and tanks. Even in May there were fewer tanks available than the German armies had mustered for the invasion of France four years before, and shortages of spare parts and repair facilities reduced operational effectiveness. The bulk of divisions made up the static formations, dug into bunkers and resistance nests in a long thin line along the coast. Because it proved impossible to get accurate intelligence on Allied intentions, Rommel had no choice but to spread his forces across the whole of northern France. By June, 7th army had fourteen divisions, six of them static; in the area where the main thrust of the invasion was expected there were twenty divisions, fourteen of them static. There were no reserves to speak of.[50]

Given the limited resources of manpower and equipment, a great deal rested upon how effectively they were deployed. Every military instinct demanded mobility and concentration of effort. These were the trademarks of German success. Rommel's static defensive crust was foreign to the battlefield traditions he had previously helped to sustain. His insistence on a shoreline battle provoked a bitter strategic argument about how best to meet the invasion when it came. The Commander-in-Chief West, von Rundstedt, was convinced that the outer shell would be easily pierced whatever Rommel did, and he deplored the idea of scattering German forces in a 'thin-water soup' along the coast. He was strongly supported by the commander of the Panzer forces in France, General Geyr von Schweppenburg. Both men wanted to absorb the initial invasion and then counter-strike with a powerful mobile reserve that would smash the enemy beachhead on ground more suitable for armoured warfare. They feared that armoured divisions placed piecemeal along the coast would be obliterated by Allied naval fire and air power and would be a wasted asset as a result. Moreover both men were convinced that the Allies would land at Calais, where the terrain favoured German tank armies. It made little strategic sense to them to place German divisions out of reach of the main battle when speed of response to the invasion was of the essence.[51]

Rommel was convinced that the invaders were at their most vulnerable at the very early stages of the operation. No mobile reserve would arrive at the beaches in time to have much impact, especially as the Allied bombers and the French Resistance between them made travel by rail a hazardous business. On the issue of air power he knew that the Allies would have air superiority, but having fought in Africa for months without effective air cover he was willing to run the risk in France.[52]

The argument could only be resolved by Hitler, who imposed a compromise solution that satisfied nobody. Hitler, like Rommel, arrived at the conclusion that a diversionary assault on Normandy or Cherbourg was a strong possibility, and agreed with Rommel's distribution of forces along the whole coastline. But he could also see the force of the argument about reserves. In early May he divided the armoured divisions between the two protagonists. A reserve under his direct control was set up under von Schweppenburg consisting of four of the divisions, held well back from the coast. The remainder were placed at intervals at or near the coast; only one, 21st Panzer, was assigned to the Normandy front, while four were scattered in southern France. Hitler's decision made a fragile defence yet feebler: there were too few mobile divisions in the reserve to have the punch effect Rundstedt wanted, and too few of them near the coast to fight the invaders on the beaches.[53] The Allies could not have disposed German forces more favourably if they had done it themselves.

German strategy in France was hamstrung by the inability to predict where the main weight of Allied attack would come. Neither was there any firm information about the precise date of attack. 'I know *nothing* for certain about the enemy,' complained Rommel.[54] By April there was an evident edginess among German forces in France. Constant patrols were ordered on the coast throughout day and night; sentries walked up and down for two hours at a stretch, forbidden to talk for fear of losing concentration. The pattern of Allied bombing – heavy in north-eastern France, light in Normandy – suggested an imminent invasion of the Pas de Calais in mid-May, but the crisis passed.[55] Talk at Hitler's headquarters was of invasion from mid-June, possibly not until August, There was even speculation that the whole invasion was a hoax. The one glimmer in the darkness arrived by chance. Routine interrogation of French Resistance prisoners in Brittany in October had revealed the meaning of coded messages sent through the BBC to French saboteurs. These included a verse from a poem by Paul Verlaine ('The long sobs of the violin . . .') which was to be broadcast in two parts, the first on the 1st or 15th of the month in which the invasion would take place, the second 48 hours before the invasion itself. German counterespionage succeeded in concealing their knowledge of the message, which remained in force right up to D-Day. German wireless operators on the French coast were instructed to listen out for the one real clue among the babble of signals that would tell them the Allies were on their way.[56]

*   *   *

On 15 May Eisenhower and Montgomery hosted a meeting of senior officers to outline the final plan for invasion. Churchill was invited, and King George VI. Soldiers and politicians filed into the model room of St Paul's School, where Montgomery had set up his headquarters, and sat on rows of hard wooden benches like so many eager schoolboys, listening to the commanders, who one after the other explained how Overlord would work. The plan was little different from the one proposed by Montgomery in January. Thanks to Ultra intelligence the German order of battle was known with more than the usual degree of accuracy, and this had helped development of the strategy. The Allied attack would be undertaken by five divisions in the first wave. To the west, at the eastern edge of the Cotentin peninsula, the American 1st army under General Omar Bradley would invade on two beaches codenamed 'Utah' and 'Omaha'; to the east, between Arromanches and the Orne river, the British 2nd army (with two British divisions and one Canadian) under General Miles Dempsey was to attack on three beaches, 'Gold', 'Juno' and 'Sword'. On both flanks paratroopers would be flown in during the invasion night to secure bridges and cut German communications: the British 6th airborne division on the eastern flank, the American 82nd and 101st airborne divisions in the west. The invasion had the support of over twelve thousand aircraft to ensure air superiority over the beachheads, and 1,200 naval vessels, including seven battleships and 23 cruisers.[57] Once ashore the object was to seize the city of Caen in the first 48 hours and then to use the area at the eastern end of the assault to pin down German forces in order to facilitate a long wheeling breakout by the American army in the west, through Brittany and then eastwards to Paris and the Seine. Montgomery gave his forces ninety days to reach the French capital. When the formalities were over the King gave a short exhortation, and Churchill, who only a few days before had confided to Harriman the oppressive doubts he still harboured about invasion, took the opportunity to tell the assembly that at long last, three weeks before D-Day, he 'was hardening towards this enterprise'.[58]

A few days before, Eisenhower had confirmed that 5 June was invasion day, all being well. Now that the plans were finalised and approved the operational orders were sent out on 25 May. Every one of the ships' captains received a book of seven hundred foolscap pages of instructions; the United States forces distributed 280,000 charts of the invasion area, and 65,000 operational booklets.[59] The Allies now faced a security nightmare. The anxiety made Eisenhower tetchy and nervous; one of his staff observed how 'worn and tired' he looked. Every effort was made to conceal the date from the enemy. Against vigorous protest, all diplomatic traffic was suspended on 17 April, and diplomatic correspondence censored; a 10-mile exclusion zone was set up around the British coast from Cornwall to Norfolk; from 25 May all letters from American troops were suspended for ten days. All trans-Atlantic communication by cable, radio and telephone was cut. During the second half of May troops were moved into transit camps and denied contact with the population outside. On 28 May all seamen were sealed in their vessels. The more daring stole a last visit to the cinema, or the local bars, but by early June three million service-men were locked away from a population whose realisation of imminent invasion was hardly less than that of the troops. Cities which had bustled with uniforms and lorries fell suddenly silent; throughout the night the constant hum of engines on the roads of southern England betrayed the secret mustering of forces. If Germany had possessed a single spy in the south the secret could hardly have survived.[60]

Now that the timetables had taken over, the whole Overlord force was suffused by a subdued, unbearable tension. 'I never want again to go through a time like the present one,' Brooke confided to his diary. 'The cross-Channel invasion is just eating into my heart. I wish to God we could start and have done with it.' Bored in his fictitious headquarters, Patton chafed miserably at the bit. 'This waiting is hard on the nerves,' he wrote to his wife. 'I must go and exercise!'[61] If it was hard for subordinates it was certainly excruciating for Eisenhower, who lived out the last week in a bored, fatalistic cloud. He played bridge and badminton; he suddenly took up sketching and then abandoned it. He developed a ringing in one ear, and sore eyes brought on by the strain of reading. His impatience was evident. There was little for him to do after the months of planning, except to give the final order to go. This was a responsibility that bore down on him more and more as the final days approached. No one, Eisenhower wrote in his own diary, could understand 'the intensity of the burdens' facing a supreme commander.[62]

The agony of suspense reached a cruel climax in the last few days. On 1 June the French Resistance, and Sergeant Walter Reichling of the 15th Army Communications Reconnaissance Post, heard the first stanzas of the tell-tale poem broadcast by the BBC at 9 p.m. The 15th army went over to the alert, but not the forces farther west. On 2 June Eisenhower prepared his Order for the Day from his new field headquarters just outside Portsmouth. Then, after weeks of clear, warm days, the weather sharply deteriorated on the 3rd. Eisenhower had always known that this was a risk, but there was nothing to be done about it. The outlook was gloomy. There were anti-cyclones over Greenland and the Azores; across the Atlantic, an easterly depression was already bringing rain and heavy cloud. At four o'clock in the morning on 4 June the commanders met with the senior meteorologist, John Stagg. He predicted low cloud cover and a force six wind. Eisenhower knew that the whole enterprise depended on air support and steady seas. Neither could be relied upon with this forecast. With great reluctance he postponed D-Day from the 5th to the 6th. Ships already at sea were recalled. The crews and soldiers on board now had to endure 24 hours sitting at anchor in high seas, their adrenaline evaporated.[63]

By the evening the winds were calmer and the cloud more broken, but the outlook was still poor. At 9.15 Eisenhower reassembled his commanders and the unfortunate weather man. Stagg spoke as wind and rain beat at the mess room windows. He detected an improvement, broken cloud and calmer winds for 48 hours. Eisenhower canvassed the views of his commanders one by one. The airmen were hesitant. Montgomery was keen to start. But Ramsay's voice was decisive. The navy had only another thirty minutes to give the orders, or risk a postponement for 48 hours, beyond the period of favourable tide. Eisenhower told the navy to start regardless, and asked them all to meet him again at four in the morning. He awoke at 3.30 to gales and driving rain. The group was tired and in drooping spirits. But Stagg had good news: the better weather was a certainty. As if by some divine hand, the rain outside abruptly ceased. Eisenhower thought for a few moments. The others remained silent. It was a moment of supreme drama. All at once Eisenhower spoke quietly but distinctly: 'O.K., let's go.'[64] The room cleared in seconds, leaving Eisenhower alone.

A strategy that had taken years to agree, and months to prepare, had almost been brought to naught by the elements. There was still a great risk. The following day the Supreme Commander sat down to pen in advance a communiqué announcing the failure of the whole enterprise, which he solemnly placed in his wallet. He no longer had any power to halt the task force. From all around the coast 2,700 ships moved to pre-arranged assembly points in the Channel, from where they were guided along five pathways cleared through the minefields. Though it was not an order, the men spoke in whispers. While the ships ploughed through heavy seas the paratroop divisions prepared to fly out. For want of anything else to do, Eisenhower rode out to visit the 101st airborne division as it embarked in the transport aircraft. He waited until the last plane took off, then returned to bed. For the first hours of D-Day Eisenhower slept.

The gloomy weather had an unexpected silver lining. German commanders relaxed at the sight of heavy rain and cloud. On 4 June the naval command in Paris reported that invasion could not be considered imminent. The regular naval patrols that might have betrayed increased Allied activity were cancelled on the 5th, so poor was the weather. Rommel took the opportunity for a short rest. On the 4th he drove back to Germany, where he arrived in time for his wife's birthday, on the 6th. The junior commanders of the 7th army in Normandy were on their way to an exercise arranged at Rennes on the same date.[65] Only Sergeant Reichling stood between the Allies and complete surprise. On the evening of the 5th he duly intercepted the final lines of Verlaine. His seniors contacted Field Marshal Rundstedt's headquarters in Paris, only to be told that no enemy would be so foolish as to broadcast invasion over the radio. Fifteenth army stayed on alert, but the rest of Army Group B, including the six divisions directly in the line of Allied fire, received no warning until paratroopers began their graceful descent on to French soil in the early hours of the morning of 6 June.

It still took time to realise what was happening. Seventh army headquarters had to piece together numerous reports of airborne landings to get any coherent picture. Among local German garrisons there was fear bordering on panic. The fighting was confused and sporadic. At the vital Pegasus bridge straddling the Caen canal British forces achieved complete surprise; the sentry stood still, struck literally dumb with terror.[66] Not until three o'clock in the morning was 7th army put on alert, and not until 3.09 did German radar begin to pick up the sea and air activity in the Channel. At five o'clock the first shore batteries opened fire on the approaching fleet. At dawn the invaders began a fearsome fusillade. The first assault came with 1,056 British and 1,630 American heavy-bombers which blanketed the coastal defences with high explosive, except for the Omaha beach, where cloud cover led the bombers to drop their loads 3 miles inland. There then followed a ferocious naval bombardment directed at the big coastal guns and the concrete emplacements too thick for aerial bombing. The shells roared over the heads of the soldiers slowly wending their way through the minefields, packed into landing craft. The naval gunfire was devastating. It did not destroy many of the fortified bunkers and gun casements, but it left their defenders stunned and deafened. As the naval bombardment eased up, swarms of fighter-bombers and medium-bombers flew in low to attack German defensive positions, while from the sea came yet another barrage, this time from thousands of rockets launched from batteries in converted landing craft which floated in on the flanks of the troop carriers. The pall of acrid smoke hanging over the beaches was so thick that the coastline at Utah beach was obscured entirely, and as a result American troops landed over a mile south of the beach they were supposed to storm.[67]

Behind the bombs and guns came a swarm of smaller craft, carrying men, supplies and tanks specially converted to an amphibious role to provide firepower in the first stages of assault. Some of the landing craft had to travel up to 11 miles from the mother vessel to the shore in rough seas. Seasick and cold, packed shoulder by shoulder, the soldiers gradually drew level with the beach defences, now lapped by the incoming tide. As the leading boats reached shallow water the loading ramps were dropped and the first troops slid and jumped down them. Some had 100 feet of water to stumble through; some just a few yards. Many radio operators, weighed down with additional equipment, sank beneath the water. Raked by small arms fire, a bewildering mêlée of floating tanks, landing craft and soldiers struggled shoreward. Unable to see the Allied wounded, tanks rolled on to the beaches over the bodies of soldiers too injured to move. From the mixture of noise, smoke and death it was impossible for each successive wave of troops to make out what progress had been made by the men in front. The beaches were consumed in the most literal sense by the fog of war.

As the morning wore on it proved possible to form a clearer picture. At all except Omaha beach the Allies had achieved the initial foothold. The airborne troops on both flanks had secured the most vital roads and bridges. On the British sector German resistance was light after the shuddering bombardment from the sea. The static divisions put up little opposition. Sword beach was cleared by mid-morning; the Canadian 3rd division on Juno captured the beach by 10 a.m. and began to move inland towards Caen. On Gold beach there were more defences to neutralise, but by midday the British 50th division was on the move, and by the end of the day it had established a bridgehead 3 miles wide and 2½ miles deep. By the evening all three divisions were well inland, but failed to capture either Caen or Bayeux as Montgomery had hoped. Not until the late afternoon did they meet serious opposition, when Rommel was finally able to order 21st Panzer division to attack west of Caen to prevent British forces from capturing the city. The effort to move some of von Schweppenburg's reserve Panzer forces towards Caen broke down, as Rommel had argued it would all along, when cloud lifted and Allied fighter-bombers attacked anything that moved on the roads.

Farther to the west American forces had more mixed fortunes. At Utah beach the failure to land at the right area was a blessing in disguise. American forces streamed ashore against the lightest beach obstacles and negligible enemy fire. By the evening they had established a beachhead 6 miles wide and 6 miles deep; total casualties were 197 from all causes. At Omaha an unpredictable set of circumstances almost produced failure. The bombers missed the shore defences entirely; the naval bombardment was short and, against high bluffs, was difficult to direct with accuracy; the sea was rougher than expected, and losses of tanks and landing craft were correspondingly higher. German defences were stronger in this sector than in any other, and troops had recently been reinforced, unknown to Allied intelligence, by a fresh German division. The troops were pinned down on the beaches by a withering fire; ahead

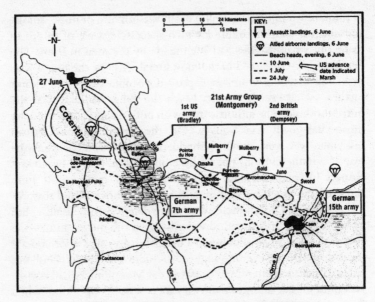

10 Battle for Normandy, 6 June to 24 July 1944

of them were cliffs broken by narrow defiles, easy to defend. American officers persuaded the men sheltering among the wreckage on the beach that there was little choice but to fight their way up the cliffs or be slaughtered where they lay. Yard by yard they battled forward until they reached higher ground, supported by accurate fire from destroyers lying close to shore. By the end of the day the American 5th corps had secured a narrow bridgehead but had been unable to clear all German resistance from the coast, or to make contact with the forces from Utah farther to the west.

The successful landings confounded Rommel's shoreline strategy. Over the next four days Allied troops and supplies poured in to stiffen the bridgeheads. By 11 June 326,000 men, 54,000 vehicles and 104,000 tons of supplies had been landed across the beaches.[68] British and Canadian forces pushed towards Caen to reach within a mile or two of the city. On 7 June Bayeux was captured and British and American forces linked up. That day Montgomery moved his headquarters to France. Eisenhower, increasingly heartened by the success ashore, crossed the Channel in a fast minelayer to see for himself the progress of battle. From the deck he watched the trails of men and lorries wending their way from the beaches through the rich farmland to the front-line. Though still well short of the initial objectives, the lodgement was firm. After four more days of fighting a continuous front was established from the Orne river to Montebourg, 10 miles across the neck of the Cherbourg peninsula.

The failure of the German shoreline battle had many causes. The German navy was swept aside by overwhelming Allied naval power. German torpedo boats and submarines inflicted only slight damage to the vast armada astride the Channel – three small ships in convoy, and a dozen minor naval vessels were lost – and of the 43 submarines sent to attack the invading naval force, twelve were forced to return to base damaged, and eighteen were sunk.[69] German air power suffered the same fate. Thanks to the critical weakening of the German air force over Germany, the contribution of German aircraft to the invasion battle was negligible. On 6 June, against over twelve thousand Allied aircraft, including 5,600 fighters, the 3rd air fleet in northern France could muster only 170 serviceable aircraft. Attempts to reinforce the front further weakened the defence of the Reich, while contributing very little to the battle. Many of the aircraft sent were destroyed in transit as Allied aircraft smashed German airfields in France. Many German pilots were plunged into conflict before they were fully trained; inadequately prepared for the invasion, many lost their way. The reinforcements were shot out of the skies. This left Allied airmen with the freedom to attack other targets. Large groups of marauding fighters and fighter-bombers attacked anything that moved on the German side, making the transfer of German ground reinforcements and supplies almost impossible by day. The bombers were free to blast German defences, to terrifying effect. Allied air supremacy proved to be a vital factor in securing and holding on to the shallow lodgement.[70]

German defences did not turn out to be as formidable as had been feared. The beach obstacles were cleared in a matter of hours thanks to careful Allied preparation and reconnaissance. The element of surprise was sufficient to allow Allied forces to storm ashore before adequate German reinforcements could be rushed to the front, and the initial German intention to build a second defensive line 3 miles inland from the shore never materialised. The great expectations of the Atlantic Wall were never realised. None the less, Rommel did have 34 divisions under his command to face the five British and American divisions in Normandy. The Allies' early success owed a great deal to the confused and hesitant response of their enemy as a result of the FUSAG deception, which continued to mesmerise the German high command long after the invasion.

There were hints of imminent invasion relayed to Hitler on the evening of 5 June. Goebbels found him 'unperturbed'. In poor health, Hitler was resting at the Berghof, high in the Bavarian Alps. At ten on the morning of 6 June the first reports of Allied action came in. Hitler was asleep and had to be roused by his valet. When he was finally told the news he was briefly exultant ('right in the place where we expected them') but he still regarded Normandy as a mere diversion from the main invasion across the Dover-Calais strait. So confused were the reports coming in from the west that Goebbels thought the invasion was taking place from Dieppe to Dunkirk, echoing Hitler's conviction that Normandy was a 'decoy'.[71] Only late in the afternoon of the 6th was Rommel allowed to take forces from the reserve to oppose the landings, too late to prevent British and Canadian troops from digging in, and now in the face of relentless air attack. But there was no question of moving the forces of 15th army away from the Calais area to help the beleaguered forces in Normandy. As long as the German high command believed that FUSAG was still poised to strike it was strategic suicide to move everything into western France and risk an annihilating pincer attack. Evidence from radio intercepts and from double-agents regarded as particularly reliable all indicated an Allied strategy of double-barrelled invasion. Everything pointed to 15 June as the critical day for the new assault, and Dieppe as its nodal point. At Supreme Headquarters it was decided to hold on at all costs in Normandy, while every effort would be strained to throw FUSAG back into the sea when it came. Against even the wildest expectations the deception lived on; late in June the BBC relayed bogus messages alerting the Resistance movements of Belgium and north-eastern France. The fifteenth of July was now regarded as invasion day. So convinced was Hitler that his enemy would do what he would have done that not until 7 August, on the very eve of German defeat, did he finally accept that FUSAG was not coming, and order 15th army westward.[72]

The failure to commit more forces early on was Rommel's undoing. Even then the margin on D-Day was slimmer than Montgomery would have liked. A week later he told Field Marshal Brooke with unusual candour that his forces could well have been defeated, for all the Allies' advantages, if Rommel had been able to mount strong assaults from midday against forces still in a confused state on the beaches.[73] Fortunately for the Allies, Rommel's reinforcements came piece-meal, tanks without their fuel, men without their lorries or horses. In the first four days German forces did all they could to establish a firm line 8–10 miles from the coast. From then on the battle entered a new phase. Montgomery had two immediate objectives: the capture of the port of Cherbourg at the end of the Cotentin peninsula, and the capture of Caen at the eastern end of the lodgement area. He knew that German forces would concentrate against the threat to Caen, because it was perceived to be the gateway to the Seine and Paris, and beyond. This would free American forces to exploit German weakness in western France and begin the long encirclement eastwards. But progress towards this stage was slow, well behind the schedule drawn up in the Overlord plan. In the western sector Allied forces could not bring their greater mobility and firepower to bear because of the nature of the terrain. They were fighting in the bocage, a farming country of narrow roads and high banks topped with thick hedgerows. This was ideal for defensive warfare. Though weakly held, the German line was difficult to penetrate. Progress was measured in yards. Morale was dented by fear of snipers, who were armed with everything from rifles to anti-tank guns. Tanks were ambushed and disabled before they could even find a target at which to aim. The German forces facing the Allies now were drawn mainly from the regular field army, and they fought with a greater determination and skill than the defenders of the Atlantic Wall. Around Caen, Rommel succeeded in concentrating four Panzer divisions by 13 June; though they were short of equipment, and constantly harassed from the air, they prevented a quick capture of the city. The front settled down to a slow battle of attrition.

As the front solidified, Allied forces became hostage to the rate at which they could be supplied with fresh troops and equipment. Within days the two Mulberries were complete enough to be operational. By mid-June the Allies had moved nineteen divisions into the bridgehead – more than half a million men, and 77,000 vehicles. Then once again the weather let the Allies down. On 18 June the day was fine and warm. The following day the Channel was suddenly and unexpectedly hit by the worst gales of the century. A force eight wind and driving rain whipped the waters into waves 6 feet high; by the 20th nothing could sail across. The shrieking wind aad fierce seas played havoc with the small craft sheltering like so many sheep inside the pens provided by the almost-completed Mulberries. Small boats were tossed up and smashed against the shore and the walls of the harbour. After eighty hours of battering, the Mulberry off the American beaches at St Laurent began to disintegrate beyond repair. Only the 55 old merchantmen, sunk farther out to sea, afforded enough flimsy protection to prevent a complete disaster. When the storm abated on the 22nd, eight hundred small craft, including many of the essential task transporters, lay smashed and stranded on the beaches. For the German defenders the gale was like that Divine Wind, *kamikaze,* that broke up the invading Mongol fleet in the Sea of Japan seven centuries before. The supplies to the lodgement area dried up at a critical juncture of the battle. Up to the 19th of June an average of 22,000 tons of supplies had been brought in every day; over the next three days the total collapsed to a little more than 1,000 tons a day, The loss of ammunition was particularly acute. Without it neither the British nor American forces could launch the attacks they had planned. For four days the German defenders were free of air attack, and could move forward men and tanks unmolested. Not until the end of July were the Allies able to make good all the losses inflicted by the capricious elements.[74]

The storm made slow progress sluggish. The only relief for the Allies came in the Cotentin peninsula, where weak German defences were swept aside isolating the whole peninsula from the German defensive line in the south. The American commander,

General Bradley, had just enough ammunition left for his forces to attack the four retreating German divisions cut off around Cherbourg. The German defences crumbled. Within a week the American 7th corps reached the outskirts of the port; by the 22nd, as the storm subsided, the attack on Cherbourg began with a thunderous bombardment from sea and land. By the 25th American troops entered the outskirts. A day later the German commander surrendered. A few German soldiers who took to heart Hitler's order to stand and die to the last man and the last bullet held out for another five days against overwhelming odds.[75] But everywhere else along the line the Allies met stiffer resistance. Protected by cloud cover during the storm Rommel moved up four more armoured divisions. By 1 July, he was prepared to launch a strong counter-attack, with five Panzer divisions, against the British position north of Caen. The attack was repulsed by concentrated artillery fire in what proved to be the heaviest fighting since D-Day. Two days later the American 8th corps began to drive south to prepare for the breakout planned by Bradley and Montgomery which would take American troops across the German line, towards the Seine, but fierce fighting, poor weather, and the combination of swampy inlets and *bocage* bogged down the assault. Rommel now moved two Panzer divisions, 2nd SS and Panzer Lehr, to obstruct the American threat to his western flank. It took American forces six days to move 5 miles. The launch of the breakout – Operation 'Cobra' – had to be postponed in the face of German resistance. By mid-July there was every appearance of stalemate.

Eisenhower, who seemed 'buoyant and inspired' when he visited France on 12 June, returned at the beginning of July 'smouldering' at what he regarded as Montgomery's excessive timidity. The mood of optimism at Supreme Headquarters engendered by the success of the initial lodgement was dissipated by the dispiriting slogging match played out in the orchards and marshes of Normandy. When Stalin began the great Soviet offensive on 20 June, timed deliberately to coincide with the Allied pressure in the west, the contrast between the early successes of the Red Army and the slow progress of the forces under Montgomery's command was plain.[76] It is not difficult to find the source of Eisenhower's frustration. With the advantage of surprise, with overwhelming air supremacy, with the success of the deception plan, with regular and ready access to German signals, it was difficult to grasp why the limited German forces could not be pushed back swiftly and decisively. The fear at Supreme Headquarters was of a return to the trench stalemate of the First World War. Relations between Eisenhower and Montgomery deteriorated. At Eisenhower's headquarters there was talk of sackings; his deputies urged him to confront Montgomery and demand action.[77]

The conflict between the two men has often been presented as a fundamental contrast in strategy, Eisenhower's hare to Montgomery's tortoise. There is certainly something in the contrast, revealed long before D-Day, between the American doctrine of the general offensive and the British emphasis on economy of effort, with the careful husbanding and deployment of limited resources. This was a difference made explicit on the ground in Normandy. Eisenhower wanted to see vigorous action all along the front, and would have been happy with a breakout anywhere, as long as the Germans were chased out of France. Montgomery feared that such a strategy would produce diluted effort and high attrition. He was conscious of how shallow was the pool of manpower that sustained British and Canadian forces. With only 37 divisions assigned to the Overlord campaign, against a potential of over fifty German, it was essential to conserve manpower by getting the enemy to concentrate his efforts against one strongly defended sector, in order to reduce pressure along the whole front. This is what happened in the area around Caen. The failure to capture the city in the first few days was in a sense immaterial, though to Montgomery's critics, then and now, it came to symbolise the apparent collapse of the Overlord plan. The enemy regarded the front at Caen as the most important Allied objective, from where they would attempt a breakout towards the Seine, and they concentrated the bulk of German forces in that sector. By the end of June there were only half a Panzer division and 140 tanks opposing the American sector; around Caen were seven and a half armoured divisions, and over seven hundred tanks. Even when Rommel switched two divisions in early July the balance of armoured vehicles remained the same. By 25 July there were two Panzer divisions and 190 tanks in front of the American forces preparing Cobra; but at Caen there were still six armoured divisions and 645 tanks.[78] What held up the Allied advance was not a failure of Allied strategy, but the failure to predict how difficult the topography of the area would be for mobile warfare, or to anticipate Hitler's order that all forces should fight for every inch of ground rather than execute the strategic withdrawals dictated by military good sense. The battle of attrition was dictated not by Montgomery but by the desperate German defenders.

Eisenhower had little appreciation of quite how brittle the German position was. Had he realised, he might well have viewed Montgomery's efforts more kindly. The failure to throw the invader back into the sea compelled Rommel to fight the kind of contest least congenial to German forces. Unable to build up a significant strategic reserve, he was forced to plug every gap that appeared in the line, using even the valuable armoured divisions, whose tactical advantages were squandered in actions more suited to infantry. Naval gunfire and remorseless aerial bombardment inflicted further heavy losses. The movement of reinforcements and supplies across France was rendered increasingly difficult by Allied air power and French Resistance. It took longer for the 9th and 10th SS Panzer divisions transferred from the eastern front to travel from eastern France to Normandy than it did for them to move from the Soviet front to France. The chaotic transport conditions made it difficult to keep front-line troops supplied with fuel and ammunition. Divisions that arrived at the Normandy battle were thrown straight into the breach before they had had time to assemble or to collect their equipment. On 17 June Rommel and von Rundstedt met Hitler at his forward headquarters, 'Wolfs Lair II', at Margival near Soissons, whither he had flown at their request to discuss German strategy in the west. Von Rundstedt wanted a fighting withdrawal to hold a front along the Loire and Orne rivers, where a mobile reserve could be built up strong enough to resume the counter-offensive. Hitler was quite unmoved. He ordered the line to stand fast at all costs and flew back to Germany.[79] But twelve days later the two generals were back in front of Hitler at Berchtesgaden, pleading for a fighting retreat across France in the hope that Germany could reach a political settlement with the west, or achieve an armistice like 1918. Again, Hitler was unmoved. He insisted once more on holding the line, where the enemy would be worn down by the struggle, and forced back into the sea 'using every method of guerrilla warfare'.[80] Yet the twenty divisions that might have turned the tide in such a battle were kept back to check the second strike expected at any moment on the Channel coast to the north.

German commanders in France knew that Hitler was quite out of touch with reality. Hitler, for his part, saw an insidious defeatism in the suggestions of withdrawal. He sacked von Rundstedt and appointed Field Marshal von Kluge in his place. The new commander arrived confident of stabilising the situation; but in a matter of days he realised that the reports from the front had spoken the truth. Shortly after his arrival Montgomery launched the first of a number of major operations on the eastern wing of the lodgement which aimed at destroying what remained of German offensive power, and creating the conditions for an American breakout farther to the west. On 7 July began the battle for Caen. The assault started with an attack by 467 heavy-bombers which carpet-bombed the unfortunate town. When British troops reached the outskirts the following morning the streets were impassable, pitted with giant craters and blocked with piles of heavy stones from the shattered buildings. The German forces abandoned the main part of the city, but destroyed all the bridges across the river, making it impossible for Allied forces to pursue the enemy farther. South of Caen the German forces dug in. Rommel organised a defensive zone 10 miles deep to prevent an Allied breakthrough. There were five lines of defence: a light infantry cover to absorb the bombing and artillery barrage; a line of tanks positioned a short distance behind; a third area of small villages occupied by infantry dug in with anti-tank guns; a powerful gun line along the crest of the Bourguebus Ridge some 4 miles south of Caen, with 78 of the formidable 88-millimetre anti-aircraft guns that doubled as tank-busting artillery; and finally a defensive zone behind the ridge itself manned by infantry and backed by a tank reserve 5 miles to the rear. Rommel waited for what threatened to be the climax of the gruelling six weeks of attritional warfare.[81]

Urged on by the tetchy Eisenhower, and anxious about the slow pace of American preparation for the breakout, Montgomery rose to Rommel's bait. On 13 July he planned a fresh operation, 'Goodwood', to drive east of Caen and seize the defended areas to the south. His instructions issued to commanders two days later made it clear that rather than attempting a breakout himself his objectives were limited to tying down and destroying the bulk of German armour, and creating a firm hinge on which Bradley could swing when he pushed open the door in Brittany: 'all the activities on the eastern flank are designed to help the forces in the west while ensuring a firm bastion is kept in the east.'[82] Ultra intelligence showed the weak state of the German divisions, and the problems of reinforcement, but did not make clear the extent of Rommel's defensive field. Goodwood was scheduled for 18 July. The day before, Rommel made a final tour of his defensive preparations. On his way back from the headquarters of Panzer Group West his car was machine-gunned on the road by two British aircraft. There was no cover. His driver was killed outright and Rommel was hurled on to the carriageway. Severely injured, he took no further part in the struggle for Normandy. His duties were assumed by von Kluge.

On 18 July Allied bombers prefaced Montgomery's attack with the heaviest bombardment of the campaign. For three hours the German defences were pounded so thoroughly that when the British armour and infantry followed the bombers' wake they encountered almost no opposition from the first line of dazed defenders. But the line of tanks and the anti-tank gun-nests behind held fast. Montgomery's troops became locked in a fierce and prolonged struggle for the villages all along the slopes of Bourguebus Ridge. Heavy losses were sustained on both sides. For two days of bitter fighting Canadian and British forces blasted German defenders from all but one of the villages in the third line of defence, but above them on the crest of the Ridge the German gun line remained intact. Then, on the afternoon of 20 July, just as Montgomery's forces were preparing to assault the Ridge, a prolonged downpour turned the battlefield into a sea of mud. Tanks could no longer be deployed and Montgomery brought Goodwood to an end. Though the Ridge eluded him, his other aims were largely met. The battle exhausted the German armoured divisions and prevented the transfer of armour to the west. To meet the Allied assault at Caen two Panzer divisions earlier moved to the American sector were recalled. The vital task of pinning German armour to the eastern end of the lodgement was achieved. Though Rommel's defensive zone held fast, the hail of 12,000 tons of bombs, and the constant attacks by fighter-bombers, ate away at precious manpower and tanks. The day after the end of the battle von Kluge wrote a long letter to Hitler, enclosing a memorandum from Rommel, penned two days before his accident. Neither man pulled his punches. 'The force is fighting heroically,' wrote Rommel, 'but the unequal combat is nearing its end.' Von Kluge added his own voice: 'I came here with the fixed determination of making effective your order to stand fast *at any price* . . . In spite of intense efforts, the moment has drawn near when this front, already so heavily strained, will break.' Since D-Day German forces had lost 2,117 tanks and 113,000 men; in return only 10,078 men and seventeen tanks had been sent to replace them. Despite heavy Allied losses the balance of tanks lay heavily in their favour, 4,500 in late July against 850 German. Allied reinforcements since D-Day exceeded one and a half million men and 330,000 vehicles.[83]

The decision to terminate Goodwood coincided with a further delay in the west. Bradley needed more supplies and ammunition before he could launch Cobra. All of this proved too much for Eisenhower. 'Blue as indigo' at the delays, he insisted on getting across to France on 20 July in weather so bad it had grounded all other aircraft that morning. When he met Montgomery and Bradley he upbraided them for their lack of vigour and drive. He wanted offensives up and down the line, and simply could not grasp Montgomery's purpose. The following day he wrote again to his ground commander: '*Time is vital*'. He flew to watch the launch of Cobra which had been postponed to the 21st, but was disappointed again. Bad weather forced yet another delay. Relations between Supreme Headquarters and Montgomery could hardly be worse. 'Ike and I were poles apart', Montgomery wrote in his memoirs, 'when it came to the conduct of war.'[84] But Montgomery had no one but himself to blame. He exuded infallibility when a modest realism would have been more politically prudent. He isolated himself from Eisenhower's Supreme Headquarters. The two men met only nine times throughout the campaign, and on these occasions Eisenhower found it difficult to create the opportunity to confront his ebullient subordinate.

In the end Eisenhower did nothing to interfere with Montgomery's handling of the situation but the myth took root that Montgomery had failed in Normandy and had to be rescued by go-getting Americans. The truth could not have been more different. The delays to the breakout were unavoidable. Bradley could not punch decisively without supplies; once he was provisioned, nothing could be done about the weather. British and Canadian forces fought against the bulk of German armoured divisions in France, against bitter resistance and with a declining rate of reinforcement. But when the breakout finally came, the long weeks of attrition around Caen bore out Montgomery's aims. The victory in Normandy was secured in the grim, unglamorous erosion of German fighting power in June and July.

★   ★   ★

On the morning of 25 July the sun shone at last on Allied endeavour. Operation Cobra began at 9.40 in the morning. It was the fruit of much careful preparation. At Montgomery's prompting Bradley abandoned the strategy of the broad offensive, which had proved slow and wasteful of American resources, in favour of a concentrated blow with aircraft and tanks to pierce and turn the weak German line. Tactical experience had demonstrated the great advantage of rolling forward behind air bombardments laid down in box formation on the enemy defences. The real problem was still movement through the *bocage*. For this the Americans turned to the ingenious invention of Sergeant Curtis C. Culin Junior. To the front of a Sherman tank were welded eight steel teeth, 2 feet above ground level. Moving at 10–15 miles an hour, the tank could drive into the hedgerow, cut the roots and plunge through with little loss of speed. The 'Rhinoceros', as it soon became known, transformed the mobility of American armoured formations. No longer tied to the roads, they could now make rapid progress across country. During the middle weeks of July Bradley's forces were brought up to strength. He had fifteen divisions in the 1st army, reinforced by a reserve of four divisions which were to form the nucleus of a new

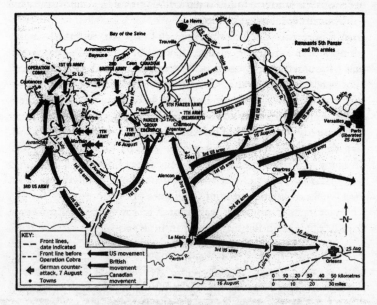

11   Breakout and pursuit in France, July–August 1944

American 3rd army under General Patton, whose FUSAG role had run its successful course. Against these forces the German 7th army could field only nine weakened divisions and 110 serviceable tanks. The stage was set for the final act.[85]

When the collapse of German resistance came it was sudden and complete. On the morning of 25 July almost 1,500 heavy-bombers laid a carpet of bombs over the Panzer Lehr division which lay directly astride the corridor that Bradley was to drive through the German line. Resistance was bludgeoned out of the forward defences. Over the next three days the forces drove on with growing momentum. While German tanks and self-propelled guns had to stay on or close to the roads, where they fell prey to fighter-bombers, the American forces developed fast mobile columns of Rhinoceros tanks, bulldozers and engineer battalions to bypass obstacles and plunge across fields and orchards. The costly advance from one hedgerow to another was abandoned. Armour pressed on, leaving pockets of German resistance to be mopped up by infantry. Each advancing column was followed by a constant patrol of Thunderbolt fighters, directed by an air controller in the front tank formations to attack strongpoints which would be marked by smoke-shells fired by the tanks. So close to the front-line did Allied aircraft operate that deaths from 'friendly fire' could not be avoided. But the effects of bombing on the enemy were severe. Against heavy bombing, the trenches, the foxholes, and even the deeper bunkers were little defence. After 45 days of severe fighting the German soldiers in front of the American attack were already weakened and demoralised. Bombing was more than many of them could stand: some went crazy, others surrendered or deserted, or drifted to the rear.[86] Under the American onslaught the last reserves of German fighting power rapidly dissolved.

In two days American forces drove 15 miles south all along the line. The armoured divisions of 8th corps reached the town of Coutances, close to the Atlantic coast, sweeping aside the shattered remnants of half a dozen German infantry divisions. While the remaining German armoured formations fought a vigorous rear-guard, the rest of the line began a disorderly retreat southwards. The American progress was relentless. The 4th Armored division raced down the coast to seize Avranches and the gateway to Brittany, driving 25 miles in 36 hours. Bradley was astonished at their progress. He had expected the usual steady steamroller, but now he had a racing-car. Aware of Montgomery's directive that 'there must be no pause', he instructed his forces to keep going. On 1 August the 3rd army was activated and General Patton was unleashed. Patton's reputation preceded him. Fiery, intemperate, vulgar, he had many vices; but he loved war. 'Made a talk,' he wrote in his diary in May, at FUSAG headquarters. 'As in all my talks I stressed fighting and killing.'[87] There was no better general on the American side to send in hot pursuit of the collapsing German armies. He took to the task with a remorseless zeal. Ignoring Bradley's precise instructions he moved seven divisions in 72 hours down the narrow coastal road and seized the whole of Brittany, save Brest and Lorient; he then wheeled his armies eastwards and began to drive towards Paris with almost no German forces in his way.

The German command was faced with a strategic nightmare.

The day after the fall of Avranches von Kluge warned Hitler's headquarters that the German left flank had collapsed, the front 'ripped open' by American armour. The choice was between holding at Caen and abandoning western France, or dividing German forces between two battles, and risking collapse in both. He was unable to retreat, at Hitler's express instruction. He decided on a compromise. He sent armoured reinforcements to the west and counted on Rommel's defensive field to hold fast. The result was predictable. Strong British and Canadian thrusts both sides of Caen immobilised the German forces and intercepted those driving towards the American front. The German effort was futile. Two months of attrition made it next to impossible to conduct the kind of mobile warfare that had taken German forces across all Europe in 1940 and 1941. There was no air support. German units fought with the frantic energy of a cornered animal, as they began to sense their imminent defeat. At last Hitler allowed forces to be released from the 15th army, but they arrived in small groups like the earlier reinforcements, unable to turn the tide of battle, and grimly aware of how hopeless the situation had become.[88]

As the German 7th army struggled to keep the rising tide of Allied forces at bay along its entire northern flank, Patton's divisions raced out of Brittany towards Le Mans and Chartres. Much of the area was already falling into the hands of armed Frenchmen. So fast was American movement that it took some time before von Kluge realised that his armies now faced the threat of complete encirclement. The natural thing to do was to withdraw quickly and in reasonably good order beyond the Seine. Though Hitler too late began to plan a new defensive line in eastern France, he was determined that von Kluge should stay where he had been ordered to stay. On 2 August he issued him with instructions to retake the western coast at Avranches and cut off Patton's forces. After von Kluge's protest that this was quite beyond his forces, Hitler sent a detailed plan on 4 August ordering 7th army and 5th Panzer army to launch a concentrated armoured offensive towards Avranches with all the Panzer divisions he could spare. Disobedience was hardly an option. Only two weeks before, on 20 July, a group of high-ranking army conspirators had failed to assassinate Hitler at his headquarters. Many of von Kluge's comrades were now in prison awaiting trial, and Hitler had sent a personal emissary to von Kluge's headquarters on 2 August to keep close watch on what the Commander-in-Chief West was doing. Von Kluge knew that to fulfil Hitler's order to counter-attack was military suicide; but to order retreat would mean a career brought to an abrupt and dishonourable conclusion. On 3 August von Kluge, against the strident protests of his tank commanders, began to arrange what Hitler wanted. By the 6th he had gathered five weakened divisions in an area 30 miles from the sea around the town of Mortain, in a position to sever the thin neck of land connecting Patton's forces with the rest of the American front. Between them they could muster four hundred tanks, but there were no reserves.[89]

Intelligence on the counter-attack at Mortain was plentiful on the Allied side. The time when it would be launched was revealed through Ultra decrypts. Bradley was able to place forces in strongly fortified areas in front of the German threat. A little after midnight on 7 August the Panzer forces began their attack. One division under cover of darkness and early morning mist covered 10 miles. But when the mist finally cleared at midday, the German armour was subjected to an air attack of exceptional intensity. The German forces made no progress and suffered heavy losses. On that afternoon Bradley began to counter-attack. By the end of the following day, the German Panzer divisions were back where they had started, and faced irresistible pressure on either flank.[90] That same day the rest of the front began to cave in. Around Caen, the Canadian and British forces pinned down in the German defensive field took advantage of the withdrawal of the armoured divisions westwards. Under cover of darkness and an annihilating bombardment that disorganised the attackers almost, but not quite, as much as the defenders, Allied forces drove to within 7 miles of Falaise, far behind the rear of the German armies trying to sustain Hitler's counter-offensive. That day Eisenhower, who had moved his headquarters to France for the first time, met with Montgomery and Bradley to discuss how to take advantage of the long weak salient now occupied by the remnants of three German armies. It was agreed that Patton should swing north to form the southern jaw of a giant pincer to cut off German forces before they could retreat. Patton obliged with a march of 35 miles in two days that brought him by 11 August to the town of Argentan, just 20 miles from the 1st Canadian army closing on Falaise from the north.

It was now only a matter of time before the jaws closed tight. All along the shrunken German line von Kluge's forces fought with desperate tenacity to keep the narrow escape route open. Von Kluge, fearful that he would take the blame for the collapse, became 'more and more gloomy'. Without permission from Hitler he ordered 2nd Panzer division to withdraw. But frantic and insulting messages from Hitler's headquarters forced him to keep the rest of his forces where they were, pressed back on all sides. There was no hope of relief. The Allies revived the Anvil attack on southern France (now codenamed 'Dragoon') as their grip in the north tightened. On 15 August Allied troops landed in southern France and against the lightest resistance, and with much willing French participation, pushed up to meet American armies driving towards Paris. Behind the trapped German armies there were only the remnants of 15th army and little else between them and the Reich. By now Hitler suspected that von Kluge was on the point of surrender. Rather than face another Stalingrad he ordered Field Marshal Walter Model, a hero of the long fighting retreat on the eastern front, to fly at once to the western pocket and perform miracles.[91]

It took Model only a few hours to see that no miracle could save the western front. He gave the German commanders permission to withdraw through the jaws, now only 5 or 6 miles wide, and closing hourly. The costs were far greater than a surrender. The exhausted Germans who ran the gauntlet through the gap were subjected to a barrage from air and artillery. Much of the equipment was abandoned. All along the roads, made impassable by bomb craters, and blocked by wagons and lorries burnt out and blown apart, trudged small groups of soldiers, who sought any avenue out of the carnage. The withdrawal had some semblance of order on 18 August, as the last German tanks held off the furious attacks of Allied forces, but when the jaws closed the retreat became a rout. A cloud of dense smoke hung over the battered mouth of the salient. Without food or munitions, German stragglers tried to fight their way through. They were raked with fire and bombed incessantly. A great part of the army escaped, but it was no longer a fighting force. Of those who did not, 45,000 were taken prisoner, and ten thousand lay in tangled heaps, among the bloated horses and crumpled tanks, victims of a slaughter for which Hitler alone bore the blame.

For the first time German forces fled before the enemy. Because Hitler had insisted on fighting the decisive battle far forward in Normandy, there was almost nothing left of German forces in France once that battle was lost. The flight that followed little resembled the fighting withdrawal in the east, or the slow, hard-fought retreat across France in 1918. The remnants of the defeated army rushed for the Seine and beyond. Here they faced the prospect of yet another encirclement, for General Patton had got there before them. While 7th army was fighting its way out of the Falaise pocket, the vanguard of Patton's forces reached the Seine at Mantes-Gassicourt, north-west of Paris. The following day the 12th corps of Patton's 3rd army pushed south of the city, and drove the weak German defence across the river. By 25 August the American forces were 140 miles beyond Paris, and only 60 miles from the German border. Model made frantic efforts to create a stable front-line, but his exhausted troops, mostly on foot or in horsecarts, were overtaken by the Allies. They straggled to the lower reaches of the Seine as American and British forces moved in for the kill. Some escaped in small craft, or on cider barrels lashed together; yet others swam across. On the far bank Model could scrape together only the equivalent of four divisions, with 120 tanks, against an Allied force of more than forty divisions. There was little else between them and the German border. The retreat continued. 'We are gaining ground rapidly,' wrote a German soldier to his family in Germany, 'but in the wrong direction.'[92]

On 25 August Paris was liberated, after a half-hearted defence by the German garrison was swept aside by the Americans, the Free French forces and a popular uprising inside the city. The following day General Charles de Gaulle, leader of the Free French forces, entered the capital in triumph. At the Arc de Triomphe he stopped to rekindle the eternal flame on the tomb of the Unknown Soldier. At the head of his forces he walked down the Champs Elysées to hysterical cheering. When he arrived at the square in front of Notre

Dame there was a burst of gunfire. De Gaulle ignored it and strode on into the cathedral, but behind him in the parade the tanks of the 2nd French armoured division opened fire on the surrounding rooftops. Inside, de Gaulle stood quietly, until he was shot at again by an unknown gunman within the cathedral itself. This time he was hustled away by anxious aides. That evening as Parisians dined out on liberation, the German air force offered one last riposte to Allied victory. At 11.15 waves of German bombers inflicted on Paris its heaviest raid of the war. They destroyed five hundred houses and a huge wine warehouse, the Halle aux Vins, whose contents burnt with such a fierce glow that the whole of central Paris was illuminated.[93]

It took only a week to clear German forces from the rest of France. By 4 September British troops entered the Belgian port of Antwerp. During the defeat in Normandy and the long rout eastwards German forces lost some sixty divisions; 265,000 men were killed or wounded, 350,000 taken prisoner. Almost all the equipment of Germany's western army was abandoned or destroyed. So rapidly did Allied armies exploit their victory that their supply lines became stretched taut and Eisenhower insisted on a period of consolidation before the final push into the Reich. The pause gave German forces just sufficient time to regroup for the last defence of the fatherland. A series of spoiling actions, culminating in the Ardennes offensive in December, in which Hitler repeated the mistaken strategy of throwing his last reserves of tanks and aircraft forward against hopeless odds, slowed up the advance of Allied armies. But after the defeat in the west, the final defeat of Hitler's Germany was only a matter of time. Pounded relentlessly from the air, threatened by a heavily armed host from west and east, German forces fought to the bitter end. On 30 April, in his bunker in Berlin, Hitler killed himself. When Albert Speer heard the news the following day he wept uncontrollably; 'Only now was the spell broken,' he wrote in his memoirs, 'the magic extinguished.'[94] To the end Hitler blamed everyone except himself for German defeat – Jews, Bolsheviks, even the German people, who died in millions for a man so obsessively self-centred that he wanted them to 'struggle to the death', 'to hold out to the last drop of blood' to prove his racial fantasies.[95]

The German people did not commit collective suicide, though many who had served Hitler did kill themselves. Rommel and von Kluge took cyanide in the autumn of 1944 to avoid dishonour. Field Marshal Model, after disbanding his defeated Army Group in April 1945, walked into a nearby forest and shot himself rather than face Allied vengeance. After Hitler's death German surrender quickly followed. On 2 May German forces capitulated in Italy. After a confused period of negotiation in northern Germany, as individual German commanders tried to surrender, the German Supreme Command finally signed the act of surrender at two o'clock in the morning of 7 May. The terms were to come into effect on all warring fronts at midnight 8 May. A formal surrender ceremony was held in Berlin on the night of 8/9 May. Air Marshal Tedder signed as Eisenhower's deputy; the hero of Stalingrad and Kursk, Marshal Zhukov, signed for the Soviet armies. In Britain the 8th and 9th of May were declared public holidays. At midnight on 7 May all the ships, large and small, around Britain's coastline set off their sirens and whistles in noisy celebration. In the Soviet Union news of the surrender was only announced on 9 May; the authorities waited another two days before allowing celebrations, for fear that German resistance might flare up again from the embers. That night a thousand guns fired a salute in Moscow, and hundreds of aircraft released red, gold and violet flares over the city. Hung from balloons, a giant red flag .hovered over the Kremlin. On the radio Stalin, who had never allowed Soviet casualties to be announced, spoke in a faltering voice of 'countless losses'.[96]

★   ★   ★

When Eisenhower placed the note in his wallet announcing the failure of Overlord he did so not from misplaced modesty, but from quite genuine anxiety as to the outcome. He found the note again on 11 July, five weeks after the secure establishment of the bridgehead. His naval aide asked if he could keep it, and Eisenhower reluctantly agreed, conscious perhaps that he might still have need of it. He was not the only one to express doubts. Although Moscow was reported to be 'awash in boozy good feeling' at news of the Normandy landings, the night before D-Day Stalin was scathing about his western allies: 'Until now there was always something that interfered . . . Maybe they'll meet up with some Germans! What if they meet up with some Germans! Maybe there won't be a landing then, but just promises as usual.'[97] Only weeks before the invasion Churchill confided to Eisenhower his nightmare of the beaches of Normandy 'choked with the flower of American and British youth'. On 5 June Brooke still believed the invasion could be 'the most ghastly disaster of the whole war'.[98]

Victory in France when it came was both sudden and complete, but it was by no means preordained. The balance of technology was at best even, though German heavy tanks, the Panthers and Tigers, outgunned those of their opponents. Until well into June the manpower of 7th army and von Schweppenburg's Panzer group outnumbered the invading force, and possessed a much greater density of both manpower and firepower than German armies in the east. Moreover the German forces had in general much tougher battle experience than their enemy, and displayed a greater willingness to fight stubbornly, hand to hand if need be, without the help of air power or the artillery barrage. Too little allowance is made in criticisms of Montgomery for the quality and tenacity of the German defence, and the inexperience of his own forces. There were points in the invasion where things could have gone badly wrong. If at any stage in the first four or five weeks the twenty divisions of 15th army had all been moved to the Normandy front the balance would have been very different. The weather was not ideal on D-Day, but if Eisenhower had decided at that critical moment to wait for the next brief period when the moon and tides held good the invaders would have been swallowed up by the great gale, which coincided exactly with the next block of favourable dates, 19–21 June. 'Thank the gods of war', Eisenhower wrote on the meteorologist's report of the storm, 'we went when we did!'[99]

Victory depended on a great many things: the prodigious organisation of supplies by sea and air, meticulous planning, the solid virtues of civilian management. But two explanations stand out above the rest. The first is the inestimable value of air power to the invading armies in every field of war – supply, reconnaissance, battlefield support and the bombing of enemy supplies and communications. The impact of Allied aircraft was magnified many times by the successful defeat of the Luftwaffe over Germany in the first half of 1944, and the destruction of German oil supplies. Aircraft gave the Allies a striking power on the battlefield that artillery alone could not supply. One German commander calculated that 50 per cent of his losses were caused by bombing. German troops were worn down by shock and fire and were demoralised by the overwhelming air superiority of the enemy. Air attack prevented the German forces from ever seizing the initiative and dictated German front-line tactics.

The second factor was deception. No doubt German forces would have remained dispersed even if more had been known about the Normandy plan, so certain were the senior German commanders that the northern Channel thrust made most operational and strategic sense. But the careful preparation of the FUSAG deception, and the exceptional good fortune over months of anxious subterfuge that shielded the Overlord plan from German eyes, ensured that the balance of forces, potentially so favourable to the defender, was evened up. German commanders were com- pelled by their uncertainty to do the very opposite of everything they had been trained to do: to dilute their forces rather than concentrate them. By the time they reverted to concentration at Mortain, the damage had been done and the Allies enjoyed local superiority in numbers strong enough to deliver the final, annihilating blow.

The defeat of Hitler's entire western army between June and August 1944 certainly did not win the war on its own, but it ended once and for all any fantasy Hitler might have harboured about stalling the western front in order to win in the east. However short the straws at which he grasped in 1944, there remained a slender chance of revival. New technology was in the offing – jet aircraft, ground-to-air missiles, rockets, long-range submarines undetect- able by Allied radar. Allied victory in France put paid to any prospect that Germany could avoid defeat. 'The war was won', Eisenhower concluded in his final report as Supreme Commander, 'before the Rhine was crossed.'[100]

6

A GENIUS FOR MASS-PRODUCTION
Economies at War

*'With our national resources, our productive
capacity, and the genius of our people for
mass-production we will . . . outstrip the
Axis powers in munitions of war.'*

F.D. Roosevelt, Message to Congress, 10 June 1941

THE SOVIET AIRCRAFT designer, Alexander Yakovlev, set
down in his memoirs one of the few first-hand accounts of
that most remarkable of Soviet wartime achievements, the
evacuation of hundreds of factories and thousands of workers from
under the very noses of the approaching German armies. In
September 1941, with the enemy only 150 miles from the Soviet
capital, the Moscow factories producing the Yak-1 fighter were
ordered to the east. Under heavy air attack, the machines were run
until they had completed every part for the aircraft on the assembly
line. Outside the factory Soviet pilots waited to fly the planes
straight into combat. Then the plant was dismantled piece by
piece.

Yakovlev's design bureau was scheduled to follow the machines.
He drove to the station at Khimki to watch the equipment leave.
He found a scene of complete confusion. Hundreds of people were
milling about on the wooden platforms. A continuous line of
battered trucks brought machinery of all kinds to be loaded on to
flat railway cars. Trains with forty cars stood on the track. One left
every eight hours for Siberia. Yakovlev found his own workers in
the mêlée. Men, women and children were loaded into box cars
hastily converted for the long journey. Each one had a number of
double bunks, an iron stove in the middle and a paraffin lamp. Each
train was placed under the command of one of the workshop
superintendents, whose job it was to supervise the loading and
unloading of their invaluable cargo.

The Yak plant was destined for western Siberia. Crude wooden
barracks were set up to welcome the industrial refugees, while local
workers prepared the supplies of electricity, water and fuel. But as

with so many of the factories shifted eastwards, there were long delays through Russia's overstrained rail system. The trains did not arrive until after the frosts had taken grip. In sub-zero temperatures the workers struggled to reassemble the workshops. Within six days of arrival production was restarted. After three months, more fighters were turned out each week than the plant had produced in Moscow. The figures may owe a good deal to the distorting effects of Stalinist economic heroism, but even after such allowance is made there is little dispute that the evacuation saved the Soviet war effort in 1942 from certain disaster.[1]

What Yakovlev witnessed was a small part of a vast exodus. Between July and December 1941 1,523 enterprises, the great bulk of them iron, steel and engineering plants, were moved to the Urals, to the Volga region, to Kazakhstan in central Asia, and to eastern Siberia. One and a half million wagon-loads were carried eastwards on the Soviet rail network. An estimated 16 million Soviet citizens escaped the German net, many of them factory workers, engineers, plant managers, all needed to keep the uprooted industries going.[2] The whole process was a messy, improvised affair. Workers arrived without their machines; equipment without its workforce. So short of rolling-stock was the rail system that car-loads of machinery were dumped beside the track in the Russian interior so that trains could return to the battlefront in the west to pick up further cargoes. The transplanted enterprises were destined for the most inhospitable regions of the vast Soviet Union. Workers struggled to assemble their new premises in temperatures forty below, while the machines, thick with hoar frost, had to be revived with braziers. There are too many accounts of work restarting on frozen earth floors in buildings with no roofs for this to be mere legend.[3]

What the Soviet people could not evacuate they destroyed. Thousands of mines, steelworks and engineering plants were abandoned. Food that could not be transported was torched. Yet for all the exceptional and desperate measures adopted, by the end of 1941 Soviet production sank to a mere fraction of the level before the German invasion. The overall levels of output were never restored throughout the conflict, but the war effort was sustained on the remarkable expansion of armaments and heavy-industrial output in the Urals and beyond. By 1942 the eastern zones supplied three-quarters of all Soviet weapons and almost all the iron and steel. The restoration of economic order out of the chaos and confusion caused by the German assault was as remarkable as the revival of Red Army fortunes after Stalingrad, and just as essential to the Allied cause.[4]

It is often forgotten that in the critical middle years of the war the balance of economic resources was not yet weighted heavily in the Allies' favour. Until the German invasion of the Soviet Union Britain and her Empire were overshadowed by the economic potential of the European Axis states and their conquered territories. Alone, it is unlikely that Britain would have survived. After the invasion of the Soviet Union the balance improved until German armies swept through the rich iron, coal and steel regions of western Russia and the Ukraine, depriving Soviet industry of two-thirds of its coal and steel. In the year leading to the siege of Stalingrad Germany produced four times as much steel as the Soviet Union. During the crisis months of 1942 and 1943, when the tide was turned on the eastern front, the balance of resources and weapons had not yet moved in favour of the Red Army. Even with the entry of the United States the situation was not transformed at once. The Battle of Midway was won with overwhelming naval strength on the Japanese side. After years of disarmament and isolation, the United States was not a major land power, and German diplomats reported to Berlin their conviction that it would take years for America's economic potential to be realised in large, well-armed forces. In reality the transformation took only a matter of months; so rapidly was America's consumer economy mobilised that by 1943 a substantial transfer of resources became possible from the United States to the two major allies.

By 1944 the balance of weapons did swing massively in the Allies' favour. But this widening gap was not a result simply of the possession of greater quantities of manpower and raw materials. In the Soviet case 8 million tons of steel and 90 million tons of coal in 1943 were translated into 48,000 heavy artillery pieces and 24,000 tanks; Germany in the same year turned 30 million tons of steel and 340 million tons of coal into 27,000 heavy guns and 17,000 tanks.[5] If the Soviet Union made the most of its attenuated resources, the

Left: Hitler as the beast from the abyss, devouring one country after another. This 1941 cartoon formed part of a collection presented by Stalin to Lord Beaverbrook.

Below: The end of old-fashioned sea power: a US Dauntless dive-bomber attacks Japanese carriers during the Battle of Midway.

Above: American submarines strangled Japanese commerce in the Pacific. This merchant ship was photographed through the periscope of the submarine, June 1945.

Left: American opinion was shocked by Japan's attack on Pearl Harbor. The thirst for revenge is strikingly conveyed in this poster from 1942.

Some sense of the scale and ferocity of the Soviet-German war is conveyed here as German soldiers cross a smashed bridge over the Don river in July 1942 through a tangle of broken vehicles.

Right: The performance of the T-34 medium tank was marred by the lack of radios and a hatch which prevented all-round vision, clearly visible here. By the time of the Battle of Kursk in 1943, these deficiencies had been made good.

Far right: The Soviet counter-offensive relied on large numbers of rugged, easily maintained weapons. Here the SU-76 self-propelled gun is in action in the advance into Prussia in 1945.

German forces called Stalingrad 'the cauldron'.

Above: The city ablaze at night under the impact of air and artillery fire from both sides.

Below: German soldiers were poorly prepared for the cold. Here an igloo has been built around a German truck to prevent the engine from seizing up in the strong icy winds.

BRITAIN'S MIGHTIEST BATTLESHIP— EXCLUSIVE DIAGRAMS AND PICTURES

12th JULY, 1940

THE **WAR**

No. 38

3D WEEKLY

Incorporating WAR PICTORIAL

OUR BOMBERS BATTER HITLER'S FACTORIES

Ever since Hitler began his Western offensive on May 10, the R.A.F. has carried out daily raids on Germany—on aircraft, chemical, armament and munitions works, petrol, oil and ammunition stores, railway marshalling yards and other military objectives. This is our artist's impression of Blenheim bombers—which fly in formations of three by day—scoring direct hits on a German chemical works during a daylight raid on an industrial area.

Artistic licence: in this drawing from the summer of 1940 there is no hint of how ineffective British bombing actually was. Factories could barely be located, let alone hit, and the slow Blenheim bombers pictured above were easy prey for German defences.

Top: In July 1943 British and American bombers destroyed 70 per cent of Hamburg. All the RAF bombers could see of the city at night were smoke, flames and the flash of anti-aircraft fire.

Above: The Kammhuber Line: across German-held Europe a wall of radar, searchlights and anti-aircraft guns was set up to combat bombing. These 88-millimetre guns were among 55,000 anti-aircraft artillery in place in 1944.

Top: Counting the cost: over 600,000 Germans lost their lives through bombing. Here the dead from a raid in Berlin in December 1943 have been laid out in a gymnasium decorated with Christmas trees.

Above: Tourist trap: the ruins of Hiroshima became a sight-seeing attraction for Allied forces in 1945. Sailors of the Indian Navy have made the 15-mile trip from the naval base at Kure to see for themselves what atomic power can do.

Top: The assault across the beaches on 6 June 1944 left well over three thousand dead in the first hours. The fragile beachhead was secured by the end of the day, helped by amphibious tanks visible at the top of the picture.

Above: A group of British artillerymen facing the German front at Caen. For many Allied soldiers war was waged at one remove, with artillery and aircraft neutralising the enemy at long range before the infantry moved forward.

Top: The two artificial ports or 'Mulberries' set up off the Normandy coast were vital to the supply of Allied forces. The American port pictured here was rendered inoperable in the 'Great Gale' of 19–22 June 1944.

Above: The mass-produced ship: here sections of Liberty cargo ships wait to be lifted by crane and welded into position. Each metal plate was clearly marked to ensure an exact match with adjacent sections.

When German forces seized the steel-producing regions of the Ukraine in 1941, Magnitogorsk in the Urals was developed to become the new centre of the Soviet heavy industrial economy.

The 'largest room in the world': the main assembly hall at Willow Run, Michigan, where Ford produced B-24 Liberator bombers like cars.

Der Gefechtstroß folgt der vorrückenden Truppe mit Munition, Verpflegung und Gepäck. Geländehindernisse halten ihn nicht fest. Auch der von den letzten Regengüssen angeschwollene Bach hemmt seinen Marsch nicht

PK-Aufnahme, Kriegsberichter Bergmann (Sch.)

This could almost be a scene from the American Civil War. It is the German army deep in Soviet territory and mud in the late summer of 1942. German forces remained heavily reliant on horses and wagons throughout the campaign.

Japanese soldiers in World War II trained in archery according to the *samurai* tradition. Armed with rifles they were judged poor shots by their enemies.

A nuclear bomb of the type used against Nagasaki, nicknamed 'Fat Man'. The bomb was 60 inches in diameter and 10 feet long, weighed 10,000 pounds, but had a yield equivalent to 20,000 tons of high explosive.

During the first summit talks of the three Allied leaders, Stalin was presented with the Sword of Stalingrad. Foreground, left to right: Stalin, Voroshilov (holding the sword), Roosevelt and Churchill.

War by committee: British and American military chiefs thrashed out policy for Overlord at the Quebec conference in October 1943. General Brooke (standing) and General Marshall, fourth from right, dominated the proceedings.

With Catholics on both warring sides, Pope
Pius XII (seated) broadcast to Washington in the winter
of 1939 the diplomatic view that the war was caused by
materialism.

Above: Atrocity pictures were used in Allied propaganda
to reinforce the popular image of a barbarous enemy.
In this picture, published in Britain in 1944, German
officers carry out the execution of two Soviet peasants.

'**These are Japs.** The would-be overlords of Asia are represented by these low-browed, puzzled-looking little men with whom American captors share food in Bataan. They seem surprised that the torture they had been expecting was not meted out to them.'

Above: The Allies in the Second World War distinguished in their propaganda between the European and the Asian enemy. The Germans were regarded as corrupted barbarians, but the Japanese were presented as primitive racial inferiors – evident in this image published in Britain in 1944, reproduced here with its original caption.

Left: Allied propaganda de-humanised the enemy. This Soviet cartoon ridiculing Nazi race theory encourages the reader to see the enemy in animal terms.

Saving Mother Russia: here Soviet propaganda married together the drive against the German invader in 1942 and the legendary defeat of the Teutonic Knights in the thirteenth century by Prince Alexander Nevsky.

When the German war effort turned sour in 1943, Goebbels played upon popular fears of Bolshevism – portrayed here with a Jewish face – to keep the population fighting.

new German empire failed to make the most of its economic advantages. This was partly a result of economic warfare. The Allies tried to deny their enemies the use of critical resources, oil in particular. Bombing seriously restricted the scope of Axis production from 1943. Unconstrained by enemy action, Axis production in 1944 and 1945 would certainly have been higher.

The key years were earlier in the war. It was in 1942–3 that the disparity between the two sides was created by Soviet industrial revival and American rearmament. Neither could be taken for granted. Success on these two economic battlefronts shaped the military successes, at Stalingrad and Kursk, in the Atlantic Battle, and in France.

*   *   *

In 1941 the Soviet economy was threatened with complete collapse. In a matter of months German forces conquered the main industrial and agricultural regions of the Soviet economy. The rich grain lands of the west, the Soviet 'breadbasket', passed over to the enemy. In 1942 grain supplies fell by half for the 130 million Soviet citizens in the unoccupied zones; meat production fell by more than half.[6] One-third of the Soviet rail network was lost, and 40 per cent of electricity generating capacity. The lifeblood of modern industry – the supply of iron ore, coal and steel – was cut by three-quarters. The availability of vital resources for modern weapons – aluminium, manganese, copper – fell by two-thirds or more.[7]

The Soviet state was for the moment reduced from being the world's third largest industrial economy, behind the United States and Germany, to the rank of smaller economies, such as France, Italy and Japan. Formerly resource-rich, the Soviet economy was now poor in almost everything except oil, timber and lead. Another government might have given up the struggle there and then, or, like the Tsarist economy in World War I, limped on from one disaster to the next. The figures alone can scarcely convey the extent of the catastrophe, as a confused mass of officials, workers and equipment fled before the invading armies into the vast, predominantly rural hinterland, there, with a shrunken fuel supply, deteriorating transport and a hungry workforce, to rebuild Soviet industry and keep over two hundred Soviet divisions in the field.

Against every reasonable expectation, the Soviet economy repaired the fractured web of industry, transport and resources and in 1942 produced more weapons than a year before, and more weapons than the enemy. Moreover many of these weapons were of improved quality, reversing the uneven technical confrontation of 1941. It proved impossible to make good the losses of coal, iron and steel, but Soviet factories were able to use the iron and steel that could be produced solely for the most urgent war production. In 1943 the gap between Soviet and German production widened further. In the middle years of the war Soviet factories produced three aircraft for every two German, and almost double the number of tanks. The balance of heavy artillery was three to one. The Soviet economy outproduced the German economy throughout the war from a resource base a good deal smaller and with a workforce far less skilled.[8]

This was a remarkable achievement by any standard, but it is easier to describe than to explain. The simple answer might be that the Soviet Union operated a command economy, directed by the state and centrally planned. There is certainly something in this. The Soviet authorities did not have to collaborate with private capitalist interests, or reach compromises with labour. The economy was governed by decree and reluctant workers or incompetent managers filled the swelling population of the Gulag camps, where they worked for the war effort behind barbed wire.[9] But Stalin could not simply command the economy to produce, any more than Canute could stop the tide. Coercion was never enough on its own to distil weapons from the disordered industrial rump left in 1942. Soviet economic achievements owed more to planning.

The Soviet Union epitomised the cult of planning that gripped a whole generation of Europeans and Americans after the Great War. Soviet planning was introduced from the mid-1920s to replace the threatened revival of a semi-capitalist economy with an economy based entirely on social ownership and modern systems of production. Planning was perceived to be the only way in which the economic and social backwardness of the Soviet Union could be overcome quickly. The pre-war Five-Year Plans, begun in 1928, transformed the Soviet Union in ten years into a major industrial player and turned Soviet society upside down. The peasantry was regimented and stripped of its private property; thirty million moved from the countryside to the cities in under a decade. On paper the Soviet Union by 1940 was the world's largest military power, and the second largest economy. Centrally planned industrial development, on greenfield sites, was nothing new to Soviet officials when the war came. The large-scale movement of population to staff the new factories, and the training needed to give them even rudimentary skills, was a familiar experience. The planning process itself was based on the crude matching of resources to plans for physical output, means to ends; though the system worked imperfectly, with much creative improvisation and a heavy cost in state terror, it gave economists, engineers and managers just those skills most in demand for organising a war economy.[10]

There was a solid base from which to work in 1941, though nothing could have prepared Soviet planners for the nightmare they faced as one after the other key elements of the planned economy fell under enemy control. Central planning became impossible and the government was forced to introduce what they called the 'regime of emergency measures'.[11] In July 1941 the head of the Soviet economic planning agency Gosplan, Nikolai Voznesensky, was instructed to draw up a regional plan for a war economy based on the Urals-Volga-Siberia heartland. He moved from Moscow to the provincial capital at Kuibyshev, on the Volga river north of Stalingrad, with the Ural mountains visible to the east. Here he drew together the Soviet commissariats responsible for military production – Aviaprom for aircraft, Tankprom for tanks – to form an industrial cabinet away from the front-line. For a year they struggled from one emergency to another, improvising, hustling, bullying. Where there was no coal they ordered factories to burn wood or peat. Shortages of rail transport compelled firms to become self-sufficient, supplying all their own parts and services, and making do without the usual ring of small contractors surrounding every armaments complex. Geologists scoured the Siberian countryside for new supplies of vital minerals. At

Sverdlovsk hundreds of scientists and technologists were sent from the Moscow Academies to spend all their time working out solutions to the thousands of technical problems thrown up by improvised, poorly-resourced production.[12]

The emergency programmes were bound together by a single, common theme: the necessity of using anything and anyone to promote war production at the expense of everything else. This was a great simplifier, but it brought its own problems. Voznesensky had to fight off military demands for valuable manpower. Gradually a set of priorities was worked out which exempted skilled men from the call-up. In the wave of fear and panic that gripped the Soviet Union in 1942, the NKVD stepped up its search for scapegoats – workers guilty of 'economic misconduct', saboteurs, spies. As German armies neared the Volga in the late summer of 1942 a second wave of evacuation began; in some instances factories were moved a second time. Improvisation yielded enough dividends for the Red Army to survive, but by the autumn of 1942 widespread confusion and inefficiency accentuated the need to return to greater central planning. In November a centralised Manpower Committee took over the allocation of labour supplies; on 8 December Voznesensky was appointed head of national planning once again, and in addition the planners were given real teeth for forcing the plans through. In 1943 a single national plan could once again be drawn up.[13]

The great strength of Soviet planning lay in the scale and simplicity of the goals set. There was nothing sophisticated about Soviet ambitions. The war required large numbers of weapons produced as simply and quickly as possible. In November 1941 the State Defence Committee ordered the production of 22,000 aircraft and 22,000 tanks for 1942. These were the benchmarks. It was up to the engineers, managers and workers to fulfil the requirements as best they might. The Soviet Union could not afford the luxury of employing a wide range of different types of weapon; there were two main tank types, the T-34 and the KV-I, and five main aircraft types, three fighters, one bomber, one fighter-bomber.[14] Technical development was confined to improving chosen models, which enabled Soviet engineers to catch up with German standards very quickly. Modifications were kept to the most essential; crude mass-production ensured large numbers and robust construction. During 1942 Soviet industry in the Urals-Volga-Siberia region supplied not 22,000 tanks and 22,000 aircraft, but 25,000 of each.

Soviet factories displayed the same priorities of scale and simplicity. When the industry of the Urals region was first developed in the 1920s, Soviet planners, uninhibited by the usual pressures of the free market, were able to build giant factories and furnaces, huge complexes producing everything from the molten metal to the finished machines and tools. The model for Stalinist industrialisation, and the later heart of the eastern war economy, was the new industrial city of Magnitogorsk, on the eastern slopes of the Ural mountains, 300 miles east of Kuibyshev. The project begun there in 1928 was to build a single steelworks capable of producing more than the whole pre-1917 Tsarist empire: 5 million tons of steel, 4.5 million tons of iron. The city took its name from the high-grade magnetic iron ores on the surrounding ironfield. It was built with the help of American engineers, and was equipped with American and German machinery.[15] By the time the war started the steelworks had six blast furnaces and twenty open-hearth furnaces, and it was expanded with the addition of evacuated equipment which was appended untidily during 1942. The whole vast complex employed 45,000 people.

Magnitogorsk was dominated by the steelworks; a perpetual dark haze hung over the city. Around the barracks and houses of the workers lay a patchwork of small allotments where they grew potatoes to supplement the meagre meals served in the vast works canteen. By American or even European standards, the plant was not very productive. It was untidy and dangerous. When the president of the American Chamber of Commerce, Eric Johnston, visited it in 1944 he found a vast inferno, filled with choking fumes, with canals of unprotected molten metal, and piles of slag and iron scrap cluttering the roadways between each workshop. In the part of the plant making shells the absence of a moving conveyor belt was compensated for by the use of long inclined wooden racks down which the shells were rolled on their way along the production line. But throughout the plant, directed by a 35-year-old blacksmith's son, Gregor Nesov, the Americans found a constant bustle and drive. Clean premises were not a priority for the war effort.[16] Magnitogorsk concentrated everything on production. Over the whole course of the war vast, dirty, ill-lit plants all over central Russia were worked day and night using standard equipment and simple procedures. While the rest of the economy remained at the crisis point reached in 1941, the output of each worker in Soviet war industry increased two- or three-fold over the course of the war.[17]

In the other Ural cities around Magnitogorsk Soviet planners sited other giant factories. At the Ural capital of Sverdlovsk, set amid rich pine forests which provided much of the fuel and building materials, there grew up a major machine tool production centre, known in Sovietspeak as 'Uralmash'. Home to a million people by the time of the war, the city became a major centre for tank and artillery production. With an increase in the size of the workforce in the gun plant from six thousand to ten thousand, the output of heavy artillery increased six-fold between 1941 and 1944. At the neighbouring city of Chelyabinsk, a vast tractor works set up during the collectivisation drives of the 1930s was turned over to the mass production of T-34 tanks. Equipment from the tank factories at Leningrad and Kharkov was married to the existing plant. The whole complex became known as 'Tankograd', 'tank city'. In three giant plants in the east two-thirds of all Soviet tanks were produced. Improvements in the production process – the tank turrets were stamped out by huge presses, rather than cast, and automatic welding replaced handwork – reduced the man-hours on each tank from eight thousand in 1941 to 3,700 by 1943.[18] Wherever possible scarce skilled labour was replaced by machines, and production standardised and simplified. The resulting weapons looked rough-hewn by western standards but refinements were an unmerited luxury. Mass-production, borrowed from American practice in the 1920s, developed in the 1930s to accelerate Soviet development, was the key to the wartime production record.

The real heroes of the Soviet Union's economic revival were the Soviet people themselves, managers, workers and farmers. The war made quite exceptional demands on the civilian population. The great majority of men between 18 and 50 were conscripted into the armed forces during the war. One million Soviet women joined them in uniform. This left a workforce made up of women, old men and teenagers. By 1943 women made up just over half the industrial labour force. On the collective farms their share was almost three-quarters. By 1944 able-bodied males made up only 14 per cent of the workers on state farms. There was nothing new about female labour in the Soviet Union – women made up two-fifths of the labour force in 1940 – but what was new were the appalling conditions in which all workers laboured in wartime Russia.[19]

The gruelling regime of work imposed on the Soviet people was not deliberately inflicted but was the product of the sudden crisis following the invasion. All holidays and leave for workers was cancelled indefinitely for the duration. Hours worked were fixed at twelve to sixteen per day; three hours compulsory overtime was introduced. On Stalin's exhortation to turn the Soviet Union into 'a single war camp', large sections of the workforce were placed under military law, first the country's building workers, then munitions workers, and, finally, in April 1943, the railway workers.[20] When the cavalcade of limousines drove the director of the Magnitogorsk plant and his American guests through the 'teeming, unpainted slums' of the city, they passed a long column of workers, marching four abreast towards the steelworks. At the front and sides of the column were military guards with fixed bayonets. The workers turned out to be hundreds of 'ragged women in makeshift sandals'.[21] For these women and for millions of other Soviet workers, the factory became a battlefield. Absenteeism and lateness were treated like desertion. Repeated offences meant the labour camp, though the conditions of everyday life were so drear for most workers that life in the camps and outside them became increasingly difficult to distinguish.

The greatest source of hardship was the food supply. Though the authorities managed to organise a nationwide rationing system, it did little more than impose malnutrition equitably across the working population. For the miners and metalworkers, engaged on heavy war work, there were 2 pounds of bread a day, eked out with a meagre 5 pounds of meat, a pound of sugar and a pound of fats each month. For the average consumer a pound of bread a day, and scraps of fat and meat, were all that could be spared. This was a quarter of German rations, one-fifth of British, for a work day that was longer and harder to endure. Hunger was averted only by sowing every garden with vegetables and potatoes. By 1943 some seven million cottage gardens kept the Soviet workforce going.[22] For those who produced the food on the collective farm the situation was worse. Rural labourers got only half a pound of bread and a potato or two each day. Farm workers were forced to work long hours. Most of the horses and a great many tractors were taken away by the army. Women workers were trained to use what machinery remained, and by 1944 over 80 per cent of tractor drivers were female. But for many farmers ploughing was done with oxen, or even, in extremes, by teams of women and youths who pulled the ploughs themselves. After hours in the fields, tired and in bitter weather, they took their turn at felling timber to feed the local factories with power.[23]

How Soviet workers kept going, month after month, exhausted, hungry, terrified that any slip or dereliction might be classified as sabotage, defies belief. No other population was asked to make this level of sacrifice; it is unlikely that any western workforce would have tolerated conditions so debilitating. The story of the Soviet people is one of epic endurance, that needs no embellishment of Soviet propaganda to make it convincing. Why did they do it? This might have seemed a curious question to the war-workers, not just because they were the victims of state conscription which limited freedom of choice, but also because the expectations, outlook and experiences of Soviet workers and farmers were entirely different from those of their more privileged western cousins. The characteristics of the war economy – large, crudely constructed factories, a harsh barrack room existence, poor food and tough discipline – were the features of Russian working life since the late nineteenth century when the Tsars began to modernise the Imperial economy. They persisted in the four years of civil war following the Russian Revolution in 1917; the forced industrial-isation of the Five-Year Plans was an upheaval every bit as exceptional and distorting as the rush for war production in 1942. Russians of sixty or more would have seen it all. This tough history did not make wartime conditions any easier to bear, but it helps in understanding why they were borne.

There are other less speculative explanations. The workforce was subjected to a barrage of propaganda that generated an ethos of struggle and commitment. In the 1930s the enemy was Soviet backwardness, socialist progress the goal; in the war the enemy was the Fascist Beast, 'Everything for the Front' the aim. Every workshop was decorated with banners promoting economic heroism and posters advertising the names of plant workers who exceeded their work norms. In a society where material incentives made little sense, since there was almost nothing to buy in the shops, there developed a popular culture of achievement; workers vied with each other to perform extraordinary feats of labour, to be rewarded with public praise and the occasional medal. Every factory had its small assembly place, the wooden stage and the dais where the plant managers announced the roll of honour for the over-achievers and castigated slackers in front of their workmates.[24] Every worker knew the name of the legendary steel-worker Bosyi who arrived in Nizhny Tagil from Leningrad in 1941 where he proceeded to produce his five-month quota in fifteen days. Bosyi got the State Prize. The spirit of what was known as 'socialist emulation' spread throughout the economy, echoing the Stakhanovite movement of the 1930s when incentives had been given to workers to exceed work norms by a wide margin. The women tractor drivers got their own hero, driver Garmash, whose team completed a year's ploughing by June through the simple expedient of keeping the tractor shifts going for over twenty hours a day.[25]

No doubt the competitive pride that fuelled the emulation programme owed a good deal to official encouragement. Other incentives were developed to keep the workforce going. Large factories had kindergartens and schools, and canteens in which the workers could get three meals a day for 5 roubles, outside the regular rationing system. Since cash had little value even on the black market, where goods were beyond the reach of the most prosperous worker by 1943, factories began to reward their workers in kind, with additional food and fuel, and to punish them by withholding food.[26] Work took on a new meaning once it was linked directly to the food supply. For a great many Soviet citizens the workplace became the literal source of sustenance, of food, warmth and companionship. Wartime collectivism, like wartime planning, was more than just a party slogan.

In the end the motives that kept workers at their ploughs and lathes through years of profound suffering and physical exhaustion can only be guessed at. Few families were unaffected by losses on the battlefield. The refugees had the bitterness of enforced exile and the wild stories of German atrocity to fire their efforts. There was no shortage of ideological enthusiasm, however naive it now seems. Teams of energetic Young Communists acted as economic shocktroops, shaming their elders into greater efforts, and spreading the socialist gospel. For those resistant to such blandishments there was Soviet patriotism, simple-minded perhaps, unduly credulous, but real enough. To attribute the suffering and efforts of the Soviet workforce simply to a sullen acceptance of coercion is to diminish both. When the American visitors to Magnitogorsk were introduced to a sour-faced young woman, an exceptional over-achiever, they asked her why she did it. Instead of a stock Marxist-Leninist answer, she explained that she worked from hatred, born of the death of her parents under German rule.[27]

Planning, mass production and mass mobilisation were the pillars of Soviet survival and subsequent revival. The Soviet Union was turned into Stalin's 'single war camp'. The costs were high for the Soviet people as they struggled to come to terms with life in an economy where there was little left over for civilians once the forces were equipped and fed. Theirs was an exceptional, brutal form of total war. The resilience of the Soviet people in the face of the German assault and the remorseless demands of their own regime needs a Tolstoy or a Dostoevsky to do it justice. Here was Ilya Ehrenburg's 'deep war', sustained, he later recalled, by an 'unobtrusive day-to-day heroism'.[28]

*   *   *

The situation facing the United States economy could hardly have been more different. There was no doubt that America had the resources for a prodigious war effort. British leaders had longed before Pearl Harbor to have that abundance at their disposal. 'There is one way, and one way only', the British economist Sir William Layton told an audience of American industrialists in October 1940, 'in which the three to one ratio of Germany's steel output can be overwhelmed and that is by the 50 to 60 million ingot tons of the United States.'[29] In 1941 America produced more steel, aluminium, oil and motor vehicles than all the other major states together.

The problem was how to turn this abundance from the purposes of peace to those of war. The United States had no tradition of military industry. Intervention in the First World War came too late to build up war production of any real size. The 'military-industrial' complex was the product of a later age. By the 1930s twenty years of disarmament and detachment left the world's richest economy with an army ranked eighteenth in size in the world, and an air force of 1,700 largely obsolescent aircraft and a mere twenty thousand men.[30] In 1940 military expenditure made up just 2 per cent of America's national product. Military weakness was a consequence of an isolation both geographical and political. The American public displayed a deep-felt hostility to war and militarism. In 1937 comprehensive neutrality legislation passed through Congress, intended to keep the United States out of other people's wars, and to limit the trade and production of armaments. Many Americans regretted the intervention of 1917 and were determined not to make the same mistake twice.

This was not the only political issue. America's was a free market economy, emerging in the late 1930s from a decade of economic hardship into the full glare of a consumer boom. The American government could not simply suppress its people's expectations and turn butter into guns. Both business and labour distrusted state power, no more so than when the state was planning to spend their money on arms. The effort to increase federal responsibilities under Roosevelt's 'New Deal' for economic recovery in the 1930s brought bitter disputes. Unlike in Germany or the Soviet Union, the growth of the military economy depended on reaching a broad consensus across the whole political spectrum, from hard-nosed Republican bosses to tough-minded Democrat unions. This remained the case even after Pearl Harbor gave Roosevelt the perfect opportunity to cut across all the objections to economic mobilisation at a stroke. The American people reacted with a fiery indignation to the Japanese attack in the early months of 1942, but the United States itself was not threatened with invasion; no bombs fell on American cities. The conflicts were an ocean away, and sustaining a popular commitment to production and economic sacrifice was an altogether different issue from how it appeared in Britain or the Soviet Union.

For all these reasons American rearmament was slow to materialise before 1942. Roosevelt was able to win additional budget funds for the navy only on the grounds that the navy was a genuine instrument of defence rather than offence. The tiny American military aircraft industry was stimulated from 1939 onwards by the demand for aviation equipment from Britain and France. Only in 1941, with a third Presidential election successfully behind him, did Roosevelt feel confident enough of popular support to begin to rearm more earnestly. On 9 July 1941 he asked the army and navy to draw up a comprehensive plan of all the resources they would need to defeat America's potential enemies. The programme became known informally as the 'Victory Programme'. The results were presented to the President in September, but a final estimate could not be prepared until British and Soviet requests for military aid were drawn up and approved. As a result a final programme for war production was barely ready before the United States found themselves at war with Japan and Germany in December. Until a clear general picture of the scale of America's planned rearmament was available, the President could do little more than authorise temporary and uncoordinated contracts, many of them to feed the needs of the other warring states for whom he had promised in December 1940 to make the United States the 'arsenal of democracy'. In the whole of 1941 military expenditure was just 4 per cent of the amount America spent between 1941 and 1945.[31]

When war broke out the United States had still a predominantly civilian economy, with a small apparatus of state, low taxes, and a military establishment that had only reached the foothills of re-equipment. America faced states which had been arming heavily for eight or nine years and now had more than half their national product devoted to the waging of war. American leaders were conscious of how much there was to catch up. The giant plans approved by Roosevelt and Congress in the first weeks of war did not just result from America's great wealth of resources, but reflected a genuine fear of military inferiority. In four years these plans turned America from military weakling to military super-power. American industry provided almost two-thirds of all the Allied military equipment produced during the war: 297,000 aircraft, 193,000 artillery pieces, 86,000 tanks, 2 million army trucks. In four years American industrial production, already the world's largest, doubled in size. The output of the machine tools to make weapons trebled in three years. The balance between the United States and her enemies changed almost overnight. Where every other major state took four or five years to develop a sizeable military economy, it took America a year. In 1942, long before her enemies had believed it possible, America already outproduced the Axis states together, 47,000 aircraft to 27,000, 24,000 tanks to 11,000, six times as many heavy guns.[32] In the naval war the figures were more remarkable still: 8,800 naval vessels and 87,000 landing craft in four years. For every one major naval vessel constructed in Japanese shipyards, America produced sixteen.[33]

Production on this scale made Allied victory a possibility, though it did not make victory in any sense automatic. For all the obvious advantages of resources and remoteness from the field of battle, the arming of America on this scale and with such speed could not be taken for granted. In the first weeks of war the administration struggled to produce a coherent picture of where and when military goods could be supplied. On 5 January 1942 the auto-mobile industrialist, William Knudsen, appointed by Roosevelt to run the pre-war rearmament agency, the Office of Production Management, resorted to the extreme of calling together a room full of businessmen where he read out a long list of military products and simply asked for volunteers to produce them.[34] For all its curious informality this was an approach more calculated to work with an industrial community that disliked taking orders and thrived on technical challenges. The urgency of mobilisation left the government with little choice but to rely on the initiative and technical flair of American business. The strengths of the American industrial tradition – the widespread experience with mass-production, the great depth of technical and organisational skill, the willingness to 'think big', the ethos of hustling competition – were just the characteristics needed to transform American production in a hurry.

Even before the outbreak of war Roosevelt had begun to mend the bridges between his Democrat administration and the largely Republican business elite. The liberal, pro-labour stance of Roosevelt's government was played down. He needed the political cooperation of business, for he knew that he could not just impose a state-run war economy. When war broke out he sought their support by building a structure for wartime planning and supervision largely run by business recruits. This made practical sense. Corporate bosses had as much, if not more, experience of the kind of planning and coordination needed in a wartime economy than did government officials, whose only real experience was the ill-starred New Deal. They preferred a strategy where business was given a good deal of responsibility to get on with the job. The new agencies – the War Production Board under the Sears-Roebuck director, Donald Nelson, the Controlled Materials plan, the Manpower Commission – dealt only with those issues that the marketplace could imperfectly direct in wartime.[35]

There was a scramble of volunteers for war contracts, in which the largest corporations were in a strong position, not least because co-opted directors sat side by side with the officials placing the orders. By the time the first rush was over four-fifths of all war orders had been given to the country's hundred largest businesses. American industrial plants eclipsed in size even the gigantic factories of the Urals. Some of them were so large that they were able to undertake war tasks on a scale no other economy could match. The General Motors Corporation alone supplied one-tenth of all American war production, and hired three-quarters of a million new workers during the war to produce it.[36] The great scale of American pre-war output, made possible by the size and wealth of its domestic market, allowed the widespread use of the most modern techniques of mass-production. Though there existed some scepticism in military circles about the feasibility of manufacturing technically complex weapons with the methods used for Cadillacs, even the largest equipment, heavy-bombers and shipping, ultimately proved amenable. No story better exhibits that 'genius for mass-production' summoned up by Roosevelt than the story of the Liberty Ship.

In 1940 the British government ordered sixty cargo vessels from

American dockyards to make good losses from submarine warfare. Working from the original British designs, American shipbuilders came up with a standard vessel, 420 feet in length, capable of carrying 10,000 tons of cargo at a plodding 10 knots. It was a plain, workmanlike vessel. Roosevelt thought it a 'dreadful-looking object'; *Time* magazine christened it the 'Ugly Duckling', and the name stuck until the US Maritime Commission, which was in charge of the building programme, insisted that what they were making amounted to nothing less than a Liberty Fleet. When the first ship was launched in September 1941 from the brand-new Bethlehem-Fairfield shipyard in Baltimore, the President himself lent dignity to the occasion. It was a gala celebration; the ugly duckling became overnight the Liberty Ship.[37]

Over the next three years the initial order for sixty ships swelled into a programme for 2,700. Each new demand placed an ever greater burden on the overstretched shipbuilding industry, short of skilled labour and berths. At first sight the ships did not lend themselves easily to mass-production, but at the west coast shipyards of Henry J. Kaiser the old principles of shipbuilding were overturned in 1942. Kaiser was new to shipwork. He began life running a photographer's shop in New York, moved into the gravel business, and ended up in California running a multi-million dollar construction company that built the Hoover Dam and the Bay Bridge. He had a reputation for tackling the impossible. When the shipbuilding programme started his initial involvement was the construction of four of the new yards on the west coast, but he then began to produce the ships as well. At his Permanente Metals Yards No. 1 and No. 2 at Richmond, on the northern edge of San Francisco Bay, the young Kaiser manager, Clay Bedford, set out literally to mass-produce ships.

The secret of the new method was to build much of the ship in large prefabricated sections. Instead of setting down a keel on the slipway, and slowly building the ship from the hull upwards, riveting one piece of steel to the next, the Liberty Ships were built in parts, away from the slipway, and then assembled there by modern welding methods. The shipyards were designed like a long production line, stretching back from the coast. A mile from the shore vast assembly sheds and storehouses were built, where the superstructure of three ships at a time was built up along a moving line. At points along the line, components and sub-assemblies were fed in from the storehouses by conveyors and overhead pulleys. Each job was broken down into a series of simple tasks which could be mastered by even the most briefly trained worker. Outside the sheds, in the open air, were 80-foot conveyors on which the completed superstructure and bulkheads, complete with plumbing and wiring, were moved to the berths, to be lifted by four great cranes on to the welded hull. The whole complex of conveyors, rail lines, cranes, the piles of preassembled, standardised parts, the army of hurriedly-trained workers distributed by time-and-motion experts along the line, the factory banners exhorting workers to build 'ships for victory', became an unexpected monument to that American obsession, rationalisation.[38] At the start of the programme ships took 1.4 million man-hours and 355 days to build. By 1943 the figure was under 500,000 man-hours and an average of 41 days. At Richmond No. 2 the Liberty Ship *Robert E. Peary* was launched in just eight days. The methods gradually filtered out to other shipyards. Over the war years productivity in the shipbuilding industry increased by 25 per cent a year. 'We are more nearly approximating the automobile industry than anything else,' the head of the Maritime Commission told a Congressional hearing in 1942.[39]

To any American audience the motor industry was an obvious benchmark. It had pioneered modern methods of production; its commercial history was one of the distinctive success stories of American enterprise. If the rationalisation of shipbuilding was a surprising bonus, the motor industry, because of its core of mass-production giants – Ford, General Motors, Chrysler – was expected to play a major role in the efficient production of war equipment from the start. American vehicle manufacture grew up in the mid-western states, on the edge of the Great Lakes. Here, in the American equivalent of the Urals-Volga heartland, was concentrated the largest manufacturing complex in the world. In 1941 over three and a half million passenger cars were produced there. During the war production dropped to the extraordinary figure of just 139 cars.[40] This decline freed enormous industrial capacity for the war effort. By 1945 the industry supplied one-fifth of all the country's military equipment, including almost all the vehicles and tanks, one-third of the machine guns and almost two-fifths of aviation supplies. The Ford company alone produced more army equipment during the war than Italy.[41]

The conversion of an industry of this size presented all kinds of problems. Until the very eve of war the powerful car-makers resisted efforts to reduce civilian output; 1941 was a record year for the motor industry. When war came they were quite unprepared for the changeover. The last day for civilian output was fixed for 10 February 1942. Small ceremonies were held in the car plants as the last chassis came down the line. Then, under the energetic super-vision of a former Ford manager, Ernest Kanzler, the machinery was ripped out and new tools were installed to begin the mass-production of weapons. This was a daunting task. Although the authorities assumed that the car plants could be turned off and on again like a tap, the manufacture of armaments on a large scale was very different from assembling cars. Weapons exhibited a greater degree of complexity and required higher precision; they needed periodic updating and improvement, which made long production runs hard to sustain. That the motor industry did adapt so success-fully owed a good deal to the character of the industry. Annual model changes accustomed managers and workers to regular large-scale adjustments on the factory floor; the large companies were used to a wide product range; the workforce was compelled by the nature of the manufacturing process to be flexible and adaptive. Car manufacture was in most cases a job of assembling parts and equip-ment provided by outside firms, and the practice of employing specialist sub-contractors – General Motors used nineteen thousand of them during the war – was carried over to the production of tanks, aircraft and engines. Once the conversion was completed the industry began to overfulfil its orders. As early as the autumn of 1942 the motor industry could provide enough vehicles, ordnance and equipment to provision America's new armies throughout 1943, and to do so with weapons of high quality and standard construction.[42]

The ultimate challenge was to produce an aircraft by the same method used to mass-produce a car. The consensus was that it could not be done. This was a technical challenge too great to resist for that apostle of progress, Henry Ford. He was the archetype of the heroic unschooled entrepreneur of American legend, a man whose faith in the possibilities of the machine age was boundless and uncritical. Early in 1941 the Ford company was invited to produce parts for the new B-24 Liberator bomber being built by the Consolidated aircraft plant in San Diego. When Ford's general manager, Charles Sorensen, visited Consolidated he was appalled at the modesty of their plans and the primitive methods of construction. That evening, 8 January, Sorensen sketched out the plan of a plant to mass-produce bombers. It was the start of one of the most famous projects of the American war effort. The army was lukewarm about the idea when Sorensen and Ford suggested it a few days later, but Ford persisted: he would either produce the whole aircraft in a purpose-built plant, or nothing at all. The army reluctantly concurred and by March 1941 work began on the factory foundations.[43]

The site chosen was dictated by the sheer dimensions of the plan and the need to have an adjacent airfield for aircraft testing. In the open country south of Detroit Ford had bought some flat, tree-covered farmland. A small creek meandered across it till it hit the river Huron. It was known as Willow Run. The name was adopted for the project itself. Within months the quiet rural vista was transformed into a sprawling, noisy building site over 900 acres in extent. The centrepiece was the main assembly hall, 'the most enormous room in the history of man', a vast L-shaped construction that eventually housed an assembly line 5,450 feet long, and covered an area of 67 acres. Ford's aim was to break down the construction of the aircraft in such a way that the components could be fed into a continuous moving assembly line. With a car this could be done relatively easily, for it averaged fifteen thousand parts; but the B-24 had thirty thousand different parts, and a total of 1,550,000 parts in all. To mass-produce something so complex was to push mechanised production to its very limits.[44]

The project was so difficult that it almost collapsed. Constant delays in the supply of tools and labour, pressure from the army to stick to building the components, the erratic intervention of Ford himself who insisted on re-siting the whole plant when he discovered that the boundary of a Democrat county ran through the complex, all conspired to hold up the promised supply of one bomber an hour. While Ford struggled with Willow Run, the 'enormous room' was even eclipsed by a room yet bigger, a vast plant for Oerlikon anti-aircraft guns built by the Chrysler Corporation in Chicago. But Ford's technical persistence paid off. By 1943 over ten bombers a day were delivered; during 1944 over five thousand were produced, reaching a rate of one bomber every 63 minutes. Willow Run, for all its problems, came to symbolise all the self-confidence and drive of American industry. Looking across the giant assembly hall was like gazing at a scene from Fritz Lang's futuristic movie, *Metropolis*. The glare of the machinery and polished aluminium and the clouds of dust made it impossible to see its whole length. At one end four moving lines carried the core of the aircraft; the four merged to two as the aircraft took shape; 1 mile from the start there was just one line of completed bombers, fed out through a cavernous opening on to the airfield beyond. It was, observed the aviation hero, Charles Lindbergh, 'the Grand Canyon of the mechanised world'.[45]

Long before Willow Run redeemed its author's ambitions the rest of American industry turned to modern methods. Productivity in the aircraft industry doubled between 1941 and 1944. By 1944 each American aircraft worker produced more than twice his German counterpart, four times the output of a Japanese worker. The war revived America's flagging enterprise culture. After a decade of depression and high unemployment, both business and labour profited from war. The contrast not only with the Soviet Union but with every other warring population was striking. In America there was enough left over from the booming war effort to provide civilians with supplies of consumer goods and food in generous quantities. Rationing was limited and loosely applied, except for petrol. Wages boomed, with an average increase, even allowing for price rises, of 70 per cent across the war years. As the nine million unemployed were absorbed into the workforce, and women – fourteen million in all – took up paid work, family earnings rose even faster. For millions of Americans who had lived on state relief and handouts in the 1930s the wartime economy was a windfall. The lure of high wages drew almost four million workers from the poorer south to the boom cities of the west coast,

Michigan and the north-east. That haunting photographic image of the 1930s, the Madonna of the Depression, gave way to the propaganda ideal of war-working womanhood, Rosie the Riveter (though given the changes in work practices Wanda the Welder might have been more appropriate). In reality the high demand for labour and the unwillingness of the government to regiment the workforce led to high labour turnover, and a strike movement that produced almost five thousand strikes in 1944.[46] The migration of labour exacerbated racial tension, and strained the housing market. But compared with the hugely deprived Soviet workforce and the bombed and bullied workforces in Germany and Japan, American labour was able to work without danger, well-fed and well-shod. This may have made it harder to exact a higher level of effort or self-sacrifice, but it certainly fostered a productive workforce.

Without the wartime opportunities the American population would presumably have done its patriotic duty with more or less enthusiasm. But can it be doubted that economic opportunism spurred on the war effort as much as raw desperation fired the Soviet population? Over the war more than half a million new businesses were started up in America. Like the Soviet over-achievers, the new entrepreneurs were hailed as examples of economic individualism at work. But in wartime American workers really could become bosses overnight. The son of a Syrian immigrant, Tom Saffody, worked as a machinist at the start of the war for a Detroit company. During the course of his work he invented a machine for metal measurement in his home garage. He left his job, built a factory using secondhand piping welded together, and by 1943 ran a business with an annual turnover of 4.5 million dollars.[47] The entrepreneurial ideal did not explain America's wartime economic performance any more than socialist emulation explained that of the Soviet Union, but it would be wrong to ignore its motivating power entirely. Much of America's economic effort was voluntary; it was therefore all the more important that it should be prosperous. 'If you are going to try to go to war or prepare for war in a capitalist country,' wrote Henry Stimson, Roosevelt's Secretary for War, with unusual frankness in his diary, 'you have got to let business make money out of the process, or business won't work.'[48]

★   ★   ★

By 1943 there existed a decisive disparity between the quantity of weapons and equipment available to the Allies and the output of their enemies, Italy, Germany and Japan: more than three to one in aircraft and tanks, four to one in heavy guns. Of the three Axis states only Germany had the economic resources, the depth of skilled manpower and technical expertise, and the industrial capacity to support a war effort on the scale of the Soviet Union and the United States. The German economy was regarded in the west as a formidable source of military power, which is why so much effort was devoted to destroying it from the air. On the success of the German war economy hinged the more general success of the Axis states.

There is little doubt that throughout the war the German economy produced far fewer weapons than its raw resources of materials, manpower, scientific skill and factory floorspace could have made possible. The disparity between the two warring sides was always wider than the crude balance of resources. Indeed up to 1943 the much smaller British economy outproduced Germany and its new European empire in almost all major classes of weapon, while the Soviet Union, in 1942 temporarily reduced to an economy even smaller than Britain's, produced half as much again as the German empire between 1942 and 1945. However much the statistics may mask differences in policy and circumstances, this is still a significant contrast. Had the situation been otherwise German fighting power might well have avoided the remorseless attrition which set in in 1944.

The irony in all this is that no state so actively worked to prepare its economy for war as Germany did in the 1930s, nor saw so clearly the intimate connection between industrial and military might. The lessons of military defeat were not lost on Germany's leaders. It was branded deep in the German public mind that blockade, hunger and material weakness sapped the will and ability to carry on fighting in 1918. The very idea of 'total war', a contest to the death between whole societies – armies, economies, peoples – was a German one, originally coined by General Erich Ludendorff to describe the bitter experiences of the Great War.[49] Hitler, who had marched next to Ludendorff in November 1923 in the failed Nazi attempt to seize power in Munich, shared entirely the view that to win wars 'the whole strength of the people' – material, moral, military – had to be thrown into the scales.

During the 1930s the sinews of the militarised economy gradually took shape. By 1939 Germany devoted almost a quarter of its national product, and more than a quarter of its entire industrial workforce to military production. From the mid-1930s the state sponsored a colossal programme of capital investment – more than two-thirds of all German industrial investment – to build a solid foundation of material resources on which to raise the later superstructure of military output. Across Germany sprang up the world's largest aluminium industry; a new iron and steel complex, planned on a scale to eclipse Magnitogorsk, was built from scratch on the large ore-fields of Brunswick in central Germany; the chemical industry began to construct whole new plants for the synthetic production of oil and rubber, resources vital for mechanised warfare, but controlled on world markets in their natural form by Germany's potential adversaries. By 1939 Germany possessed the kind of military-industrial complex that would characterise the two incipient super-powers, the United States and the Soviet Union, after 1945. Its productive potential was enormous, though reliant on the supply of raw materials and oil from outside German borders. To compensate Germany drew into its orbit the resources of central and eastern Europe, partly by conquest and occupation, as in the case of Austria, Czechoslovakia and Poland, partly through aggressive trade treaties that gave Germany privileged access to raw maerials, as in the case of Hungary, Yugoslavia and Romania (and between 1939 and 1941, the Soviet Union). These regions, described by German economists as the 'Great Economic Area' (*Grossraumwirtschaft*), supplied coal, iron ore, machinery capacity, oil, bauxite, lignite (for synthetic oil production) and a host of other valuable resources. Hitler recognised that Germany would not win a future war if, as in 1914, she had to draw just on her own supplies.[50]

On this foundation Hitler hoped to build a military power that could face the prospect of war in the 1940s with every chance of success. In 1938 he approved the weapons that would make

Germany the first super-power: an air force of twelve thousand modern front-line aircraft; a huge new battlefleet to replace the one unceremoniously scuttled in Scapa Flow in 1919; a quantity of explosives almost three times that produced at the height of Germany's war effort in 1918. The development of rockets, jets and new forms of nerve gas was well advanced. The army in 1939 evaluated the possibility of developing atomic weapons. The war that broke out in 1939 cut across all these preparations. Neither foundation nor superstructure was complete; despite the growth of state planning and control, the transformation of the German economy into the instrument of super-power status was slower than expected. If war had not started until the mid-1940s Germany might well have proved unstoppable. In 1939 the whole military-industrial complex was still in the throes of expensive and lengthy construction.

When war broke out the economy was conscripted at once. Strict rationing was imposed on the population. Inessential businesses were closed down; those that could be converted to war purposes were ordered to do so. A world of shortages, queuing, ration coupons and a dreary, monotonous diet was imposed overnight. Life in Berlin, recalled an American correspondent in 1939, was 'Spartan throughout'.[51] An army of bureaucrats, military and civilian, descended on the German people, requisitioning cars and lorries, horses and tractors, issuing permits for travel, directing workers to the arms factories. Already by the summer of 1941 almost two-thirds of the industrial workforce laboured on military orders. (Even at the height of the German war effort in 1944 the proportion was not much greater.) It was a level of commitment to war production higher than the British in 1941, and higher than that of the United States throughout the war.[52]

On any reasonable expectation Germany should have won substantial dividends from a mobilisation on this scale. By 1941 the economy was supplemented by the rich resources conquered in Belgium, France and Luxembourg, which gave Germany access to almost the whole European coal and steel industry, a large additional labour force and trainloads of looted machines and metals. Instead Germany's industrial effort stagnated; 8,000 aircraft in 1939, 10,000 in 1940, only 11,000 in 1941. The output of weapons and equipment was not much greater after two years of war than it had been after one. A great chasm opened up between Hitler's plans and the material reality. There were not many more tanks and aircraft available to attack the Soviet Union than Germany had used to defeat Britain and France, a factor that played an important part in Soviet survival in the first critical six months of the Barbarossa campaign.

There is no easy explanation of this paradox. Hitler's Germany was an authoritarian single-party state, dominated by the personal dictatorship of a man who was not only head of state, head of the party but titular Supreme Commander-in-Chief as well. Hitler had all the powers necessary to order the weapons he wanted. It has sometimes been argued that he chose instead to launch short, limited wars to reduce the pressure on civilian living-standards by producing limited quantities of armaments; this political preference explains the so-called 'peace-like war economy', until Germany was forced to switch to a 'total war' economy in 1942. But this argument sits ill with Hitler's own demands for enormous increases in output, formulated in the year from late 1938 until the winter of 1939/40. Nor does it fit with the statistical picture which shows a rapid conversion to the purposes of war on such a scale that by 1941 almost two-thirds of the industrial workforce was committed to wartime orders (only a very little less than in 1944) while living-standards fell further and faster between 1939 and 1941 than at any other time in the war. The German problem was not one of under-mobilisation but over-mobilisation, in a context where there was constant pressure on the supply of scarce materials and no single central planning agency, as there was in the Soviet Union. Hitler's orders, though they shaped national policy, were refracted and distorted by a system that was poorly coordinated, uncooperative and obstructive. There was no straight line of command between Führer and factory and no single forum where issues of matching resources to output could be thrashed out and solved. In between there was a web of ministries, plenipotentiaries and party commissars, each with their own apparatus, ambitions and interests, producing more than the usual weight of bureaucratic inertia. Each branch planned and organised a great deal, but they did not plan together. At the end of the line was a business community most of whom remained wedded to entrepreneurial independence and resented the jumbled administration, the corrupt party hacks, and the endless form-filling, all of which stifled what voluntary efforts they might have made to transform the productive performance of the war economy.[53]

No organisation was more guilty of limiting Germany's war potential than the military. Answerable only to Hitler, the armed forces treated the German industrial economy as an annexe to the front-line. Military priorities dominated arms production, from the design and development of a weapon through to the final factory inspection. Military officials were posted to factories to monitor production. Changes to design and specification were made endlessly in response to every cry for improvement from the battlefield. Production schedules were set by military agencies; consultation with the industrialists and engineers who had to produce the goods was rare and one-sided.

The military domination of production had mixed effects. There is no dispute that Germany developed weapons of very high quality, with a level of finish and attention to detail that astonished the Allies when they inspected crashed aircraft or captured guns. By the end of the war German designers had begun to develop many of the weapons that armed NATO a decade later. But the pursuit of advanced weaponry came at a price. Instead of a core of proven designs produced on standard lines, the German forces developed a bewildering array of projects. At one point in the war there were no fewer than 425 different aircraft models and variants in production.[54] By the middle of the war the German army was equipped with 151 different makes of lorry, and 150 different motor-cycles.

With such a variety it was difficult to produce in mass. In 1942 Hitler remarked how industrialists 'were always complaining about this niggardly procedure – today an order for ten howitzers, tomorrow for two mortars and so on'.[55] But this was the nub of a system in which the military dictated the selection and development of anything that promised battlefield dividends. It produced a situation where a proper sense of priority and evaluation was replaced with a chaos of demands and programmes. 'Nobody would seriously believe', complained a group of German engineers in 1944 based at the research centre in Rechlin, 'that so much inadequacy, bungling, confusion, misplaced power, failure to recognise the truth and deviation from the reasonable could really exist.'[56] As long as the military tail wagged the industrial dog, German war production remained inflexible, unrationalised and excessively bureaucratic.

Not surprisingly the experience of mass-production in Germany was very different from that of the Soviet Union and United States. The German armed forces shared a widespread prejudice against 'American' production methods. Göring's jibe that all America could produce was razor-blades was not an isolated belief. Mass-production was associated with cheap consumer goods and shoddy standards. The German military preferred to establish close links with smaller firms with traditions of skilled craftsmanship, which would be sensitive to frequent design changes and produce a sophisticated custom-built weapon. The great strengths of the German industrial economy had always been high quality, skilled workmanship, the conquest of technical complexity. German weapons were very good, but very expensive – in skilled man-power, time and materials.

Nothing more clearly reveals these preferences than the story of the German car industry, which during the 1930s became the largest manufacturing sector in Germany. The major names – Adam Opel, Ford (in Cologne), Auto-Union, Daimler-Benz – slowly adopted American production methods. Hitler was an enthusiast for the motor-car, so much so that he dreamed of making cheap motoring available for the masses. In 1933 Hitler met the designer Ferdinand Porsche. He asked him to try to develop a low-priced family car 'in which one could go for weekend trips . . . a car for the people'.[57] The design that Porsche came up with was the prototype of one of the century's most famous cars, the Volkswagen. In 1938 work was started at Fallersleben in the Brunswick countryside, not far from the vast new iron and steel complex at Saizgitter; around the Volkswagen plant was to be built one of the new model cities of the Third Reich, Wolfsburg. The factories were planned on a gigantic scale, together capable of turning out half a million cars a year, rising eventually to one and a half million. The central workshop boasted the largest metal press in the world, capable of punching out whole car bodies at a time. By the beginning of the war the installation of machine tools for the production of 200,000 cars was completed. Yet the contribution to the German war effort of the plant designed as one of the largest and most modern mass-production factories in the world was absurdly small. Only one-fifth of its capacity was ever utilised, for a hotch-potch of equipment from one and a half million camp stoves to army-issue footwarmers. Only a few thousand chassis were produced for a military version of the Volkswagen which the army took with great reluctance. Otto Hochne, who worked at Wolfsburg during the war, later recalled that 'there seemed to be no plans at all'. The works were used haphazardly until they were bombed in 1943.[58]

The rest of the industry suffered something of the same fate. Postwar estimates showed that barely 50 per cent of its capacity had been utilised during the war. The largest and most modern producer, Opel, almost disappeared entirely at the start of the war. Opel was owned by the American car giant General Motors, which did not endear it to the German authorities. The directors had clashed with the army in 1936 when they closed down a small army supplier; Wilhelm Opel had earned Hitler's displeasure when he introduced his new small car at the 1937 Berlin Motor Show with the words 'This, Herr Hitler, is our Volkswagen'. When war broke out in September the military authorities wanted to break the whole complex up and parcel its tools and workforce out among other smaller producers. There were no orders for war work and no mobilisation plan. The company resisted dispersal, but not until 1942 did it become a major producer after the authorities had dithered for almost three years over what it should make.[59] During the war the productive performance of the Opel workforce dropped by almost half. The systematic conversion of the motor-cycle, motor-car and tractor plants only began in 1942 and 1943.

There was more to the problems of the German war economy than the failure to mobilise the largest mass-producing industry in the country, but this was symptomatic of a productive system in which technical sophistication was routinely preferred to turning out large quantities of standard weapons. In the end it took Hitler to break the logjam. Prompted by complaints from industrial leaders, Hitler called military and civilian leaders together at Berchtesgaden in May 1941 and berated the soldiers for burdening industry with unnecessary technical demands. He demanded 'more primitive, robust construction' and the introduction of 'crude mass-production'. The military took little notice. In December 1941 Hitler turned his request into an order, a Führer Decree on 'Simplification and Increased Efficiency in Armaments Production'. Citing the success of the Soviet factories, he ordered German industry to embark on 'mass-production on modern principles' and insisted on putting industrialists in charge.[60]

The real turning-point came a few months later when Hitler appointed as Minister of Armaments the young architect Albert Speer, a close courtier who indulged Hitler's fascination for monumental architecture. He was only 36, a man with no military experience or close knowledge of industry but a sound organiser. He quickly built up a team of young managers and engineers, almost all of them in their thirties or early forties, most of them co-opted from German industry. German industrial resources were at last centrally planned, and in response to Hitler's efficiency decree industrial rationalisation became the driving force of the new structure. A remarkable amount was achieved in three years. By 1944 the number of weapons had been reduced to a few chosen types; 42 aircraft models became five; 151 lorries gave way to just 23; a dozen anti-tank weapons were replaced by just one; and so on through the whole range of German weaponry.[61] The shift to mass-production, though far from universal, brought an instant increase in efficiency. Weapons output trebled in three years; the productivity of German industrial manpower doubled, even though by 1944 one-third of the armaments workforce were foreign forced labourers. Large factories were expanded and small ones closed down. The Messerschmitt-109 fighter was produced at the rate of a thousand a month in three giant plants in 1944, where once 180 had been produced in seven smaller plants. Industrialists now revelled in the freedom to work without the constant fear of military interference.

Under Speer the German economy at last promised to deliver the sort of numbers produced in the Soviet Union and America. It was a promise still difficult to redeem. The military continued to disrupt long production runs and standardisation if they felt military necessity required it. The army procurement officers regarded Speer as an 'inexperienced intruder'. Speer recalled ruefully in his memoirs the survival of 'excessive bureaucratisation', which he fought 'in vain'.[62] Göring kept a jealous hold on aircraft production until the spring of 1944, when real mass-production was introduced at last. The one thing that kept Speer going was Hitler's backing, but the war economy always threatened to return to the unrationalised, disordered mêlée of its early years. For all his enthusiasm and sense of urgency, Speer was unable to reap an early harvest. Not until the summer of 1943 did the reforms begin to bear real fruit, long after the Soviet Union and the United States had mass-production established. At just the moment when German potential was on the point of being realised in the large, mechanised, centralised assembly plants, bombing began in earnest.

Bombing was the enemy of rationalisation. Speer's deputy responsible for tank production explained to his postwar interrogators that bombing forced measures that contradicted mass-production: 'the breaking down and dispersal of plants, starting up factories on account of their geographical position instead of their technical capacity . . .'[63] As German factories moved into smaller, camouflaged premises, into woods or even underground, it became progressively more difficult to expand production. There was enough momentum in the Speer reforms to carry German industry to a peak in September 1944, but bombing made it impossible for managers and workers alike to achieve the maximum. By the autumn of 1944 the war industries were living off accumulated stocks of materials and components. Under the impact of bombing, conditions on the home front deteriorated rapidly. More and more of the workforce was made up of unwilling labourers forced from their homes across Europe to fill the German workhalls. By 1944 seven million of them – a quarter of the workforce – lived and worked in squalid conditions on low pay, regimented by the Nazi authorities, bullied and victimised by German workers whose own conditions in the industrial regions of the Reich were growing steadily worse. Food supplies declined and urban amenities were strained to breaking point; millions of Germans were made homeless and could not be satisfactorily rehabilitated. Hundreds of hours were spent huddled in shelters and absenteeism rates soared. The more bombing affected the willingness to work, the more the regime resorted to draconian methods to extract the labour. The SS mobilised the population of its empire of concentration and exter- mination camps, which was literally worked to death. Workers caught pilfering or slacking were taken into the camps or sent on 'work education weekends' organised by the Gestapo; even industrialists who displayed defeatism or obstructed the growing number of SS officials recruited to run the bombed economy could follow their employees behind barbed wire. Work was kept going on a basis of fear: dread of the Soviet enemy as the Red Army neared the borders of the Reich and dread of the rule of terror at home.[64]

The German wartime economy was a paradox. Germany possessed a wealth of resources, a large class of competent entrepreneurs and engineers and a highly skilled workforce, all at the disposal of an authoritarian system that brooked no opposition, led by a dictator with delusions of super-power grandeur. This was a rich mixture, which promised a great deal more than it could deliver. The German economy fell between two stools. It was not enough of a command economy to do what the Soviet system could do; yet it was not capitalist enough to rely, as America did, on the recruitment of private enterprise. Only too late was an effort made to do both these things, to coerce more ruthlessly and at the same time to give industry more responsibility for production. Before that German mobilisation was hostage to the ambitions of a highly professional and exclusive military elite which saw war in all its elements as a military affair and clung on to that prerogative with stifling effects on Germany's industrial effort.

*   *   *

Hitler had little respect for American economic power. 'What is America', he asked, 'but millionaires, beauty queens, stupid records and Hollywood?' He had even less for the Soviet Union. On the eve of Barbarossa he told Goebbels that between German and Soviet strength there was no comparison; 'Bolshevism will collapse like a pack of cards,' wrote Goebbels gleefully in his diary.[65] These turned out to be profound misjudgements, though who, in the summer of 1941, could have clearly foreseen how rapidly and on what a scale America and the Soviet Union would arm themselves? Two years of production turned both states into the super-powers Hitler's Germany hankered to become.

What Hitler failed to see was how central industry was to the Allied view of warfare. 'Modern war is waged with steel,' Churchill told Hopkins when he weighed up the imbalance between American and Japanese economic strength. Stalin's view of the war was entirely conditioned by economics, as befitted any disciple of Marx: 'The war will be won by industrial production,' he told an American delegation in October 1941. Roosevelt's opinion of American power was every bit as determinist. In planning the Victory Programme in 1941 he told War Secretary Stimson to work on the assumption 'that the reservoir of munitions power available to the United States and her friends is sufficiently superior to that available to the Axis powers to insure defeat of the latter.'[66] Though Hitler was the inspiration behind the German adoption of mass-production in 1941, he did not consider economics as central to the war effort. Rather, he stuck to the view that racial character – willpower, resolve, endurance – was the prime mover; weapons mattered only to the extent that they could be married to the moral qualities of the fighting man.

There was much in common between the Soviet and American experience. One of the correspondents who made the trip to Magnitogorsk observed that the Russian people were 'in many ways like Americans . . . they have a fresh and unspoiled outlook which is close to our own.'[67] These were sentiments that pre-dated the Cold War, but they contained a grain of truth. In both countries mobilisation was a hustling, improvised affair; technical tasks were tackled head-on and quickly; production was big in scale and easily standardised; engineers and managers were given wide scope to solve problems themselves. Both economies had a good deal of central planning, but here the similarity ends, for the one enjoyed a supervised abundance, the other a regimented scarcity. Both countries shared the sudden shock of unprovoked aggression, which gave a genuine urgency to economic planning, and forced their economies to give priority to finding a cluster of advanced weapons on which to concentrate production energies.

The situation facing Germany was very different. There was no direct threat to the homeland until the onset of serious bombing. German planners and designers had almost two years at war before conflict with the Soviet Union and America. During that time there was little pressure to mass-produce, even if the military had approved it, or to concentrate on a narrow band of designs. The military slowly built up their version of a heavily bureaucratic command economy, which displayed a ponderous inflexibility beside the enemy. A few hours before the attempt to assassinate Hitler on 20 July 1944, Speer wrote to him that the great strength of the American and Soviet systems was their ability to use 'organisationally simple methods'. He drew a contrast between Germany's 'overbred organisation' and the 'art of improvisation' on the other side.[68] Speer was a man constantly banging his head against officialdom. Posterity, he warned Hitler, would judge that Germany lost the struggle by clinging on to an 'arthritic organisational system'. Posterity might find this view a little harsh, for Speer himself had given the system a good shaking. But the contrast between American and Soviet productionism and Germany's bureaucratised economy was more than superficial. No war was more industrialised than the Second World War. Factory for factory, the Allies made better use of their industry than their enemy.

7

A WAR OF ENGINES
Technology and Military Power

*'. . . modern imperialist war
is a war of engines
– engines in the air and
engines on the land.'*

N. Voznesensky, Chairman
of the Soviet State Planning Commission, 1940

ONE OF THE most famous of the German armoured divisions, the Panzer Lehr division commanded by General Fritz Bayerlein, experienced in 1944, at the height of the struggle for Nomandy, the full weight of Allied technical might. The hapless division, reduced after 49 days of continuous fighting to a mere 2,200 men and 45 serviceable tanks, held a stretch of French countryside 3 miles wide south of the town of St Lô. It sat right in the path of overwhelming American forces poised for the breakthrough – Operation Cobra – which destroyed the German front in France.

On the morning of 25 July waves of American Thunderbolt fighter-bombers swept over the division, every two minutes, fifty at a time. They dropped high explosive bombs and napalm incendiaries. They were followed by four hundred medium-bombers carrying 500-pound bombs. Then from the north came the sound every German soldier dreaded, the heavy drone of the big bombers – 1,500 Flying Fortresses and Liberators. From their swollen bomb-bays 3,300 tons of bombs obliterated almost everything on the ground. Finally, the German line, or what was left of it, was pounded by three hundred Lightnings carrying fragmentation bombs and more of the new incendiaries. It was an awesome display of power, numbing and terrifying for the soldiers and French civilians who lay under it. One survivor remembered that everything shook so much 'it was like being at sea in a force 10 gale'.[1]

Almost half the Panzer Lehr remnant perished in the bombardment. Hundreds more were killed or blasted by the barrels of ten thousand American guns that opened up as the last aircraft disappeared from view. Bayerlein later told his captors that the experience of that morning was the worst he ever saw in battle. In front of him stretched a moon-like landscape, so filled with dust that German artillerymen were forced to fire blind at the oncoming enemy. All communications were destroyed. He could see no human life left where the bombs had fallen. To the astonishment of the Americans the bemused and wretched dregs of Panzer Lehr fought on, until they were swept aside by American armour the following day. Bayerlein reported that evening the utter destruction of his division. The same night the American General Hobbs, leading the assault, reported his success: 'This thing has busted wide open.'[2] Panzer Lehr's few survivors joined the general retreat across France. When they were regrouped beyond the Seine they could find only twenty tanks and six self-propelled guns and a handful of lorries. From then until the end of the war this elite Panzer division averaged a strength of only 22 tanks instead of the 74 it should have had. Most of those who joined the division in 1944 soon became casualties, victims of an exceptional disparity in firepower and mobility between Germany and the western Allies.[3]

Here was a complete reversal of fortune. Four years before it was German armour and aircraft that tore the Allied front to shreds and sped almost unopposed across French soil; the combination of tank and aircraft proved irresistible. German forces were regarded as the most modern in the world, the product of years of frantic mechanisation and technical improvement. That apparent modernity gave the German armed forces in 1941 what one senior staff general, Gunther Blumentritt, called 'the reputation and aura of invincibility'.[4] Something of that same aura touched the Japanese when they swept aside western forces in southern Asia and the Pacific. The fighting skills of both peoples contrasted starkly with their enemies' feeble efforts to obstruct them. Some of that reputation, though not all, was hastily constructed myth, serving to explain Allied humiliations. But about one thing there was no doubt in 1940 and 1941: the German use of air power and ground mobility set their armed forces apart from every other major state. Why that lead was lost, and then overturned, why Panzer Lehr was obliterated in Normandy, is a central explanation of Allied victory.

The answer lay in imitation of German practice. By the end of the war America, Britain and the Soviet Union possessed very large and increasingly effective mechanised forces; American and British Commonwealth armies were fully motorised; each of the Allies developed sophisticated systems of tactical air support, marrying together air and ground power. Given the extraordinary success of German arms, the Allied learning curve had to be short. In the Soviet Union, against which the bulk of German armour and air power was concentrated in 1941 and 1942, Stalin grasped the obvious lesson. In his speech to mark the 24th anniversary of the revolution on 6 November 1941 he pinpointed Soviet weaknesses in tanks and aircraft as the critical difference between the two sides: 'In modern warfare it is very difficult for the infantry to fight without tanks and without adequate protection from the air.'[5] Priority in 1942 went to the production of tanks and planes, with revolutionary changes in the way the Red Army deployed them. Britain and America were forced to build up large modernised forces almost from scratch in the early 1940s to counter the substantial operational skills displayed by the enemy, and both adapted German practice to their particular circumstances.

By contrast the military effectiveness of both German and Japanese forces first stagnated, then went into decline. Both had been flattered at the beginning of the war by the weaknesses of Allied forces, which served to exaggerate their military prowess. American servicemen were issued with pamphlets 'Exploding the Japanese "Superman" Myth' to counter the image of Japanese superiority. In reality Japanese forces were poorly armed, a situation masked by high levels of training and endurance. The technological balance between Japan and her enemies was greatly in the latter's favour, and the underdeveloped state of the Japanese economy made extensive modernisation of the armed forces difficult to achieve. For every American soldier in the Pacific war there were 4 tons of supplies; for every Japanese a mere 2 pounds.[6] German forces were also under-armed for much of the war. Modernisation was concentrated on a small portion of the army; German air power, for all its successes, remained limited in role and in size and was all but eliminated by 1944 from the skies of battle. By 1944 a great portion of the German ground forces relied on horse-transport. Neither the quality nor quantity of German technical resources was sufficient to prevent the dissipation of the myth of German invincibility.

★   ★   ★

The core armoury of offensive warfare in the Second World War consisted of aircraft, tanks and trucks. The effectiveness of these weapons in German hands depended on their use in combination, concentrated in great number at the decisive point of battle. Operational success also relied on communication. Radio played a vital role in linking tank to tank, and tank to aircraft. Good communications enhanced the flexibility of armoured forces and helped to concentrate its firepower. They were essential for units that were effectively self-contained divisions, operating with their own motorised infantry – some in trucks, some in armoured carriers – engineers, artillery and anti-aircraft batteries. The German Panzer forces did not have to wait for the mass of ordinary infantry divisions to catch up by rail and by foot but could fight at the pace of the tank, punching holes in the enemy front to outflank and envelop his forces. Panzer divisions were organised by the German army as battle winners, designed to 'strike decisively'.[7] They worked at their best on flat open plain, less well in built-up areas, and with least effect in mountainous terrain. The wide even grasslands of the Soviet Union showed German armoured forces at their most deadly in the summer of 1941. Against overwhelming odds – some 3,648 tanks against an estimated fifteen thousand Soviet tanks – the Panzer armies cut swathe after swathe through the Soviet defences, virtually destroyed the Soviet tank and air arm, and brought the Soviet Union almost to the point of collapse.[8]

Though the Soviet tank and air forces were numerically large, the German attack exposed fundamental weaknesses in the way they were used. The key to Soviet revival in 1942 and 1943 lay in improving the quality of both forces and in the transformation of the way in which they were used tactically on the battlefield. The eastern front was vital to Allied survival. It was here that German armour and air power were blunted. More than two-thirds of

German tank casualties were suffered in the east. Later on at Kursk Soviet armies won the largest tank battle of the war.

The almost complete destruction of the Soviet mechanised corps in 1941 allowed the Red Army to start from scratch in the spring of 1942. The basic organisational unit became the tank corps, equipped with 168 tanks, supported by anti-tank guns, a battalion of Katyusha rockets and anti-aircraft artillery. Two tank corps and a rifle division were put together to form tank armies, combining tanks, infantry and supporting arms and services, like the German Panzer armies. The shortage of purpose-built armoured carriers for the infantry, which helped German soldiers to keep up with the tanks, was solved by the simple but dangerous expedient of welding hand rails on to Soviet tanks, each of which then advanced into battle with a dozen riflemen clinging precariously to its hull. In September 1942 the Red Army began to organise mechanised corps, with a higher proportion of infantry and fewer tanks. The infantry were heavily armed and more mobile than the bulk of Soviet forces, particularly when, from December 1942, they began to receive the first self-propelled artillery units. During 1943 both tank and mechanised corps were strengthened and the modern-isation of their equipment was accelerated. By the autumn of 1943 each tank corps boasted 195 of the most modern tanks, each tank army about four hundred. Between 1942 and 1945 43 tank corps were activated, and 22 mechanised corps. Overall Soviet tank strength climbed steadily from the low point of November 1941; by the end of 1942 it was more than double German strength, and by the autumn of 1943 more than three times greater.[9] The new tank and mechanised forces were designed to do what German armour did, to punch hard at the weak points of the enemy line, to pierce it and to exploit that penetration in sweeping pincer movements. It was first used with real success at Stalingrad, and repeated devastatingly at Kursk, when other improvements in equipment and training began to bear fruit. The new Soviet tank armies were equipped almost entirely with the new generation of tanks, the versatile, hard-hitting T-34, and the heavier KV-1. The T-34 provided mobility, the slower KV-1 firepower against enemy strongholds or tanks. During 1942 they were converted to diesel engines, which improved the tanks' radius of action by a factor of three; they had wide tracks, unlike most German tanks, to cope with the autumn muds and the winter snow; and during 1943 the greatest defect of the Soviet tank arm was made good with the gradual introduction of two-way radio. At a stroke this improved the battlefield performance of the tank arm. Air support could be called up, and tank commanders could communicate with each other quickly to bring armour to bear where it was needed. The use of just two tank types at any one time throughout the war (the KV-1 was phased out as too slow in 1943 and replaced with the 'Iosef Stalin' IS-2 in 1944) simplified problems of repair and maintenance. Teams of engineers accompanied the armoured corps, travelling with lorries full of equipment and spare parts. When Rotmistrov's 5th Guards tank army drove 150 miles in 48 hours to help at the Battle of Kursk along roads so choked with dust that the red circle of the sun was barely visible through the shroud, mechanics performed miracles keeping almost all the tanks going. In two days of heavy fighting the Guards lost four hundred tanks, but 112 were returned by the repair crews within hours.[10] The Red Army learned quickly from its mistakes; critical improvements were made. In 1941 and 1942 Soviet armoured forces lost six or seven vehicles to every one German; by the autumn of 1944 the ratio was down to one to one.

The modernisation of the Soviet air force was just as essential. The reform of the tank arm would have achieved considerably less without air support. In 1941 Soviet aircraft suffered the same fate as Soviet armour. Most of the estimated eight thousand aircraft facing the German assault were wiped out, either where they stood, invitingly unconcealed, on open airfields, or in hundreds of uneven tournaments in the air between Soviet I-15 and I-16 fighters and German Me-109s. Soviet air tactics were stuck in the First World War. Fighters flew in tight formations of three aircraft abreast. They were no match for German pilots who operated in pairs or fours, flying in loose vertical formation with good air-to-air communication. The feeble fighter shield left Soviet ground forces open to the same demoralising pounding from German medium-bombers and Stuka dive-bombers that had helped to break Polish and French resistance earlier in the war.[11]

Air force reform ran alongside the reform of ground forces. The Soviet authorities rightly saw concentration of air attack and close support for ground operations as the core of German success. This persuaded them to put their effort into winning local air supremacy over the battlefield, and using aircraft as part of the battering ram to penetrate the enemy front. The vast area of the eastern front, and the short combat range of most fighter aircraft, made it imperative for them to concentrate air forces at the critical point of attack, rather than dispersing the aircraft all along the front as they had done in 1941. The Soviet air force borrowed and refined the German strategy with striking results. During 1943 air superiority over the battlefield passed to the Soviet air force.

The architect of Soviet revival in the air was a little-known general, Alexander Novikov. A veteran of the Civil War, and a Communist Party member, he survived the army purges of the 1930s to become head of the Leningrad District air forces in 1941. Like his chief, Zhukov, he won promotion on his performance in battle. In February and March 1942 he organised the air defence of Moscow and Leningrad, but he did so by grouping all his aircraft together into a powerful offensive instrument to carry out the first concerted massed air strikes against German targets. A month later his success was rewarded with nothing less than his appointment as Commander-in-Chief of all Soviet air forces. At 42 he was four years Zhukov's junior, but looked younger. Handsome, forthright, energetic, he came to his new office with a self-driven mission to modernise Soviet aviation. His abilities brought him appointment to a new rank of Marshal of Aviation in 1943, but like Zhukov he fell foul of the renewed wave of Stalinist purges after the war. Incarcerated in 1946, he was released and had his honours restored only on Stalin's death, in March 1953.[12]

Novikov's reforms were comprehensive. Starting at the top the air force was placed under central direction, instead of being parcelled out to each front-line army or division. Aircraft were organised like German air fleets into large 'air armies' made up of fighters, bombers and ground-attack aircraft. The air armies were then assigned to critical points of the front, where they came under direct command of local front commanders in order to achieve the closest possible collaboration between army and air forces. The air armies, like the tank armies, were set up to follow the strategy of concentrated offensive. Their task was to attack enemy air power and smash a path forward for the armoured forces below. To enhance this striking power the Supreme Command built up a strategic reserve of air corps which could be sent to critical points of the battle to reinforce existing air armies. Though the strategy of reserve-building, insisted on by Stalin himself, sometimes left parts of the front denuded of air support, it did allow the Red air force to impose irresistible force where it mattered most, at the front of the main axes of advance.[13] During 1942 each of the seventeen air armies could muster only 800-900 aircraft; but by 1943 they averaged 1,500 aircraft. By the end of the war air armies of 2,500–3,000 aircraft were built up. The Soviet Union ended the war with over eleven thousand front-line aircraft.

There were a host of smaller changes which together transformed the fighting power of Soviet airmen. A new generation of simple but technically advanced aircraft was introduced – the Yak-9, Lagg-3 and La-5 fighters, and the formidable Sturmovik Il-2 dive-bomber, with its tank-busting rockets. They proved capable of regular upgrading, which avoided disruption to the assembly lines, and by 1944 they could outperform the latest Me-109 and Fw-190 fighters. The models were few in number, and relatively easy to maintain. Novikov overhauled the rear services for aviation, providing teams of mobile repair units, improving the building and organisation of a host of small, temporary airfields, and insisting on high standards of camouflage and deception, so that Soviet aircraft would never experience again the pre-emptive destruction of 1941. Radar was gradually introduced; dummy aerodromes were constructed; aircraft in forward areas were hidden in woods and farm buildings. The Soviet air force of 1943 was an effective fighting force, unrecognisable from the clumsy and incompetent instrument of 1941.[14] By the time of the Battle of Kursk the reforms were largely completed. After a week or so of fighting, Soviet aircraft built up a superiority of ten to one on some parts of the battlefield. For the first time it was the turn of German soldiers to scan the skies in desperate search of friendly craft.

The pace of modernisation never slackened thereafter. The Soviet Union emerged in 1945 with the sinews of a military superpower in place. Some of this was the fruit of Allied assistance. The United States motorised not only its own army but the Red Army too. Under 'Lend-Lease' war aid agreements America supplied over half a million vehicles – 77,900 jeeps, 151,000 light trucks, and over 200,000 Studebaker army trucks, the backbone of the Soviet motorised supply system.[15] American aid also made possible the revolution in radio communications by supplying 956,000 miles of telephone cable, 35,000 radio stations and 380,000 field telephones.[16] During 1943 the Soviet air force concentrated on improving radio control of its aircraft, and radio communication between ground and air forces. Radio control centres were set up about a mile and a half behind the front, enabling aircraft to be directed quickly and flexibly to help the ground forces, or to attack enemy aircraft and air bases. In the main, modernisation and reform were driven by the sharp spur of necessity. The Soviet solution to the crisis of 1941 was deceptively easy: tanks and aircraft were organised into large, flexible offensive forces, backed by huge reserves of robust and effective weapons. Mistakes were still made, and Soviet losses remained high throughout the war, but enough was done during 1942 to erode the margin between the two sides in military performance and military technology.

The German experience on the eastern front was almost exactly the reverse. German forces started the war on the Soviet Union with over 3,600 tanks; by the summer of 1943 they had only two-thirds this number. In 1941 the Luftwaffe fielded 2,500 aircraft; in the summer of 1944 there were 1,700 aircraft along the entire eastern front, including a mere three hundred bombers and three hundred fighters.[17] The vaunted Panzer divisions became primarily defensive units: the number of tanks and vehicles in each division declined steadily over the course of the conflict, though German production of tanks in 1944 was more than ten times higher than in the first year of war. Panzer divisions started the war each with an established strength of 328 tanks; at Kursk they started the battle with an average of 73; in March 1945 their average was 54.[18] Though German forces fought with high tactical skills throughout the war, they experienced a progressive 'demodernisation'.

There were extenuating circumstances. The slow development of mass-production has already been observed. By 1944 German forces had to be spread thinly over three main fronts. Bombing contributed a good deal to the dislocation of the German military supply system. But the process of demodernisation was already evident in 1942, when the main theatre of operation was the eastern front, and bombing had barely begun. The explanation for the growing technical weakness of German forces lies with Germany rather than the Allies, and it goes back to the summer of 1941 when Panzer forces raced with apparent ease across the Soviet steppe.

The army that Hitler assembled to crush Bolshevism was essentially two armies, one small modernised army based on tanks and trucks, and a vast old-fashioned army still reliant on rail and horse. The first contained an expanded number of Panzer divisions; 21 of them had been created by parcelling out the stock of tanks in smaller packets than before. The second comprised 119 divisions moving at the pace of the foot-soldier and the horse and cart. This was the result of a conscious decision taken in the 1930s to neglect the modernisation of a large part of the army in favour of developing a hard-core of heavily-armed mobile divisions. The split was maintained in the Barbarossa plan. The Panzer armies were established as entirely self-contained forces, carrying the fuel and ammunition required to take them 400 miles into the Soviet Union. This strategy avoided trying to supply the mobile forces through a poor road and rail system clogged with the slow infantry army behind. It was managed only by starving the war effort elsewhere of fuel and vehicles. Some 600,000 vehicles were mobilised for the invasion, many of them in the Panzer armies; the rest of the army made do with some 700,000 horses.[19]

In reality even the modernised core of the army was weaker than it looked. To assemble the required number of armoured divisions, the number of tanks in each had been reduced to 150. Of these tanks half were of poor quality – the weakly armed Panzer-II and the captured Czech TNHP-38 tank which was now becoming obsolescent. There were a thousand of the Panzer-III, and 479 of the Panzer-IV, but neither was a match for the T-34.[20] Much of the Soviet armour destroyed in the summer attack was the victim not of German tanks but of German artillery. The incompetence of Soviet forces in 1941 allowed the Panzer armies to penetrate far and fast, but by the autumn the toll was very great. The losses sustained in combat and from wear saw the modern army crumble away; when the front stabilised in late 1941 the infantry mass had caught up to support it. By August tank strength was down to half what it had been in June 1941; by November there were only 75,000 vehicles still working out of the half-million committed in June. By December the Panzer armies were using horses again. These were rates of loss never anticipated by German leaders. Little thought or preparation had gone into the question of what to do if the quick campaign of annihilation failed. The German army too needed to modernise in 1942.

The collapse of the air-tank offensive in 1941 was compounded with problems of geography and terrain. When the autumn rains came German vehicles were brought to a standstill on roads that were little more than sand-tracks. The supply of fuel oil and lubricants for the front broke down; tanks that could cruise on a pint of oil for 60 miles on flat summer roads now consumed over 4 gallons to cover the same distance, creeping forward in first gear. The coming of cold winter weather gave some respite, but with their narrow tracks German tanks were unsuited to traversing frozen mud, and the sub-zero temperatures brought new problems. Turrets and gun-sights froze solid. Rubber seals became brittle. Efforts to warm up engines with small braziers produced explosions. On open grassy airstrips mechanics literally froze to the machines they worked on. The supply system, overstretched already by the rapid advance, threatened to snap entirely.[21] It was rescued by horses, which pulled tanks and guns when engines failed, and dragged rows of covered wagons full of supplies and ammunition. Even this brought difficulties, for the large European horses suffered from frostbite and malnutrition. In the first months of the campaign there were over a quarter of a million horse casualties. Their place was taken by the smaller Russian *panje* horse and sled, which became the primary form of transport for much of the German army. During 1942 another 400,000 horses were brought to Russia from German-dominated Europe; the same year German industry turned out just 59,000 trucks for an army of three million men.[22]

The growing reliance on horses was not due only to climate and terrain. Great difficulties were also experienced in servicing and repairing motor vehicles, a problem that persisted in the German forces for the rest of the war. The German army, unlike its Soviet enemy, had a bewildering array of different vehicles and engines. In the effort to provide lorries and fighting vehicles for the Panzer armies Europe had been scoured for motor vehicles of all kinds. There were two thousand different types of vehicle in the assault on the Soviet Union. Army Group Centre alone had to carry over a million spare parts. One armoured division went into battle with 96 types of personnel carrier, 111 types of truck, and 37 different motor-cycles.[23] Repair work became a nightmare, all the more so as the poor terrain and harsh combat produced exceptionally high levels of wear and tear. By November 1941 the same armoured division reported only 12 per cent of its entire stock of vehicles as roadworthy. Understandably, vehicle spares were not top priority in the strained supply organisation. Nor were they a production priority. In the Reich much more emphasis was placed on producing the whole product than a satisfactory ratio of spares and components, particularly engines. During 1942 and 1943 efforts were made to produce more spares, but the absence of standard-isation and the poor state of the long supply routes into Russia left a great many vehicles immobilised for months or weeks, forcing the German armed forces to rely increasingly on horse-drawn artillery and long marches on foot.[24]

During the winter of 1941–2 the German army hoped to make good the collapse of the armoured force by developing new tanks that could both outgun the Soviet T-34, and remain immune to anti-tank fire with greatly strengthened armour. Hitler took a leading role in planning them, but instead of developing a tank that was easy to maintain and to produce in quantity, he demanded large, technically complex tanks of very great weight. The result was the 'Tiger' and the 'Panther'. Though they could deliver the enhanced firepower, they were slow and unmanoeuvrable, a liability on poor ground. When Hitler insisted on throwing the first six Tigers into battle in the early summer of 1942 the result was a fiasco. On a road fringed by marshland the tanks were ambushed by Soviet troops. The first and last tank were hit in the poorly protected side and rear, immobilising the remaining four which were destroyed one by one.[25] The Panther's baptism of fire was no more propitious. The first batch was sent into battle before the development work was complete. All 325 had to be returned to Berlin for modification of the steering and control mechanism; the

Maybach engine proved inadequate and had to be improved, making an already complex piece of engineering yet more sophisticated.[26] The level of technical quality of the new tanks was such that they were difficult to produce and maintain. There were too few to turn the armoured battle in Germany's favour in 1943; they were too heavy and cumbersome to be used on anything but flat summer ground; and like other German vehicles they were difficult to repair. The Tiger tank required a small army of mechanics to keep it in the field, while the ratio of spares produced was derisory. For every ten Tiger tanks only one spare engine and one transmission were produced. The new heavy tanks were supposed to be repaired at the front-line, but the absence of heavy repair equipment or adequate stocks of spare parts meant a long trail of tank transporters carrying damaged vehicles back to repair depots in Germany, 2,000 miles away.[27]

Even with the introduction of higher quality tanks, the slow attrition of German technical resources in the east could not be reversed. Both the Red Army and the German army relied largely on horses, but the Soviet stock of motor-vehicles steadily expanded while the German stock declined by almost 50 per cent from a peak reached in January 1943. The number of half-track transporters fell from 28,000 in 1943 to eleven thousand at the end of 1944. As German forces fell back the effort to maintain mobility and levels of mechanisation was abandoned. By 1944 still only one-tenth of the German army was mechanised. Instead the production effort was diverted to heavy infantry weapons necessary to fight the long retreat. Even the Tiger and Panther, and the lumbering 'King Tiger' and 'Mouse' that followed, became tools of defence, mobile anti-tank platforms to blunt enemy armoured attacks. Their gradual switch from offensive to defensive firepower helps to explain why German troops remained an effective fighting force, even when vastly outnumbered in tanks and aircraft. But for the bulk of the German army during 1943 and 1944 there was a progressive decline in the number of tanks and supporting aircraft, of lorries and half-tracks to move the infantry and supplies, and even of artillery pieces.[28] The German soldier got used to fighting without a German aircraft in sight, with dwindling supplies of fuel and ammunition, and trucks and tanks short of tyres, spares and lubricants, and long marches on foot or by sled. The small core of high quality weaponry sustained pockets of a modernised army; the rest of the army fought a refined version of the infantry and artillery battles of 1918.

The German air force in the east suffered the same fate, though for rather different reasons. There were the same problems of supply and maintenance over long distances, the severe shortages of spares and replacement engines, the problems of flying in cold weather for which German aircraft had not been properly equipped. The use of rough grass airfields in forward areas increased the losses of aircraft and crew through accidents. But the real failure in 1942 and 1943 was in not matching either the quantity or the quality of Soviet aviation. Luftwaffe leaders, no less than the army, failed to anticipate the high levels of attrition that would be experienced even against a much weaker enemy. Before losses could be made good that enemy began to fight with large numbers of aircraft of much higher quality, making it difficult for German air units to avoid the attrition cycle. The German effort to introduce a new generation of advanced aircraft in 1942 to restore the technical lead enjoyed at the start of the war failed disastrously.

In 1941 responsibility both for the production and development of aircraft in Germany lay with Colonel-General Ernst Udet. It was a post for which he was utterly unsuited. He was appointed in 1935 at Hitler's suggestion. A veteran fighter-pilot, Udet made his reputation in the 1920s as a stunt-man of the silent screen. He was a notorious *bon viveur* and womaniser, who narrowly survived a knife attack by one of his mistresses. He was a talented cartoonist; he ate only meat, a habit that left him in chronic ill health when he held office; he loved hunting. For the job of chief technical director of the German air force he had almost no credentials whatsoever.[29] All he had wanted to be was a test-pilot, and even in his new office he still flew dangerously to get the feel of new aircraft. He was fully aware of his wide limitations. His only contribution to air force development was to insist that all bomber aircraft, even large four-engined craft, should have a dive-bombing capability, a requirement that set German bomber development years behind that of the Allies.

Under his wayward stewardship the Luftwaffe lost its direction.

The real architect of German air power in the 1930s was Erhard Milch, an ex-director of Lufthansa, and Göring's deputy. But in 1939, jealous of his successful subordinate, Göring excluded Milch from the realm of technical development and production altogether, and left the work to Udet. Between 1939 and 1942 aircraft production stagnated. Udet lacked the organisational skills and strength of character to impose what views he might have had on the scientists and entrepreneurs under his control. Unable to reach clear decisions about the development of new aircraft models he became the victim of rivalry between aircraft designers and the butt of every complaint from bureaucrats and airmen. Except for the Focke-Wulf 190, which proved to be a fighter aircraft of high quality when it was introduced in the autumn of 1941, every one of the new aircraft Udet selected was a failure. The He-177 long-range bomber, which Göring and Hitler wanted for attacks on Soviet industry and the Atlantic trade routes, was plagued with technical problems that stemmed from Udet's dive-bombing order, and was never produced in any quantity.[30] The new generation of medium-bombers and heavy fighters, the Junkers Ju-288 and the Me-210, were technical flops and were scrapped, but only after a heavy investment of money and production effort. The replacement for the ageing Stuka dive-bomber, the Hs-129, which was expected to match the Soviet Sturmovik, had to be withdrawn because its engines caught fire too easily, and were so susceptible to the dust of the southern Russian steppe that serviceability could not be maintained.[31]

The failure of almost the whole range of new air designs forced the Luftwaffe to stick with older, proven models. But the damage was done. Production of older models was run down to make room for the new; a great deal of modern factory space and skilled labour was allocated to production that never materialised. The prospect of developing German air strategy with long-range bombing and enhanced battlefield firepower evaporated. It proved impossible to expand numbers to meet all Germany's many air fronts; air reserves, which for the Soviet Union comprised almost half of all combat aircraft, could not be built up. Udet did not survive long enough to see the chickens home to roost. On 17 November 1941, unable to stand the strain of office any longer, he drank two bottles of brandy and shot himself through the mouth. Milch was his successor. The unravelling of Udet's legacy took almost two years, by which time German aircraft production was dwarfed by that of the Allies. Priority switched to defending the Reich during 1943, while air power on the eastern front declined. The last attempt to mass German aircraft offensively in the east came at the Battle of Kursk, when over a thousand planes supported the armoured thrusts. By this time the gap between the two air forces had all but disappeared; German losses at Kursk amounted to 1,030 aircraft. Air superiority passed to the Red air force. As more and more aircraft were sucked into the defence of the Reich, numbers in the east collapsed. German air forces mustered little more than three hundred fighters and three hundred bombers along the whole front during 1944, manned by crews whose levels of training and tactical skill were often a fraction of what had been the norm in 1941.[32]

*   *   *

It was Germany's misfortune to be allied in the Second World War with two states whose ability to produce and deploy the new technologies of war was limited in the extreme. Neither Italy nor Japan developed large armoured forces; neither was effectively motorised; Japanese aircraft production, like Italian, was restricted by shortages of raw materials and industrial capacity. Though both states could produce designs of high quality, they lacked the technical means to turn them into large numbers of battlefront weapons that could compete on equal terms with the enemy.

The mechanisation of ground forces made scant progress by Allied standards. Italian military leaders well understood the principles behind modern armoured warfare. The Italian army actually organised the first armoured corps in 1938, and used an armoured division for the invasion of Albania in 1939. And there it stopped. Italy fought the war with the three mechanised divisions she possessed in 1939. They were equipped with tanks that were between ten and twenty years old, with small-calibre armament. A new tank, the MI3, was introduced into battle in the invasion of Greece in 1941, but proved easy prey for Greek artillerymen. The MII tank used in the North African desert was under-armed and poorly maintained. Two of the three armoured divisions were destroyed at the Battle of Alamein. None of them had adequate numbers of supporting vehicles. Most of Italy's 75 divisions were horse-drawn; in August 1942 two Italian cavalry squadrons made their country's last cavalry charge with sabres drawn against a Soviet infantry division. The divisions nominally designated as 'motorised' divisions had only 350 vehicles each. Tank production was too small to supply more – only 667 tanks were manufactured in the whole of 1942 and 350 in 1943. Even for these small numbers the supply of fuel and ancillary services proved difficult.[33]

Japan made no attempt to embrace the new developments. The army remained an infantry army, reliant largely on horses. Army divisions had on average three thousand mounts, but fewer than three hundred vehicles. Tanks were produced in small quantities to serve as infantry support. They were poor by wartime standards. Armour was so thin it could be penetrated by small-arms fire, while the standard heavy tank boasted only a 37-millimetre gun. Production of tanks, most of them light vehicles, declined during the war from 1,100 in 1942 to 400 in 1944, and a mere 142 in 1945. For all its postwar success, the Japanese car industry was tiny in the 1930s. In 1938 Toyota produced 458 cars, Honda 1,242 and Datsun 2,908.[34] Most of the army's stock of trucks in 1941 were imported American models, for which the supply of spare parts dried up. The entire Japanese army had approximately seventy thousand trucks and cars, and limited numbers of personnel carriers and half-tracks. This represented one truck for every 49 soldiers overseas against an American figure of one for every thirteen. During 1942 the Japanese army explored the prospect of building up self-contained armoured forces but rejected the idea as beyond Japan's capability. Tanks played only the smallest part in the conquest and defence of the southern empire.[35]

Possessing few tank forces of their own, the Japanese army failed to develop anti-tank weapons. Southern Asia and the Pacific were not ideal terrain for tanks, but both American and British forces deployed them in ever-increasing numbers. The Japanese army responded with primitive solutions. Coconuts were hollowed out and primed with explosives; soldiers stacked with explosives threw themselves under the tracks of oncoming tanks; others were positioned in foxholes in the ground holding bombs over their heads to detonate when tanks rumbled above them. Much other Japanese weaponry was obsolete. The main infantry rifle dated from 1905; it was difficult to fire rapidly, and was accurate over only relatively short distances. The heavy machine gun was a 1914 Hotchkiss model. The standard infantry artillery pieces dated from 1905 (a 75-millimetre gun) and 1922 (70-millimetre), but there was nothing to match the Allied artillery barrage. The key infantry weapon was the light machine gun, also a 1922 model. Japanese soldiers were encouraged whenever possible to use their bayonet, a 15½-inch blade kept on the rifle at all times. Officers used their swords, sharp enough to cut through a man's body at the first stroke.[36]

Once the United States brought the weight of its new technology to bear in the Pacific war the contest became very one-sided. On the ground, in the air and at sea America enjoyed a wide lead in both quality and quantity. Japanese industry provided obsolescent material in substantial quantities until the blockade began to bite in 1944, but it suffered exceptional levels of attrition, leaving Japanese forces short of air cover, naval equipment and fuel. The demodernisation of Japanese forces threw them back on the one resource they had left: exceptional levels of endurance and moral commitment. Training and discipline in the Japanese forces was harsh by western standards. In 1934 a young American infantry lieutenant, Harold Doud, spent a year with a Japanese regiment. He observed at once a regime of strict obedience and elaborate protocol; the most minor transgression brought instant and violent punishment. Training pushed men to the very limit. Forced marches in the hottest parts of the day were supplemented by hours of bayonet practice. On an exercise he marched almost continuously for 29 hours without sleep. The men slept on long wooden platforms in barracks and lived mainly on a diet of rice and *daikon*, a pickled turnip. Each mealtime was accompanied by a long lecture on tactical or technical questions. Idle time was unknown; in the field soldiers were posted on sentry duty or patrols to keep them active. 'Holidays' to mark past Japanese victories were regarded as days on which training efforts were redoubled in honour of dead comrades and the Emperor.[37]

Military culture in Japan demanded the highest sacrifices from soldiers, even a willingness to kill themselves, flying aircraft packed with explosive into enemy ships, or making suicidal attacks on tanks and enemy positions. To retreat or be taken prisoner was an act of profound dishonour. Military codes were savagely enforced. The unwillingness to surrender, and the ability to survive for long periods of time in conditions of appalling deprivation, made the Japanese soldier a difficult enemy to defeat. That the long retreat through the islands of the South Pacific was so hard fought was not explained by the firepower or technology available to the Japanese garrisons. It owed more to the value-system of the military in Japan, and to high standards of tactical training. Nevertheless, fighting to the death was a self-defeating virtue. Japanese forces were drained of their best manpower early in the conflict. Levels of attrition were far too high to build up adequate reserves, and left Japan with a low ratio of military technology to manpower. Japan began the war with 232 serviceable naval craft; by the beginning of 1944 there were 180 and by the start of the following year only 95. The loss rate of Japanese naval aircraft in the first eighteen months of the war was 96 per cent.[38] The weaknesses of the Japanese economy and the constant interruption of supply lines by enemy aircraft and submarines made it impossible to reverse the cycle of attrition, and condemned Japan to fight an old-fashioned war against an enemy whose technology and fighting power constantly expanded.

That such an imbalance should exist might be regarded as entirely predictable. Japanese fighting power reflected both the strengths and weaknesses of Japanese society. America's abundant use of military technology was in turn a reflection of a culture fascinated by technical progress and invention, with a great depth of technical skill and experience. America was the world's leader in the application of the internal combustion engine, on the ground and in the air. In a very general sense the advantages of technical familiarity were carried into America's war effort. But they needed to be translated into effective use on the battlefield. In 1940 there were no American armoured divisions; the air force had given little thought to its role in offensive warfare. American military technology compared poorly with the best available in Europe. The defeat of the Panzer Lehr division four years later depended on

America's ability to forge a modern armed force out of the raw manpower and technology to hand.

In the 1930s the only mechanised force in the American army was the 7th Cavalry Brigade. It mustered 224 light tanks and a few self-propelled guns. The success of German forces in Poland and France forced American military leaders to rethink the whole shape of their army. In July 1940 the first two armoured divisions were created under the command of a separate Armored Force; in 1941 three more were activated. The Victory Programme planned for a final force of 61 armoured divisions. The first divisions were built in what was mistakenly thought to be imitation of Panzer forces, with an overwhelming preponderance of tanks rather than a mixed force. But field exercises held in 1941 showed the disturbing evidence that modern anti-tank weaponry could destroy massed tank attacks. Soviet armies learned the same lesson the hard way in 1941. In 1942 the structure of the armoured division was over-hauled. The number of tanks was reduced to two regiments from three, and a regiment of motorised infantry with three battalions of self-propelled artillery was added. With the addition of engineers and service battalions, the new division was actually much more like its German counterpart, a self-supporting combined-arms unit with considerable tactical flexibility. However, the American division had the advantage that all its infantry was carried in armoured half-tracks, and all its artillery was motorised. In addition the unit had two Combat Commands, A and B, which allowed the division to develop smaller combined-arms combat forces capable of fighting on their own. This gave the whole structure even more flexibility. Each division had 375 tanks and 759 other tracked and armoured vehicles; the first Panzer divisions had had 328 and 97 respectively.[39]

The initial purpose was to use the armoured divisions as the decisive shock force, like the Panzer armies of 1940, but experience in the early fighting in Tunisia and Italy, where the terrain was largely unsuitable for the massed deployment of armour, brought a final reform of the way in which the US army organised its forces. Instead of fighting with two armies, one formed of the heavily armoured tank divisions, one based on infantry, the two were fused together to create, in effect, one vast Panzer army, with the resources more evenly spread. The number of armoured divisions planned was reduced from 61 to sixteen, and they were integrated with the general army command structure. They had a larger complement of tanks to allow them to be used in pursuit and exploitation, but were otherwise indistinguishable from the regular army divisions, all of which were allocated a battalion of tanks and self-propelled artillery and were fully motorised. The object was to fight in combination, tanks, artillery and infantry each supporting the other. As with the Soviet system a large reserve of tank battalions could be allocated to critical points in the battle. The breakout from St Lô was a good example of the system at work – an initial thrust by heavily armed and mobile infantry divisions to create a protected channel through which the armoured divisions could move, to exploit the fracturing of the enemy line.

The American decision to produce a completely mechanised and motorised army, rather than a heavily armed core, was made possible by the generous supply of vehicles from American industry and the very large pool of drivers and mechanics available at home. America was the most heavily motorised of the major combatants. In 1937 she produced over 4.8 million vehicles; Germany produced 331,000, Italy 71,000, Japan 26,000. For every thousand Americans there were well over two hundred motor-cars and trucks; for every thousand Germans only sixteen; for every thousand Japanese there was less than one.[40] America's motor-car culture helped to compensate for the lack of a conscript army. Both air force and tank force enjoyed high levels of serviceability. Standardisation of design – the original five tank engines were reduced to just one type for ease of maintenance – allowed interchangeability of parts and rapid repair and redeployment of damaged vehicles. American forces were backed up by a formidable supply organisation based on sturdy purpose-built army trucks. These were produced in such quantity that there was little need to resort to civilian requisitioning, whereas the German army was forced to denude the home economy of motor traffic. During 1944 America produced almost 600,000 army trucks. German industry turned out just 88,000, against losses between January and August of well over 100,000.[41]

The American army became in two years the most modern army in the world. Its air force followed the same course. Rather than concentrate on the single role of providing air support for the mechanised armies, the army air corps under General Hap Arnold began to plan in 1940 a multi-role air force, capable of strategic bombing, of large-scale air transport, of air defence and tactical air support. Aircraft were widely understood to explain Germany's success in Europe. 'Air power', announced Secretary for War Stimson at the height of the Battle of Britain, 'has decided the fate of nations; Germany with her powerful air armadas has vanquished one people after another.'[42] American air intelligence exaggerated German air strength in 1940 by a factor of ten. Roosevelt needed little persuasion that the air force should be given everything it wanted. A new generation of bombers, fighters and fighter-bombers – the Flying Fortress and Liberator heavy-bombers, the B-29 Super Fortress introduced in June 1944 in the Pacific war, the Lockheed Lightning fighter-bomber, and the Mustang and Thunderbolt fighters – were all at the forefront of aviation technology, and were constantly refined as the war went on. They were produced in greater numbers than were needed, with generous spares and additional engines, to match what it was believed Germany was producing.[43]

The development of a battlefield air force, designed to win local air superiority and to help push the ground forces forward, was a natural adjunct to the build-up of large mechanised armies. In 1941 the American air force began to experiment with close-support tactics. The results were very mixed, for there was no means for local army commanders to communicate directly with strike aircraft as there was in the German system. As a result air units responded slowly to demands for front-line assistance; messages had to be routed through major command posts, which was time-consuming and tactically ineffective. It was the system used by the RAF in France 1940 which American forces largely borrowed. When it was used in the North African campaign in 1941 and 1942 it proved a dismal failure. The solution arrived at was not dissimilar to Novikov's reform of the Soviet air force that same year. The RAF placed mixed forces of bombers, fighters and fighter-bombers under a central tactical air commander who coordinated attacks closely with the requirements of the army command. At the front-line, mobile control centres, with radar and radio communications, kept contact with aircraft in the air and the army in front. The bombers attacked rear areas and disrupted communications; fighters kept the enemy air force at arm's length; fighter-bombers attacked battlefield targets and tanks.[44] These tactical changes were sufficient to blunt Axis air power in the desert, which had never been strong. In the weeks before El Alamein Rommel had a foretaste of the devastating air power he would witness in France − 'the paralysing effect which air activity on such a scale had on motorised forces'. 'Anyone', he ruefully reflected, 'who has to fight, even with the most modern weapons, against an enemy in complete command of the air, fights like a savage against modern troops . . .'[45]

The American air forces assigned to the North African campaign were still in their tactical infancy. The cumbersome command structure that evolved in 1941 had predictable effects: army units found that air support took so long to arrive that there was little point in requesting it. Airmen and ground troops had difficulty distinguishing enemy from friendly forces. Attacks were made at the front-line, but little attempt was made to counter Axis air power in the rear, or to attack supply and reinforcement areas. The ability of ground and air to work in combination, which was central to the new form of warfare, was conspicuously absent. The humiliation in Tunisia forced a major reassessment of how American tactical air forces were employed. They were assigned to an independent central air command which worked in tandem with the army commanders. Their mission was to neutralise enemy air power first, then to attack supplies and troop movements, and finally to attack critical points on the battlefield. During the Italian campaign the system was improved with the introduction of 'Rovers' − pairs of air controllers and army liaison officers who literally roved the battlefield calling up fighter-bomber strikes − and 'Horsefly', a small light aircraft acting as a forward air controller for directing bomb attacks on enemy artillery and transport. By the time the Allies deployed overwhelming air power over Normandy they operated a sophisticated system of air-ground cooperation that could cope with everything except poor weather.[46]

By the time the western Allies brought their mechanised armies and tactical air power to bear in France in 1944 much of the damage had already been done to German forces. The bombing offensive diverted and destroyed the better part of German air power, so that the tactical air forces had little or no opposition. They were free to roam far and wide over German-held areas, destroying equipment and vehicles in order to accelerate the decline in German technical resources. Bombing also interrupted the German supply of fuel, tanks and ammunition to the French front, making a difficult supply situation worse. When it was turned on the front-line itself, as it was on the Panzer Lehr division, it could turn the tide of battle. 'Utilisation of the Anglo–American air force *is* the modern type of warfare,' commented Rommel's deputy, Friedrich Ruge. Under these circumstances the Allies could afford a higher margin of error than the enemy. The application of technological strength through air power compensated for the difficulties in turning civilians with little experience of military life into effective soldiers. Western forces preferred the bomb to the bayonet. German soldiers wrote disparagingly of the poor fighting qualities of American and British Empire troops, particularly at close quarters. The effective deployment of modern technology, against an enemy forced to fight with little air cover, few tanks and dwindling quantities of trucks and guns, made the difference between victory and defeat.

The campaign in Normandy destroyed what was left of the German mechanised army after its mauling on the eastern front. The 7th and 15th armies facing the French coast relied for their mobility on 67,000 horses. There were only 14,500 trucks for the whole front, and severe shortages of fuel and tyres. During the fighting in Normandy half the horses were lost and most of the trucks.[47] Out of 2,300 tanks only a hundred survived. The soldiers fought with declining mobility and firepower against armies overflowing with resources. 'If I did not see it with my own eyes,' wrote one German division commander, 'I would say it is impossible to give this kind of support to front-line troops so far from their bases.'[48] Hungry, exhausted, outnumbered in the air and on the ground, German forces none the less fought with all their traditional operational and tactical skill. Hitler expected them, like Japanese soldiers, to fight to the last man. Most did not, but they fought with a desperate competence as long as it was possible to do so. It is widely accepted that man for man, German forces were generally more skilled than the enemy, east or west, for most of the war. But it was a margin that narrowed each year as the Allies reformed and modernised their forces, and as the slow process of de-modernisation set in for large parts of the German army. When the Allies sat on Germany's borders in January 1945 to deliver the *coup de grâce*, the technical gap was beyond redemption. The western Allies mustered 25 armoured divisions and six thousand tanks against a thousand German tanks; on the eastern front over fourteen thousand tanks faced 4,800 tanks and self-propelled guns. Soviet forces were backed by 15,500 combat aircraft against 1,500; the United States alone had 32,000 aircraft overseas.[49] For Germany modern war had come full circle since 1939.

★ ★ ★

The war of engines, of aircraft, tanks and trucks, was a war run on oil. 'Petroleum products', wrote the American geographer C.F. Jones, in 1943, 'are the blood of battles that bring victory.'[50] The Axis states were at a profound disadvantage in fighting modern war because they largely lacked this critical resource. Over 90 per cent of the world's natural oil output was controlled by the Allies. The Axis states controlled just 3 per cent of the output and 4 per cent of the oil-refining capacity.

So vital was oil for modern industry and modern war that the Axis states were willing to fight to gain control of it. The Japanese decision to attack the United States and Britain in 1941 was forced by an oil embargo that cut off from Japan 90 per cent of her supply. The prime target of Japanese soldiers and sailors in the drive southwards was the valuable oilfields of Borneo, Java, Sumatra and Burma, which produced more than enough oil to cover all Japan's military needs. In 1942 Hitler's drive into southern Russia, against the generals' advice, was aimed at the even richer oilfields of the Caucasus which could provide four times the quantity of oil Germany had in 1941. When German soldiers reached the oil centre of Maikop, and threatened to cross the Caucasus mountains to Baku, German and Italian forces in North Africa were battling towards the Suez Canal and the oil resources of the Middle East beyond. Axis armies came close to achieving a geopolitical dream, the conquest of the oil-rich crescent from Trans-Caucasia to the Persian Gulf.

Japan conquered her oil supplies; Germany and Italy did not. Without the oil of south-east Asia the Japanese war effort would have petered out. Though retreating oilworkers blew up or ignited what they could before the Japanese arrived, the damage was not irreparable. The Japanese expected to take two years to get oil production going again, but some of the fields were working within days. During 1942 the southern region supplied almost as much oil as Japan had imported from the United States before the embargo. Most of the oil was refined on the spot and absorbed into the battles in the south Pacific. This left the rest of Japan's forces and the home economy short of oil throughout the war. The large home fishing fleet was ordered to convert to sail and abandon oil. Even naval ships were converted back to coal, though it reduced their performance. Every effort was made on the home front to cut back on civilian oil consumption, which by 1944 was a mere 4 per cent of the amount used in 1941. From 1938 Japan had pursued a programme to produce synthetic fuel from coal, a process already pioneered in Germany. Although production went ahead, precise details of how the processes worked were withheld by the Germans until January 1945. In the interim the grandiose pre-war plans to produce 14 million barrels of oil a year fizzled out in the general climate of shortages; the peak of synthetic production was reached in 1942, a mere 1.5 million barrels of poor quality fuel.[51]

The one problem to which little thought had been given was how to transport oil safely around the new imperial seaways. There was a large tanker fleet, to which 1 million tons of new tankers were added during the war. By the end of the war most of them were on the ocean bed. Knowing how much Japan depended on oil, American submarines and dive-bombers sought out the vulnerable tankers, which had virtually no protection, sailing in ones or twos. If they radioed for air protection they were more likely to be visited by American aircraft than Japanese, directed to the target by the regular interception and decoding of Japanese messages. The flow of oil to Japan's empire began to slow down early in 1943; it was no more than a dark trickle by 1945.[52] Too late, efforts were made to sail in large convoys, but even these were given little air or naval protection. On 9 January 1945 Commander Kuwahara led a slow convoy of tankers and merchantmen out of

Saigon harbour, with six armed escorts. The following day it was spotted by a B-24 bomber. On the 12th it was ambushed by 16 fighter-bombers in mid-morning; one freighter was destroyed and the convoy was scattered. As it began to regroup seventy dive-bombers appeared, circling round the ships like hungry birds of prey. They bombed and torpedoed the convoy until dusk. All the oil tankers were sunk, all the merchant ships destroyed, and three escorts lost. The remaining escorts, heavily damaged, limped into harbour, where they were attacked and damaged three more times by carrier aircraft. Kuwahara survived, but of the convoy that followed from Saigon on the 10 January not one ship got through.[53]

Japan's oil situation was critical in 1944. Imports dropped to less than one-seventh of the pre-war volume. Though enough oil had been stockpiled to provide an estimated eighteen months' supply, it was eaten away in the months of fierce fighting in the southern islands. By 1945 imports were almost zero, and stocks fell to a level so low that the fleet could no longer operate. During 1944 Japanese warships were hampered by fuel shortages from fighting effectively. They were forced to sail slowly to conserve oil, or to sail by direct routes when good sense called for evasion. Pilots could not be fully trained for want of scarce high-octane fuel, and pilot losses reached levels so high in 1945 that there was little to distinguish the regular flyers from the suicide squads. During 1945 thousands of aircraft and small suicide boats or *shinyo* – wooden vessels of 2 tons, driven by a truck engine, and primed with high explosive – were gathered in Japan for the last-ditch defence of the home islands. They needed little fuel, and there was little to be had. The last barrels of carefully hoarded oil were hidden in caves and storerooms to power the suicide forces on their one-way passage.[54]

In Germany the decision was taken long before the war to find a substitute for natural oil. Although Germany produced a small quantity of oil from her own wells – about one-fifth of German consumption in 1939 – most German oil, like Japanese, came from the New World in the 1930s. One resource Germany did have in abundance was coal. In 1909 the German chemist Friedrich Bergius discovered a way of producing oil from coal by combining it with hydrogen under pressure. The process was known as hydro-genation. Though Bergius could produce oil in the laboratory, he could not find a way to extract it in quantity. During the 1920s the German chemical giant IG Farben bought up the Bergius patents and began a long and costly attempt to produce synthetic oil commercially. Though oil was produced in quantity it proved impossible to market: the synthetic material was six times more expensive than natural oil.

The project was saved by Hitler and the German armed forces. Even before Hitler came to power in January 1933, he had met with directors from IG Farben to assure them that a Nazi government would do everything to promote hydrogenation in order to render Germany immune to the threat of embargo or blockade. The new German armed forces in the 1930s needed oil as a top priority. In December 1933 the Hitler government signed a deal with IG Farben: the state promised to subsidise synthetic oil if the company committed itself to a four-year oil project. In 1936 the regime finally set up a programme to make Germany virtually independent of external sources of oil if war broke out. The programme was boosted by the seizure of the Sudetenland from Czechoslovakia in 1938, for the Sudeten areas contained rich deposits of lignite, or brown coal, which was far more suitable for synthetic production. By 1939 Germany produced just over one-third of her oil needs; by 1943 the figure was well over 7 million tons, three-quarters of all the oil she used.[55] Almost all the requirements of high-octane aviation fuel were met from the synthetic plants. What could not be produced at home was supplied from the Romanian oilfields at Ploesti, once Romania had become a virtual satellite of the Third Reich in 1940. Small quantities – about 10 per cent of German consumption – were provided in 1940 from the Soviet Union under the terms of the Nazi-Soviet Trade Agreement of 1939.[56]

If Germany had enough oil to fight with, the margin was slim indeed. The home front was starved of oil. Cars and lorries were compulsorily converted from petrol-driven to gas-driven motors. Those still permitted for non-military use were driven through the streets with large canvas gas bags attached, like so many small Zeppelins. The horse once again was in demand to haul goods and supplies. German inventors scoured Europe's flora for fuel substitutes. The acorn harvest, traditionally fed to pigs, was commandeered to produce a usable oil which was capable of fuelling haulage vehicles. European nuts were transformed chemically into a good-quality fuel to be used in tanks and trucks, and the residue was used as cattle-feed. A rich yellow oil, good to smell and taste, was pressed from the grape waste of French vineyards and was used as a machine lubricant.[57] None of these expedients could cope with demand from the armed forces and from industry. Fuel shortages slowed up progress in Russia in 1941; Rommel in North Africa was plagued by oil supply problems, though Axis forces drove back and forth across the undiscovered oilfields of Libya. While Allied forces could draw on the oil supplies of the Middle East, delivered by pipeline to the eastern Mediterranean, Axis oil supplies had to be carried on the dangerous convoy routes to Libya and Tunisia. They then had to be transported long distances from the ports with a severe shortage of tankers and trucks, under increasing air attack. Low mobility restricted the Axis drive to capture the very oil they needed.[58]

The alternative to acorns and nuts lay in the large oilfields of southern Russia. It is arguable how important oil was when Hitler planned Barbarossa. But by the winter of 1941–2 the oil question was central. The assault on Moscow was held up for six weeks while Hitler insisted that the drive to the Caucasus had priority. In the event neither was captured, but Hitler was convinced, against the forceful protests of his generals, that the oil question could decide the war. The capture of the Caucasus would kill two birds with one stone: the Soviet armies would be deprived of the oil needed to fight, and Germany would capture the oil she required to combat Britain and the United States. Hitler's directive for the summer campaign of 1942, Operation Blue, published on 5 April, placed the main weight of the German offensive in the south, 'in order to secure the Caucasian oilfields'.[59] By July German forces had swept through the Crimea and across the Don and were within striking distance of the first oilfields and the Soviet oil lifeline up the Volga, past Stalingrad.

The oil campaign was a chapter of disasters. German forces were divided and weakly spread over a vast geographical area. Shortages of fuel meant resort once again to large numbers of horses. When the first German forces reached the oil-town of Maikop, with its

annual supply of over 2 million tons of light crude oil, they found
the wells and refineries burning and demolished. German forces
never reached Grozny but were stopped 100 miles west of it. The
annual production here exceeded all German supplies. Beyond, on
the Apsheron peninsula by the Caspian Sea, lay Baku, a city literally
floating on oil. From a forest of derricks and chimneys came
annually over 20 million tons of oil, three times Germany's yearly
consumption.[60] At Maikop the revival of oil output was beset with
difficulties. Almost nothing had been done, for all Hitler's
insistence on the economic objectives, to prepare for the revival
and exploitation of the Caucasus oil. Germany was short of drills
and oil-producing equipment. What drills she had were already in
use in Germany and Austria, in the search for new sources of
natural oil. There were few oil technicians to spare. A group of
forty experts sent to Maikop were housed in a large barrack block
with German guards at the door, but during the night Soviet
partisans broke in and slit their throats. There were severe food
shortages for the oilworkers. When the drilling equipment was
finally prepared for transit it was held up in the overtaxed rail
network and had not reached Maikop by the time Soviet armies
recaptured the town. All Germany acquired during the period of
occupation was 70 barrels of oil a day.[61]

The defeat at Stalingrad completed the rout of the oil offensive.
Germany was forced to rely on synthetic production and Romania
for the rest of the war. Oil supply from these sources reached a peak
in 1943 and then declined sharply. In 1944 Germany received less
than half the oil from Romania supplied in previous years. The
Allies had long regarded oil as the German Achilles heel. The
British and French planned to attack German sources of oil supply
from the air in 1939 and 1940 but lacked the technical means to do
so.[62] During 1944 the bombing offensive was used to impose an
aerial oil blockade on Germany. The Romanian fields at Ploesti,
one of the most heavily defended targets in Europe, were attacked
first on 5 April, and were then bombed heavily until 19 August, the
day before the Soviet invasion of Romania forced her to change
sides and cut off all supplies of oil to Germany. The Danube had
been mined by British bombers, reducing traffic along the river
from Romania by two-thirds.[63] Germany was compelled to rely

more and more on home production, but in May 1944 the Allied bomber forces began a systematic assault on synthetic oil production. The impact was immediate. In June synthetic output fell by 60 per cent; by September it had been reduced to a mere 10 per cent of the output before bombing began. The American air force mounted 127 oil attacks, the RAF some 53. From the summer of 1944 the German forces lived off reserved stocks. At the beginning of the year these had been kept at roughly a month's consumption. By September 1944 they were down to a mere 150,000 tons, less than half the monthly total needed to keep the German forces in the field.[64]

During the course of 1944 the lifeblood of German armoured and air forces drained away. It was impossible to fight modern war without fuel. German air training, like Japanese, was reduced to simulation. New pilots got their first taste of real flying when they reached the squadrons, and the high loss rates reflected it. Reports from every German front-line complained of the lack of oil. Tanks were dug in as artillery. Oil use was rationed to only the most essential operations. The shortage of aviation fuel in particular – down by September to 5 per cent of its peak production level that year – gave the Allied air forces virtual command of the skies, allowing them to attack Germany's meagre oil supply lines at will. Even had the modern weapons been available in sufficient quantity, the problem of driving and flying them was insuperable. Both Germany and Japan gambled that they could win with careful use of their existing oil supplies and the seizure of oil-producing regions, but in reality oil remained a vulnerable aspect of their war efforts, and one that limited the extent to which their forces could be armed and supplied with modern petrol-driven equipment. The oil shortage was exacerbated by enemy action. The seaborne blockade of Japan and the aerial blockade of Germany were deliberately designed to exploit this Axis vulnerability.

It would be wrong to argue that oil determined the outcome of the war on its own, though there could scarcely have been a resource more vital to waging modern combat. Because they had sufficient supplies, the Allies enjoyed a much greater flexibility both in choosing the structure of their forces and deciding how to use them. But even the Allies, for all their preponderant supply of oil resources, were not free of oil problems. Soviet production fell by almost half thanks to the disruption of the industrial economy, and the supply of high-grade aviation fuel was made possible only by importing chemicals from America to upgrade Soviet fuels. Britain fought for much of 1942 and 1943 to keep open the oil supply lines from the Americas across the Atlantic, and to guard the gateway to Middle Eastern oil in Egypt, Even the United States, endowed with more oil than the rest of the world together, could not simply turn on the tap. California alone produced more than the Soviet Union, but there were no pipelines running eastwards, and west coast oil went largely to fuel the Pacific war. After a decade of low output and prices the oil industry suddenly found itself faced with demands from the armed forces and America's allies that it could barely meet. A state oil commissioner was appointed to plan the expansion of output. The choice fell on Roosevelt's Secretary of the Interior, Harold C. Ickes. He was a politician of the New Deal, far too liberal for the oilmen. Ickes swallowed his scruples and built bridges to the industry. Government investment and subsidies poured into some of America's largest and richest corporations.[65]

There were two particular problems to be solved. The first was to find a safe way of getting oil to the main industrial areas of the east and north-east in the face of the German submarine threat to the oil supply routes along the coast. At the beginning of the German-American war tanker tonnage took heavy losses. Petrol rationing was introduced in the eastern states, and speeds restricted to 35 miles per hour. Ickes urged the industry to build new pipe-lines to deliver the oil across country. So began the construction of the 'Big Inch' pipeline, one of the largest engineering projects undertaken anywhere during the war. The pipe was in fact 24 inches in diameter, and ran from Longview in Texas to Phoenixville, Pennsylvania. It was designed to carry 15 million tons of oil a year. Pumping stations were constructed every 25 miles along its 1,380-mile course. The line was completed in 1943.[66]

The second problem was the production of high-octane fuel for aircraft. Hundred-octane fuel produced much better performance than the 87-octane fuel used by the Luftwaffe. It allowed aircraft longer range, greater manoeuvrability, and those surges of power that gave Spitfires the edge over the Messerschmitt-109 in 1940. In

1940 the United States produced fewer than 40,000 barrels of it a day. Ickes pushed hard for another great engineering effort to expand 100-octane production above all else. By 1944 the United States produced over 90 per cent of the Allies' high-octane fuel, more than half a million barrels a day.[67] American production made it possible to sustain a vast navy, the bomber fleets of both western Allies, and a fully motorised army. Though the distribution of supplies never worked perfectly – the shortage of fuel for Patton as he raced across France is the most notorious example – the oil weapon gave the Allies the means to exploit the modernisation of their forces to the full. At the dinner for Churchill's birthday in the middle of the Teheran conference Stalin stood up to propose one of many toasts: 'This is a war of engines and octanes. I drink to the American auto industry and the American oil industry.'[68]

★   ★   ★

The only way the Axis states could have reversed the technological balance decisively was by finding entirely new weapons. In effect that meant the atomic bomb. A great many technical novelties were proposed on both sides, but nothing approached the unique destructive power of atomic energy. Every combatant knew by the war that atomic physics might have a military application, but only the United States was able to turn this knowledge into a usable atomic weapon by the war's end.

Japanese scientists pursued a number of secret weapons, none more bizarre than the 'death ray'. By the end of the war they had developed an electronic device based on a high-frequency electro-magnetic wave tube emitting 13-foot waves that could stop a petrol engine at a short distance, or kill a rabbit at 10 feet from haemorrhages of brain and lungs. The plan to turn the ray on enemy bombers never materialised.[69] Atomic research got little further. Japanese physicists were well aware that a bomb was a possibility, and that their American enemy was working on just such a project, but they did not believe that any power could produce one within the likely period of conflict. The Japanese navy began exploratory experiments in 1942, working with Japan's leading atomic scientist, Yoshio Nishina, a pupil of the Danish

288 A WAR OF ENGINES

physicist, Niels Bohr. The navy was more interested in the prospects for nuclear propulsion – given the shortages of oil – but by March 1943 naval researchers could see that the likelihood was poor and abandoned atomic research. Nishina switched loyalties to the Japanese army, which would not collaborate with its sister service, and continued to pursue the separation of the vital U235 isotope of uranium which held the key to atomic fission. Neither his equipment nor method matched his ambition. By 1944 the experiments faced stalemate. The money and high-quality scientific equipment Nishina wanted could not be spared from the war effort. When he complained to his army contact that he still needed 10 kilograms of U235 to make a bomb, the general replied, 'Why not use ten kilograms of a conventional explosive?'[70] Four months after this interview B-29 bombers hit Nishina's laboratory. The wooden building burnt to the ground; Japan's atomic weapons programme was reduced to ash.

The prospects for producing atomic weapons were much greater in Germany. Much of the pioneering work in atomic physics was German. Germany was endowed with extensive scientific resources, and an outstanding scientific community, even after the emigration of distinguished scientists to the west in the 1930s. It was a German chemist, Otto Hahn, who published in January 1939 the first paper to demonstrate that the nuclear fission of uranium was a possibility; this was the process at the heart of nuclear power. The fruits of overseas research were freely available in the scientific periodicals. In the spring of 1939 German physicists grasped that the energy released in the fission of uranium would be sufficient not only to produce a new source of fuel, but also to create an explosion that would dwarf those produced by conventional weapons. On 24 April 1939 the Hamburg chemist Paul Harteck, a Nazi Party member who had worked in Cambridge in the 1920s with Ernest Rutherford, the New Zealand physicist who first split the atom, wrote to the Army Ordnance Office with news of the potential new weapon: 'That country', he warned, 'which first makes use of it has an unsurpassable advantage over the others.'[71] The same month the German Education Ministry set up a high level nuclear research team; in September many of its resources were absorbed by an even larger project funded by the German

army. In December 1939 Werner Heisenberg, Germany's outstanding theoretical physicist, submitted to the research team a
paper that outlined in detail how a nuclear reactor worked, and the
possibility of an atomic bomb.[72]

Over the next two years the research teams pursued the practical
issue of turning these theoretical insights into a usable product.
Nuclear fission required processes of exceptional complexity and
expense. Uranium, the heaviest known element in 1939, contains
two isotopes. The first, U238, makes up 99.3 per cent of the metal;
the other, U235, a mere 0.7 per cent. The preponderant isotope
absorbs free neutrons in the element and keeps it stable, but
neutrons can divide U235 and create a chain reaction, which
releases enormous kinetic energy. To create a bomb it was
necessary to increase the number of unstable isotopes until a
sufficient mass of fissionable material was available to cause an
explosion. Uranium, when bombarded with neutrons, also
produced a new artificial element made up of U239 isotopes. This
new element, numbered 94, was christened plutonium. It was
found to be as fissile as U235, and promised an alternative source of
material for bomb-making. Both the production of additional U235
isotopes and of plutonium were examined, but the army opted for
enhanced uranium, and so this was where most of the German
research effort was concentrated until 1945.

There were several ways of producing more U235. It could be
done electro-magnetically, as it was at the Oak Ridge facility in the
United States later in the war. Or it could be produced by adding
a moderating material which could slow down the neutrons in
uranium without absorbing them. This allowed the neutrons to
avoid the U238 isotopes, which absorbed them only when they
moved at higher speeds, and to divide the U235 isotopes instead.
Heisenberg suggested a number of moderators: pure graphite was
one, and another was so-called heavy water (D_2O or deuterium
oxide), produced by removing hydrogen atoms from ordinary
water by electrolytic conversion. Graphite was required for the
army's rocket research programme, and so by default the choice fell
on heavy water. The more effective electro-magnetic method was
ignored. Europe's only source of heavy water lay in Norway, at the
Norsk Hydro plant at Vermork. Production was time-consuming

and costly. In February 1943 the plant was successfully sabotaged by the Norwegian resistance. By the end of the war German scientists had 2½ tons of heavy water, only half the quantity needed even to begin serious production of fissile material.[73]

In early 1942 the army gave up on the project. The prospect of producing a weapon in time for it to be used seemed remote. The Education Ministry once again assumed responsibility; atomic physicists were ordered to use their time for more immediate war-work, and neither a nuclear reactor nor adequate quantities of enhanced uranium were ever produced. Research work continued, interrupted by bombing and evacuation, but when Allied intelligence teams scoured Germany in 1945 for scientists and laboratories they found that Germany was still years from producing an atomic weapon. Some German scientists blamed the system with its excessive compartmentalisation and political interference; others, Heisenberg among them, argued that they had deliberately dragged their feet to prevent nuclear weapons falling into Hitler's hands.

The truth is more complicated. Whatever the views of Germany's scientists, Hitler remained hostile to the whole project. Without the wholehearted support of the political leadership the atomic programme was unable to generate the huge resources of labour, materials and brain power necessary. Though Hitler saw himself as an expert on guns and tanks, he found the principles of modern physics hard to grasp and disliked discussing them. Party scientists branded much of the new work as non-Aryan, 'Jewish physics'. When Speer tried to talk to him about the research Hitler condemned it as 'a spawn of Jewish pseudo-science'. He retained the fear that nuclear explosions would prove incapable of control, and might burn up the hydrogen in the atmosphere, destroying the globe. Nothing that was presented to him during the war carried sufficient conviction about the short-term feasibility of atomic weapons to disperse that scepticism.

Hitler did not bring nuclear physics to a halt, but the research lacked the urgency of other more pressing armament projects. The potential scale and expense of the task, given the uncertainties still evident in atomic physics, weighed heavily against committing resources to it. Though German scientists knew of American research they remained reasonably confident that they were ahead in the atomic race throughout the war, despite its low priority. This arrogant assumption, shared, it should be added, by the anxious enemy, was the product of a science establishment heavily weighted towards pure theory. When the theoreticians turned to the practical development of the ideas they ran into difficulties. The experimental physicists were looked down upon, and the practical problems in producing atomic energy were neglected. Lacking a solid experimental foundation, Heisenberg undermined the whole development by insisting that a uranium bomb needed only a small amount of U235 to achieve critical mass (the American project required 30 pounds), and by arguing that a chain reaction could not be controlled. There were practical researchers who by the middle of the war had discovered that one of the best ways of producing fissile material was to employ extremely low temperatures (at Oak Ridge the Americans used liquid helium to achieve this). Harteck realised in 1941 that low-temperature reactors would solve the production problems, but was denied the uranium to verify his findings. Another researcher, Baron Manfred von Ardenne, who worked, improbably, for the German Post Office, successfully developed the method of electro-magnetic isotope separation.[74] Neither idea was taken up by the elite of theorists who dominated the field. Was this a form of resistance, or sabotage? or was it simply that they failed to grasp their significance? Did it reflect professional self-interest? or a genuine belief that a workable bomb was many years off? No doubt moral scruple played a part with some of Germany's science community, but there were enough scientists who supported the war effort and understood the nature of the atomic bomb to have produced a nuclear capability if the regime had been enthusiastic enough. As it was, the scepticism was borne out by events; even the Americans, with vast resources, 150,000 workers, and top priority for the project, failed to produce a bomb until the war in Europe was over.

Just what a difference Hitler made in the field of research and development was evident from the fate of the German rocket programme. Rocket research dated back to the 1920s, fired by the seminal work of a German-speaking Transylvanian, Hermann Oberth. His book *The Rocket into Interplanetary Space*, first published in 1923, provided the theoretical frame; but when Oberth tried to build a liquid-fuelled rocket in 1929 as publicity for the Fritz Lang film *The Woman in the Moon* the project was a technical disaster. The furore over Oberth's rocket attracted the attention of the German army, which began a programme of rocket development initially with Oberth's assistance. In 1932 the army decided the project was so important that it set up its own top-secret laboratory. The first employee was a twenty-year-old student at the Berlin Technical High School, Wernher Freiherr von Braun. Despite his youth, von Braun had grasped the principles of rocket propulsion, and led the way in developing the first generation of army rockets. By 1935 enough was achieved to secure generous funding and the cooperation of the fledgling Luftwaffe. The two services combined their research on a remote and barren stretch of the Baltic coast, at Peenemünde. Here laboratories, production facilities and giant rocket launchpads were built under the leadership of an ex-captain of artillery, Walther Dornberger.[75]

The army leadership wanted to be sure that a viable weapon would result from their investment. In 1936 Dornberger promised them a missile that could deliver a 1-ton warhead twice as far as the famous Paris Gun, the Krupp cannon used in the Great War. For the next six years von Braun's team of engineers wrestled with the technical difficulty of turning their early experiments into a real weapon. The Dornberger rocket, the A4, suffered from the immature state of research into engines fuelled with liquid oxygen, and persistent problems in the guidance and control systems. Hitler, though aware of the research, did not give it a particular priority, until Dornberger and von Braun visited him in August 1941 to discuss their work. Rocket technology did not alienate Hitler like nuclear fission did. He hailed the new missile as 'revolutionary for the conduct of warfare' and promised more money. By June of 1942 the first rocket was ready for launching. Hermann Göring was the guest of honour. Among the sober team of engineers and soldiers, Göring arrived dressed in bright red riding boots with silver spurs, and a vast opossum-skin coat, his hands covered in ruby-encrusted rings. In front of the baroque visitor the research team watched as the 13-ton rocket blasted from its pad and roared into the clouds. A minute and a half later the roar began to get louder again. To the horror of the onlookers the missile plunged into the ground only 800 yards away. The guidance system was still insufficiently developed.

On 14 October 1942 the first successful launch was carried out. The rocket travelled 120 miles and landed within 2½ miles of the target. This was hardly long-range precision, but Hitler was impressed enough to call for production of five thousand. Technical problems continued to plague the rocket, however, and by the summer of 1943 there were still no finished weapons. By that time the Luftwaffe had developed its own pilotless missile, the Fieseler-103 'flying bomb'. Both projects emerged as technical possibilities at just the time that Hitler began to search for some means of retaliation against Allied bombing, or some wonder-weapon that might turn the tide of war. On 7 July 1943 Dornberger and von Braun visited Hitler again to press their case for developing the rocket. After a brief discussion, Hitler was invited to the cinema at his headquarters, where von Braun showed a colour film of the successful launch of a rocket. It was an ideal tactic. Hitler was overcome with excitement. He told Speer that the A4 was the 'decisive weapon of the war'; he agreed to appoint von Braun a professor on the spot, and ordered immediate mass production.[76]

The new weapons, the air force pilotless bomb and the rocket, were nicknamed '*Vergeltungswaffen*', 'weapons of revenge', which gave rise to the popular abbreviations V1 and V2 respectively. Hitler wanted thousands of them, there and then, to shower down on England and the preparations for invasion. In reality things moved slowly; neither weapon was yet at the stage of development suitable for mass production. The A4 became bogged down in a further round of design changes to cope with the unexpected distortions created by the addition of a warhead. In August 1943 the RAF destroyed the Peenemünde rocket-research base, as news of the secret weapons leaked to the west. Despite this setback Hitler determinedly persisted with the 'wonder-weapon' strategy. The whole project was taken from the army and given to the SS. Himmler built a production complex for the rocket deep in the Harz mountains, and promised to fulfil Hitler's demand for five thousand missiles a month. Thirty thousand concentration camp labourers slaved in the underground caverns in unspeakable conditions. When Speer visited the plant he was met by lines of exhausted, grimy men with 'expressionless faces, dull eyes'. The brightly lit tunnels, 2,800 yards long, were full of stale air and the stench of excrement. Speer felt numb and dizzy. When he returned above ground he took a stiff drink.[77]

In the subterranean nightmare of Himmler's '*Mittelwerk*', the slaves installed the machinery, the rail track and equipment. One-tenth or more of their number perished every month. By January 1944 production began, with an output of fifty rockets rather than five thousand. Constant technical changes held progress up. Over the whole year only 4,120 were produced. Despite Dornberger's warning that the rocket was still not fully developed, Hitler and Himmler pushed on with the 'decisive weapon'. The first fying-bombs were despatched on 13 June, but only ten could be launched, and only five reached London. The first rocket was launched on 8 September. Out of the six thousand produced only 1,403 were fired at Britain, and only 517 actually landed on the capital. Some 5,800 flying-bombs were eventually launched, 2,420 of which reached the capital, out of a total production of thirty thousand. The weapons of revenge killed nine thousand Londoners, but the total tonnage of explosives they contained – approximately 2,500 tons over nine months – represented just 0.23 per cent of the tonnage dropped on Germany by the Allied air force over the same period.[78]

Neither rockets nor flying-bombs were war winners. The technology was immature, the warheads, even with the high velocity of the rocket impact, which magnified the effects of a ton of high explosive, too small to achieve any major damage, and too imprecise to target particular installations. The weapons ignited the persistent fantasy among Nazi leaders that the fortunes of war could be turned at the last hour by an unforeseen technological break-through. The V-weapons programme cost over 5 billion marks, and absorbed tens of thousands of workers. The resources that went to build them could, according to the American Bombing Survey, have produced an additional 24,000 aircraft. Himmler's brutal efforts to squeeze out the manpower and materials compromised the rest of the war economy, while the research teams were forced to tackle the A4 and the flying-bomb instead of concentrating on projects with greater strategic worth. One of the programme casualties was a ground-to-air missile, *'Wasserfall'*, developed at Peenemünde, which was closer to mass-production and less expensive than the A4. Its strategic value in 1944, as a weapon capable of destroying the heavy-bombers, greatly outweighed the A4, but there were not enough resources to go round and the latter enjoyed Hitler's personal support. Only a different warhead would have made the A4 worth the effort. Much recent research has attempted to demonstrate that such a warhead did in fact exist, but the evidence is not entirely clear cut. Himmler is supposed to have authorised development of so-called 'dirty bombs' using the radioactive waste generated by the nuclear research programme. Small spherical bombs 'the size of a pumpkin' were stored in the tunnels of the *Mittelwerke*, where the rockets were produced. A detonation with conventional explosive could have produced a device capable of irradiating an area of a square mile; but German engineers failed to produce a fuse that could have detonated the dirty bomb in the atmosphere, even if Hitler had approved their use. It has also been claimed that small tactical nuclear weapons were developed out of the research into nuclear energy, but the precise nature of the bomb, if it existed, cannot now be verified.[79]

There is a profound irony here, for it was fear of Germany's technological prowess that prompted first the British and then the American government to pursue their own atomic programme with their collective scientific energy. British scientists faced the same scepticism as did their German counterparts when the prospect of an atomic bomb was first suggested in 1940. But Britain faced a situation of desperate isolation, even defeat. In April 1940 a committee of top British scientists was set up to evaluate and report on the possibility of producing an atomic bomb within the likely timescale of the war. The 'Maud' committee, as it was known (the word was plucked from a cryptic telegram sent by Niels Bohr, and turned out to be the name of his former English governess), reported in July 1941 that it was possible to produce a bomb either from enriched uranium or from plutonium produced in a reactor. A great deal of experimental work supported their conclusions. They spelt out the industrial costs and problems, even down to the price of producing one bomb per week. On 17 October the decision was taken to proceed. That autumn two American physicists were sent to Britain to examine the atomic programme; they reported back to Washington that the bomb was a realistic project. The day before Pearl Harbor the Office of Scientific Research and Development approved the research for an atomic bomb.[80]

In June 1942 the British finally recognised that the industrial and research effort needed to make a bomb quickly was beyond them. America took over the whole project. British scientists were coopted to three 'secret cities', one at Oak Ridge, Tennessee, where the uranium separation plants were built, another at Hanford, for the production of plutonium, and the third in the New Mexico desert, at Los Alamos, where the laboratories were sited. In August 1942 the bomb project – codenamed 'Manhattan' or 'DSM' (development of substitute materials) – was launched. Enriched uranium was produced by electro-magnetic process; plutonium was developed in reactors cooled by liquid helium; graphite, the second of Heisenberg's moderators, proved the most successful. There were still innumerable technical hurdles to surmount. British scientists had forecast in 1941 that a bomb could be produced by the end of 1943. But even with strong political and military backing, generous resources, a large international team of scientists and engineers and no disruptions from bombing or sabotage, there was still insufficient uranium to produce a bomb by the time the war in Europe was over.

Not until 16 July 1945 was it possible to test the new weapon. Both a plutonium bomb and an enriched uranium bomb were planned, but the test used a plutonium warhead, which generated less fall-out. Early in the morning the Los Alamos scientists made their way to a position some 10 miles from the Alamogordo Air Base site where the detonation was to take place. The weather was poor, and the test, planned for 2 a.m., had to be postponed for three and a half hours. All personnel were instructed to lie face down on the ground with their hands over their eyes until the flash of the bomb was past. At 5.30 a.m. the bomb went off. Otto Frisch, the German emigré who in 1939 helped to calculate the necessary critical mass of U235 to cause an explosion, stood 20 miles away, behind a radio truck for fear of ultra-violet rays. The hills around him were flecked with faint dawn. Suddenly, soundlessly, the hills were glaringly illuminated 'as if somebody had turned the sun on with a switch'.[81] When he looked over the truck he saw a 'pretty red ball, about as big as the sun' connected to the earth by a grey stem. One of the technicians who had been placed in a dug-out 5 miles closer to the explosion, arrived back at the main group. He said simply: 'The war is over.' General Groves, the military head of the project, replied: 'Yes, after we drop two bombs on Japan.'[82]

Both were wrong. Atomic weapons did not win the war, for they came far too late to affect the outcome. Japan was on the point of surrender by the time the two available bombs were used. The technical triumph helped to illustrate how intimately technology and war-making were linked in the western war effort. The war accelerated the technical threshold, and brought the weapons of the Cold War within reach, but no state, even the most richly endowed, was able to achieve a radical transformation of military technology before 1945. The war was won with tanks, aircraft, artillery and submarines, the weapons with which it was begun.

*   *   *

The fate of German military technology provides one of the central paradoxes of the war. Germany was without doubt a modern state by the standards of the 1940s, but her forces were stage by stage deprived of the modern weapons they needed. While German scientists pioneered the world's most advanced weapons – rockets, jets, atomic weapons – German forces lacked adequate quantities of the more humdrum petrol-driven equipment. Only later in the war was an effort made to change this balance, but by then the attrition ratio was too far in favour of the Allies to be reversed. Billions of marks were spent on projects at the very frontiers of military science which brought almost no strategic advantage whatsoever.

The paradox can be explained in part by the warped outlook of Germany's leaders, who persuaded themselves as the war began to turn against them that German science could conjure up a new generation of fantastic weaponry that could reverse the war's course at a stroke. 'Retribution is at hand,' Goebbels told Party leaders in February 1944. 'It will take a form hitherto unknown in warfare, a form the enemy will find impossible to bear.'[83] Towards the end of

1944 Himmler tried to revive atomic bomb research, and to develop a new chemical weapon, the new incendiary 'N-material' that was literally inextinguishable. All these projects absorbed men and materials that would have been much better spent in conventional fields, but the fantasy of secret weapons helped to cushion the reality of impending defeat.

There is a second explanation: the German armed forces pursued technical excellence for its own sake. By the late 1930s they had developed the weapons that would in the main be used to fight the Second World War. They were now keen to move on to the next technical threshold to keep ahead in the arms race. At the outbreak of the war they were already at the very starting edge of the world of jets and missiles. When the war came they tried to speed the process of development up, to win the war with the weapons of the 1950s. The result was a technical disaster: shortages of resources, constant political interference, the inherent difficulty of accelerating research work at the forefront of science, all meant that German forces got little operational payback from the new weapons to match the great expense of producing them. The Allies – except for the Manhattan atomic project – stuck with the weapons of the late 1930s, and pushed them successfully to their limits, in most cases overtaking the performance of Germany's more conventional weaponry. When after the war they came to develop missiles, jets, advanced submarine technology, and a host of other vanguard equipment, they simply took German scientists and blueprints.

By contrast with Germany, the Soviet Union was far less modern, though the difference is often prone to exaggeration. But over the course of the war very great strides were made in all fields of advanced technology. Hundreds of thousands of poorly-educated, technically illiterate Soviet citizens were trained to drive trucks and tanks, or fly some of the world's fastest aircraft. The Soviet Union had nothing to lose and everything to gain by upgrading her manpower and equipment. For twenty years Soviet communism had preached the virtues of modernisation as the answer to all the country's problems. Now the gospel was needed more than ever. The struggles of collectivisation and industrial transformation became the struggle for military progress. Soviet soldiers were tough, used to harsh conditions of life and few luxuries. When that hardiness was married to modern technology the mix proved much more effective than many in the west had thought possible. When John Erickson, later a distinguished historian of that modernisation, first confronted the Red Army as a young infantry sergeant in 1945 he was struck by the curious blend he beheld: 'an army of unwashed, uncouth, lithe Ukrainians, squat riflemen from the central Asian republics, combat medals a-jingle, cradling superb self-loading rifles – but, above all, the tanks in their fungicidal green colouring, the paint just slapped on over those powerful turreted guns'.[84] In the end the Soviet army proved just modern enough; but German forces were too modern for their own good.

8

IMPOSSIBLE UNITY . . .
Allies and Leaders in War

*'Our greatest triumph lies in the
fact that we achieved the impossible,
Allied military unity of action'.*

General George C. Marshall, 1945

THE FIRST TIME that the three Allied leaders came face to face
was at the Teheran conference in November 1943. At four
o'clock in the afternoon of the 28th, they were to assemble at a
small conference room in the Soviet Embassy. Stalin was already
waiting for them. Roosevelt was pushed in his wheelchair from the
small building in the grounds of the Embassy where the American
delegation had been persuaded to stay. Dressed in a blue business
suit, he arrived half an hour before Churchill. Churchill had already
met Stalin, but to the President the Soviet dictator was an unknown
quantity. When Roosevelt entered Stalin ambled towards him,
walking clumsily 'like a small bear'.[1] A little later Churchill left the
British compound with his retinue to walk the few hundred yards
to the conference hall. Soviet secret policemen peered from behind
every tree. When he arrived he greeted Stalin cordially and shook
Roosevelt's hand. 'Everything was so relaxed,' recalled the
American interpreter. 'It did not seem possible that the three most
powerful men in the world were about to make decisions involving
the lives and fortunes of millions of people.'[2]

Teheran was not an ideal location. Roosevelt had travelled 7,000
miles to get there, Churchill 4,000. Roosevelt had suggested the
Bering Straits, Khartoum, Cairo, even Asmara in Somaliland, but
Stalin rejected them all.[3] Stalin chose Teheran only because he said
he wanted to stay in close communication with the front in an area
where Soviet security could operate freely. No one ran any risks.
The city was patrolled everywhere by Soviet agents, guns bulging
ostentatiously beneath their jackets; there were 2,500 American
troops on hand; Churchill was protected by a whole regiment of
Indian Sikhs. The conference room was comfortably furnished,
with a large green-covered table in the centre. The leaders, with

their interpreters and chief advisers – all except General Marshall, Roosevelt's Chief-of-Staff, who was out sightseeing because, incredibly, nobody had told him about the meeting – sat in a circle to begin the formal proceedings.[4]

Their opening remarks were entirely in character. Roosevelt began lightly: as he was the youngest of the three he welcomed his elders to the table. Churchill was almost absurdly grandiloquent: 'In our hands we have the future of mankind.' Stalin, speaking softly, almost indistinctly, as he did for much of the conference, said simply: 'Now let us get down to business . . .'[5] There were other contrasts. Stalin's habit was to sit very still, to make few gestures, and to speak only when necessary. He occupied himself by doodling with a thick blue pencil, which he carried around with him all the time. Roosevelt was affable and talkative. He acted as an informal chairman, removing his pince-nez and waving them in the air to underline his remarks. Churchill was ill for much of the conference, and obviously ill-at-ease. He fidgeted at the table, scribbled notes and passed them to his colleagues, and talked at such length without pausing that his interpreter was hard pressed to reproduce what he had said. The one habit they had in common was smoking, Stalin a pipe or a cigarette, Roosevelt cigarettes, and Churchill a recalcitrant cigar which he lit and re-lit throughout the discussion.

Beneath the outward show of bonhomie there were strong currents of distrust and uncertainty. The three leaders were united in what was popularly called the 'Grand Alliance', but their three states were not even allies in any formal sense. Britain and the Soviet Union had signed an alliance of cooperation in May 1942, against Churchill's better instincts, but the United States refused to enter into fixed agreements with either of their fellow combatants. The Soviet Union was not even at war with Japan. The coalition survived through a common interest in the defeat of Germany, and little else. There were differences between them even on this central ambition. The Soviet delegation arrived at Teheran determined to force the west to enter into an unambiguous commitment to open a Second Front in France, after almost two years of delays. Churchill and his staff were just as determined to assert the British preferences for further action in the Mediterranean theatre. Stalin would have preferred to meet Roosevelt on his own. He let the President know before the meeting that he and Churchill 'got into each other's hair'.[6] Only Roosevelt, who had long hoped to meet Stalin face to face, travelled to Teheran with the expectation of smoothing the rough edges of collaboration.

For all the President's notorious charm, the first contacts were awkward. He wrote to his wife of the atmosphere of 'great distrust' emanating from Stalin when they first met. Arnold, Roosevelt's air force Chief-of-Staff, was struck by Stalin's patronising treatment of the British, 'half-humorous, half-scathing'.[7] Roosevelt decided to side with Stalin in these exchanges in order to break the ice. At the plenary meeting on the second day of the conference the President began by ignoring Churchill and chatting to the Soviet delegates; he began to poke fun at the British, and the more Churchill scowled the more Stalin smiled. When Stalin finally burst out laughing at Churchill's discomfiture, the tension between the two leaders was broken. Though Roosevelt later recalled that they now talked 'like men and brothers', he strained the relationship with Churchill, who had to accept not only the jibes of his partners but their strategy as well. For at the second session Stalin at last got his commitment to the Second Front, even from Churchill.[8]

By the evening of the second day there was a more affable mood. Dinner was held at the Soviet Embassy, and was by the traditions of Soviet hospitality both lavish and long. But Stalin could not leave Churchill alone. Throughout the evening he returned to the theme that the Prime Minister wanted to treat the defeated Germans too kindly. When Stalin suggested shooting fifty thousand German officers as an example Churchill, red in the face, rose indignantly from the table and denounced his host with a passionate ill humour; Roosevelt smilingly agreed to 49,000 and with that Churchill swept out of the room. Stalin hurried after him and persuaded him to return and the baiting ended.[9]

The following day the seal was put on the coalition's strategy for the defeat of Germany. In the evening the British Legation hosted a birthday dinner for Churchill. With the Second Front secure, the Soviet delegates shed their inhibitions. Stalin refused the offer of cocktails, for which he had an unexplained distrust, but attacked the whisky, and then a great deal of champagne, which he was less used to drinking. Churchill decreed that the party would adopt the Russian habit of endless toasts and speeches. He toasted 'Stalin the Great' with more than a touch of irony, and 'Roosevelt the Man'. Stalin hailed 'my fighting friends', but added the sting: 'if it is possible for me to consider Mr Churchill my friend'. Later Stalin accused Churchill's Chief-of-Staff, General Brooke, in front of the whole dinner party of being 'unfriendly' towards the Soviet Union. Both Churchill and Brooke gave as good as they got in reply, but the aspersions gave an unfortunate edge to an occasion generally remembered by all present to have been an optimistic expression of real partnership.[10]

By the end of the evening the sour note was drowned out. Stalin was unusually intoxicated, moving from one guest to another and clinking glasses, and then forcing the disconcerted waiters to drink with him too. Roosevelt ended the proceedings by calling on the three coalition partners to work as 'a harmonious whole', with 'the traditional symbol of hope, the rainbow' before them in the skies of war. Churchill – who of all of them had the most to regret from the conference – went to bed that night satisfied that 'nothing but good had been done'.[11] The following day, 1 December, the spirit of the birthday party was embodied in a public communiqué signed by all three leaders. They announced a 'common policy' directed at 'the destruction of German forces'. The final line betrayed its American authorship: 'We leave here, friends in fact, in spirit and in purpose'.[12] Roosevelt returned to Washington confident that the foundation for a firm personal relationship with Stalin had been secured, and that this was essential for the defeat of Germany and the rebuilding of the postwar world. Neither Stalin nor Churchill saw the meeting in such rosy terms. In Moscow the commitment to a Second Front was treated sceptically right up to its final launch, on 6 June 1944. Churchill, aware of how his influence in the coalition had been diminished, drew away from the close relationship with Roosevelt established earlier in the war, and continued to distrust Stalin. None the less, the one solid achievement of the conference, a mutual commitment to the final defeat of Germany, was forcefully and publicly stated. This final coalition, which had been fragile and embryonic since 1941, was an essential condition for the eventual victory of the Allies.

★   ★   ★

The coalition was from the first a product of necessity rather than deliberate intention. The relationship between Britain and the United States was far closer throughout the war than the relationship of either with the Soviet Union. Yet even Anglo-American collaboration was difficult to secure before Pearl Harbor, and was marred by friction thereafter. The roots of western collaboration went back to the early summer of 1940 when Britain faced certain defeat in Europe, and Churchill, recently appointed Prime Minister, appealed to Roosevelt to throw American assistance into the Allied scales. American opinion was divided on the issue. The President was forced to take account of strong anti-war sentiment, and a powerful residual distrust of British imperialism. Nineteen-forty was also election year. Roosevelt was ambitious for an unprecedented third term in office, and would not run risks with public opinion. He was sympathetic to the British cause, but he was by nature, he told Averell Harriman, 'a compromiser'. He promised the American people that he would keep their country out of war, while agreeing to provide some of the material assistance Churchill wanted. In November 1940 Roosevelt was re-elected. A few weeks later, following an impassioned plea from Churchill for real aid, Roosevelt took the risky step of proffering more American goods and weapons. The scheme of provision was named Lend-Lease, to preserve some spurious notion that goods were on extended loan and might one day be returned. The British saw it as a lifeline. 'This is tantamount to a declaration of war by the United States,' Churchill declared to his private secretary. He regarded the economic alliance as 'the most important thing' next to winning the war.[13]

Lend-Lease provoked a political storm in the United States, but Roosevelt stuck steadfastly to the commitment. In March 1941 the proposal passed through Congress. Churchill now worked to turn American economic aid into active belligerency. The US navy gradually extended its activities out into the Atlantic to protect the sea-lanes from submarine attack; American servicemen were stationed in Greenland and Iceland. Beyond that Roosevelt would not go. He was, remarked the journalist Dorothy Thompson, 'attempting to win a war without fighting it'.[14] Whether Roosevelt would ever have declared war on the Axis states if American non-belligerency had not been swept aside by Pearl Harbor remains unanswerable. Churchill regarded the Japanese attack as Britain's salvation. He recalled in his memoirs the emotion he felt at hearing the news: 'We had won the war . . . Once again in our long Island history we should emerge, however mauled or mutilated, safe and victorious.'[15] Though Churchill had failed to secure America's active participation sooner, he did succeed in securing the moral support of both the President and a good fraction of American opinion. When the United States entered the war there already existed a framework for close cooperation between the two states. Unlike in the First World War, when the United States refused to integrate its war effort closely with Britain and France, the two western powers swiftly established an exceptional degree of collaboration.

It has become fashionable to see this as the time when Churchill, half-American himself, sold out to his richer cousins, leaving the future of the British Empire hostage to American wishes. Churchill recognised, however, sooner than many of his countrymen, that without American help Britain was not going to win the war. The alternative was a negotiated peace with Hitler, with the odds stacked heavily in Germany's favour. Churchill, of all people, would never have countenanced agreement with Hitler, hence his single-minded pursuit of a 'special relationship' with Roosevelt, something, he told the Commons in February 1942, 'I have dreamed of, aimed at and worked for . . .'[16] Churchill was without doubt right. The Soviet Union was an unknown quantity, and its survival in 1941 very much in doubt. For all the hollow rhetoric of English-speaking brotherhood, there did exist more in common between the two democracies than between Britain and authoritarian Europe. Roosevelt was not taken in by Churchill's sentimentality, but he understood the real danger to American interests if Britain were isolated and defeated, leaving the United States 'an island of peace in a world of brute force' facing the 'contemptuous, unpitying masters of other continents'.[17] The decision of both men to build a common cause between their two states ranks as perhaps the most important political explanation of ultimate Allied success.

The association was intimate from the start, and took its lead from the personal contacts between Churchill and Roosevelt. At regular summit meetings the two leaders argued out Allied strategy and coordinated their productive and technical efforts. By November 1942 Churchill confessed to his Foreign Secretary, Anthony Eden: 'My whole system is based upon partnership with Roosevelt.'[18] There were strong differences of opinion; neither statesman was entirely candid with the other. The warmth of the relationship is easy to exaggerate, and Roosevelt for one seems to have seen through Churchill's florid personality and mercurial intelligence. 'He has a hundred [ideas] a day, and about four of them are good,' he told his Secretary for Labour.[19] He teased Churchill ruthlessly, and remained wary of climbing into bed with a Tory grandee. Both men continued to defend national self-interest throughout the war, but the chief among those interests was the common pursuit of victory.

Anglo-American collaboration took its lead from Roosevelt and Churchill, but it soon developed a momentum of its own. At every level the two states combined their activities. In December 1941 it was agreed to hold common strategic discussions in a Combined Chiefs-of-Staff Committee. An intricate network of bodies for sharing intelligence and technical information and for pooling industrial and shipping resources was established during 1942. By the end of that year there were over nine thousand British representatives in Washington, reproducing there the pattern of Whitehall deliberations, with which the American administration was much less familiar. This produced a good deal of friction, since British officials were generally much better briefed, and much more versed in detailed committee work. General Dykes, acting for the British Chiefs-of-Staff in the American capital, found his opposite numbers '*completely* dumb and appallingly slow'.[20] American representatives found the British patronising and elusive. A Senate report in 1943 painted a lurid picture of 'smart, hard-headed' Britons, 'daily outwitting, ousting and frustrating the naive and inexperienced American officials'.[21] Even Roosevelt complained that he came away from any discussion with his ally with 20 per cent, while they kept 80. Over the course of the war American negotiating skills improved with the establishment of effective secretariats and a fuller domestic committee structure. By the time of the Teheran Summit American officials felt they were at last a match for the British.[22] By 1944 the balance in the alliance tilted more obviously towards the United States as its military power and political experience ripened.

If the two partners squabbled and bickered, their marriage survived none the less. With the Soviet Union marriage was out of the question. It was difficult to find a more unlikely associate. Diplomatic relations between the three states in the 1930s were tenuous. The legacy of the Russian Revolution threw up a barrier between communist east and capitalist west that was difficult to penetrate even in the name of expediency. When Stalin made his pact with Hitler in August 1939 the western democracies added the Soviet Union to the number of their potential enemies. The Soviet war on Finland in December 1939 provoked a 'moral embargo' on all trade with the Soviet Union. In the six months before Barbarossa relations between the United States and the Soviet Union were, according to one Soviet view, 'growing worse and worse'.[23] Stalin remained impervious to all warnings from the west of impending German attack, assuming that Britain was trying to provoke a struggle between communism and fascism, hoping to profit by it. When Germany actually invaded, Stalin was completely uncertain how Britain would react.

The immediate response from the west was, under the circumstances, surprisingly favourable. Churchill, one of the most outspoken critics of the Soviet system, who had sent British troops in 1919 to help the counter-revolution, gave an immediate pledge of support. It was a very personal gesture, made, he told his private secretary, only because his single purpose, above all others, was the destruction of Hitler. On the evening of 22 June he broadcast his decision without even showing the text of his speech to the Foreign Office. He did not disavow his unalterable hostility to communism but he spoke instead of the innocent Russian people battling against Hitlerism. He promised all the economic and technical assistance in Britain's power to provide.[24]

Roosevelt's response was just as personal. Though no friend of communism, his close associate, the wealthy lawyer Joseph E. Davies who was American ambassador in Moscow from 1936 to

1938, gave him glowing testimonials on the young Soviet society and its remarkable leader. When news of the invasion arrived Roosevelt told his cabinet that the priority was 'to give help to the Russian people'. Unlike Churchill he did not publicly announce the commitment from fear of popular hostility but he worked hard to begin an active aid programme against the strong advice of many of his close colleagues and his military leaders.[25] Without the whole-hearted support of Churchill and Roosevelt neither the economic aid nor moral comfort given to the Soviet Union would have been possible. They insisted on the commitment in the face of strong opposition. Neither the British Foreign Office nor the State Department was happy with their new co-belligerent in the early weeks of war. The British censors were instructed that the Soviet Union was to be described for the time being not as an ally, but as an associate.[26] General Pownall, chief of the Home Army, confided to his diary his view of the Russians: 'a dirty lot of murdering thieves themselves, and double crossers of the deepest dye'.[27] The prevailing view was that the Soviet Union would collapse in a matter of weeks, and that any aid the western powers sent her would quickly fall into German hands. Even Roosevelt and Churchill, for all their sudden enthusiasm for the Soviet Union, recognised that their prime interest was in the damage the Red Army could inflict on German forces, 'killing Huns' as Churchill bluntly put it.[28] Economic aid, Roosevelt told Stimson in August, should only be sent as long as the Soviet Union 'continues to fight the Axis powers effectively'.[29]

The commitment to supporting the Soviet Union was sustained by a small circle of political supporters around the Prime Minister and the President. Roosevelt's coterie included Harry Hopkins, his personal adviser, who travelled to Moscow in July to see whether the Red Army could hold out. After two days, in which he neither met Red Army leaders nor discussed military affairs in detail, he reported back to Roosevelt that 'the morale of the population is good'. Hopkins felt 'ever so confident' about the Soviet front.[30] The news tipped the scales in Washington. On 2 August Roosevelt announced publicly that although the Soviet Union would not qualify for Lend-Lease – which was technically only for democracies fighting aggression – she would be given 'all economic assistance practicable'. Two weeks later the Soviet Embassy provided a detailed list 29 pages long of everything they wanted.[31] In September a schedule of deliveries was finally agreed, and the following month Roosevelt personally pledged a billion dollars for Soviet aid. On 7 November, as German forces were advancing on Moscow, the American President finally persuaded Congress to grant the Soviet Union full Lend-Lease facilities, not because of her democratic credentials, but because Soviet survival was 'vital to the defense of the United States'.[32]

The task of helping the Soviet Union was eased by the reaction of public opinion in both states. In Britain the cause of the Red Army was taken up with enthusiasm. The great battles of the east came at a time when the British war effort was stagnating. The British labour movement now had a fellow working-class with which to identify, but even among British elites it became the fashion to hail the Soviet Union as a firm ally. On 1 January 1942 a New Year pageant was held in the Albert Hall in London for 'Empire and Allies'. When the Soviet ambassador, Ivan Maisky, was introduced he received, according to *The Times,* a tumultuous welcome; the entire audience greeted him with a 'V' for Victory sign.[33] In the United States popular opinion was more divided. Little more than half the people admitted in opinion polls to trusting the Soviet Union, but by October 1941 73 per cent in a *Fortune* poll favoured working with her, and by February 1942 the figure was 84 per cent.[34] They could see, like their President, that Soviet belligerency was good for American security.

To the Soviet authorities expressions of goodwill and solidarity were no substitute for a firm commitment to help. Stalin wanted a formal alliance between the three powers. Roosevelt would not entertain the idea. Britain accepted a limited joint declaration, signed on 10 July 1941, that both states would keep fighting against Hitler and would not make a separate peace. Beyond that, assistance was conspicuous by its absence. In September Maisky complained to Eden that the British were more like 'spectators' than allies.[35] The same month Stalin sent a desperate appeal for military help to Churchill. To Maisky, who delivered it, Churchill replied: 'I do not want to delude you; until winter we cannot give you any serious help . . . All we can give you is a drop in the ocean.'

Throughout 1941 the United States sent only 20 million dollars in aid, against more than a billion for Britain.[36]

Over the next two years the Soviet Union wanted only one thing from the west: the opening of a Second Front. The story of the western response has already been told, but the ambiguities in the western position did nothing but harm in Moscow. Roosevelt's promise of a Second Front when Soviet Foreign Minister Vyacheslav Molotov visited Washington in May 1942 was given more to keep the Soviet Union in the war than from genuine commitment. Soviet suspicions that western assistance was 'prompted by expediency rather than by friendliness' were not entirely misplaced. Averell Harriman, Roosevelt's personal representative in Moscow, recalled long after the war that he had been sent to the Soviet Union 'to keep Russia in the war and save American lives'. In November 1942 Admiral King observed 'in the last analysis' that 'Russia will do nine-tenths of the job of defeating Hitler'.[37]

The Soviet authorities did little to help their cause. They refused to supply their co-belligerents with detailed information about military strategy or economic planning. They shared almost no technical or intelligence information with the west, while expecting a great deal in return. Contact between the Soviet people and visiting foreigners was carefully limited. Soviet citizens could not fraternise without running serious risks. Journalists and diplomats were confined in Moscow except for short stage-managed visits to the front or to showpiece factories. In the capital, according to one British correspondent, 'an invisible fence sprang up around us . . . and moved with us wherever we went'.[38] Soviet officials were routinely awkward and obstructive, without explanation. General Deane, head of the American Military Mission, wrote after the war that when it was ' "kick-Americans-in-the-pants week" even the charwoman would be sour'.[39] By contrast a good deal of sensitive technical and military material was passed regularly to Moscow. Fifteen thousand Soviet experts visited factories and military installations in the United States during the war. Not until 1944, when it became clear that Soviet economic demands were now being tailored for the postwar economy, did the American authorities begin to refuse further Lend-Lease supplies. Throughout the period of the coalition relations between the three states were strained by mutual distrust and prejudice. Teheran eased the tension but it did not eradicate the deep political gulf between east and west.

The proverbial man from Mars, observing such a coalition, might well ask how he should distinguish friend from foe. The common denominator for the Allies was hostility to Hitler's Germany. Only the German threat was sufficiently powerful to hold three such unlikely partners together through so many disagreements. Each had an interest in the others' continued hostility to Germany, and each worked to prevent a separate peace. Their options were in reality exceptionally narrow. Before Barbarossa, Roosevelt interpreted America's prime interest as 'everything we can do, short of war, to keep the British Isles afloat'. After the German attack in the east his priority was to keep both Britain and the Soviet Union fighting. Stalin in his turn needed to be confident that the west was really committed to the war, and not just hoping that fascism and communism would batter each other to a standstill. He watched hungrily for any evidence of western good faith. In June 1943 the Soviet ambassador to Washington, Maxim Litvinov, sent back to the Kremlin a very full report on American attitudes to the war. Everything he had seen or heard confirmed what Stalin wanted to know: 'the most important strategic task of the United States is the struggle against Hitler.' Though the coalition fought what the British politician Stafford Cripps called 'two relatively unrelated wars', their joint commitment to the defeat of the Axis was scarcely in doubt.[40]

The temporary confluence of interest was sustained throughout by American economic aid, which bound both Britain and the Soviet Union willy-nilly to the coalition. Over the course of the war the United States supplied Britain with one-fifth of all military equipment, and large quantities of food, oil and machinery. The Soviet Union received much less in the way of weapons, but was sent large amounts of vital industrial equipment and materials to enable Soviet factories to produce their own military equipment. One-fifth of supplies to Russia consisted of food, enough to provide every Soviet soldier with a daily ration. The ubiquitous tins of Lend-Lease spam were nicknamed 'Second Fronts' by the Soviet soldiery.[41] Without American aid neither Britain nor the Soviet

Union could have fought so effectively. The recipients were not particularly grateful for this wartime dependence. Churchill once remarked that he had no intention of paying it back, while the Soviet regime complained constantly at the delays in delivery or the quality of goods supplied. A generation of postwar Stalinist writing ignored altogether the role of Lend-Lease in Soviet victory. It is now known that Stalin privately viewed the aid as vital to Soviet survival. Nikita Khrushchev, his successor, recalled in taped interviews in the 1960s (released only thirty years later) that Stalin had on a number of occasions told his close circle that without Lend-Lease the Soviet Union 'would not have been able to cope'.[42]

There was little idealism about the future of collaboration to bind the three leaders together, and what there was has long been discounted by historians as misplaced. Little of it touched Stalin, who for most of the war thought the worst of his allies but remained impressed by Hitler, 'a very able man' he informed Harry Hopkins at Teheran.[43] Churchill was quite realistic about the limitations of the coalition. To one Soviet complaint, late in 1941, he retorted: 'We in this island did not know whether you were not coming in against us on the German side . . . we never thought our survival was dependent on your action either way.'[44] Roosevelt was more of an optimist. He seemed genuinely to have hoped that cooperation with the Soviet Union would create a foundation both for greater freedom in the communist world and for international peace after the war. The close personal relationship he believed he had with Stalin mattered to him perhaps more than the certainty of Churchill's loyalty. 'Stalin hates the guts of all your top people,' he told his British ally. 'He likes me better, and I hope he will continue to do so.'[45] Roosevelt's motives in extending help to Britain and the Soviet Union stemmed from a real desire for a more hopeful international order after the war, based on a genuine warmth between the leaders of the coalition. National interest nevertheless prevailed. When the three Allies began to address the political issues that were raised by the imminent defeat of Germany and Japan in 1945, the wartime friendship rapidly evaporated, to be replaced by the incipient antagonisms of the Cold War.

*   *   *

The account of the personal efforts made by the three Allied leaders to secure and sustain a coalition that could defeat the Axis prompts the further question of how important leadership was in explaining the outcome of the war. There has never been much doubt that it mattered a great deal in Hitler's case. The defects in his leadership contributed to Germany's eventual defeat in ways that can easily be demonstrated. But in the Allied case the issue is more complex. Two of the leaders were democratically elected; Stalin was a dictator like Hitler. None of them expected to be called upon to lead his nation in war, whereas Hitler made war a central ambition of the regime he led. In the event all three Allied leaders in their different ways, and from their very different backgrounds, became wartime leaders of real substance. Hitler, on the other hand, failed the test of war.

Unlike the Allied leaders of the First World War, who in the main left the fighting to the generals, Churchill, Roosevelt and Stalin dominated the whole process of waging war. Throughout its course they were closely involved both with military operations and with the mobilisation of the home front. Though only Stalin was a dictator, all three men were able to exert considerable personal authority. Each acted as commander-in-chief of his respective forces. This was Roosevelt's prerogative as President of the United States. Stalin acquired the title on 8 August 1941, though it was not widely publicised at first because Stalin did not want his name to be identified with defeat.[46] Under the British constitution the King was formally Commander-in-Chief, but Churchill informally assumed the role. As Prime Minister, Minister of Defence and chairman of the Defence Committee he was uniquely placed to act the military chief, though it was never entirely clear whether he could constitutionally compel the armed services to obey his commands.[47] All three leaders encouraged the centralisation of the Allied war efforts around their own person. Here the similarities end. Roosevelt and Churchill had some things in common beside language, though less than the rhetoric of 'special relationship' might suggest. But with Stalin, the artisan's son from Georgia, the contrast was complete.

Any assessment of Stalin's leadership during the war has to steer between two extremes. Soviet propaganda presented Stalin uncritically as the saviour of his people and the architect of victory. In unravelling this distorted image the temptation is to highlight the weaknesses and errors in Stalin's conduct of the war, and to diminish the contribution of his personality. Neither approach does justice to its subject. Stalin's was an extraordinary story. Born in 1879 in the small Georgian town of Gori, his early life was spent in conditions of urban squalor; he was beaten mercilessly by his father, a failed and drunken cobbler. At six he survived smallpox, but he bore its facial scars for the rest of his life. An infected ulcer left him with a slightly withered arm. His tough upbringing produced in him, a childhood friend recalled, a personality both 'grim and heartless'.[48] At school he excelled, thanks to a phenomenal memory. He was transferred to a seminary in Tiflis where he first made contact with Russian Marxism. He became an active revolutionist, and moved in and out of Tsarist jails. In 1917 he was prominent in the inner circle of Bolshevik leaders working to turn the overthrow of Tsardom in February 1917 into a communist revolution. When Lenin's coup succeeded in October, Stalin was rewarded with his first taste of office, as People's Commissar for the Russian Nationalities. In 1922 Stalin was appointed General Secretary of the Party. It was a fateful appointment, for this ambitious, secretive, devious revolutionary, both more plebeian and more brutal than the committee of intellectuals who ran the Party, used the new office as the instrument to create an unassailable power-base. By the late 1920s Stalin – 'man of steel' – dominated the Soviet state apparatus and his own party.

In the decade before the war Stalin was the driving force behind the rapid state-imposed modernisation of Soviet industry and agriculture, and the build-up of the largest armed forces in the world. This 'revolution from above', with its ceaseless chaos and emergencies, permitted him to centralise power even more completely in his own hands. The apparatus of propaganda lionised Stalin as the genius of socialist reconstruction and father of his people. The reality was grotesquely different. Stalin was obsessively distrustful of those around him and used the machinery of state terror to insulate himself against enemies real or imagined. The massive problems of social dislocation thrown up by forced modernisation were viewed by Stalin as so much sabotage, to be punished with deportation or death. Stalin appeared to those who met him in public unassuming, almost timid, quiet-spoken but firm, with little of the monster about him. In private Stalin revealed a coarseness in his nature; he was vindictive and bullying, given to bouts of red-faced fury. His colleagues learned when to talk and when to stay silent. After Stalin spoke in committee all those present were said to applaud.

Stalin travelled a long road from early poverty and obscurity to the dizzy heights of state power. He was above all a survivor, who created his dictatorship through a combination of hard work, political acumen and an almost complete lack of moral scruple. Though he was hardly a charismatic figure, the possession of absolute power gave him a distinct presence in any company, enhanced by the public demeanour of imperturbability. Harsh-tongued, brutal and cynical, he was well-informed on all issues and trusted no one. Given such a personality it was all the more disconcerting that Stalin should have been caught so completely by surprise when Germany launched the Barbarossa campaign on 22 June 1941. His conviction that Hitler simply would not attack while the British Empire remained undefeated in the west became unalterable during the spring and summer of 1941 and those who tried to challenge it ran very great risks. The German attack provoked the most serious potential crisis in Stalin's whole history of dictatorship. It was once thought that Stalin had suffered a complete nervous collapse and retreated from his official duties. Witnesses certainly recall a quite different Stalin, ashen-faced, tired and irritable. But the post-glasnost revelations have shown that Stalin flung himself into an exhausting spell of activity after the invasion, scarcely sleeping for days on end. When he withdrew to his dacha at the end of June it was not to abandon his office, but to draft decrees setting up a new Soviet Defence Council and his first address to the Soviet people, which he gave on his return to Moscow, on July 3. None the less, Stalin was aware of how he had let down his people, just as he had been betrayed by Hitler. At a speech in May 1945 after the war had ended he took the curious step of thanking the Soviet people for not saying to him at the point of crisis in 1941: 'Go away, we shall install another government'.[49] On June 30, when a delegation from the Soviet government came out to see Stalin at his dacha, he may well have wondered whether this betokened something more sinister. Instead, his colleagues came to make sure that he would put himself at the head of the new defence committee, and to reassure him of his indispensability to the Soviet system.[50] That same day the State Committee for Defence was formed.

On 10 July the Supreme Headquarters was set up, based at Stalin's modest apartment in the Kremlin building. Stalin now took over running the war. He assumed the role of Commissar of Defence and appointed himself Supreme Commander-in-Chief. He had no scruples about easing his party crony Klement Voroshilov out of the key post of commander of the northern front. A veteran of the Tsaritsyn battles of the civil war, Voroshilov was utterly unqualified for the military offices he held. Khrushchev regarded him bluntly as the 'biggest bag of shit' in the army. Following the older Russian tradition of promotion on merit, Stalin selected Georgii Zhukov as his deputy, an appointment that proved to be of critical importance for the eventual revival of Red Army fortunes.[51] Under Zhukov's lead the role of the notorious 'commissars' in the armed forces – political officers who enjoyed equal rank with the military commanders – was emasculated and the armed forces were restored to military control.

From then on Stalin exercised very close supervision over the conduct of the war. He worked long hours, rising late but working until two or three o'clock the following morning. He took hardly a break over the next four years. At Teheran western officials noticed how much he had aged, his dark grey hair gone almost white, his swarthy complexion sallow. He imposed this arduous regime on his General Staff, who literally operated round the clock throughout the war, sometimes working an eighteen-hour day. Stalin expected reports three times a day from the General Staff, the last, in the evening, a personal briefing in the Kremlin. At this session Stalin would go over the maps of the front in detail, and look through the operational orders for the following day prepared by the staff. He seldom travelled far from Moscow, but he insisted that his Chief-of-Staff and his deputy, Zhukov, should make regular trips to the front-line to oversee major operations. He was methodical and rigorous, but he made no pretence that he had great operational imagination. The one thing he insisted on, that Hitler would not strike on the southern front in 1942, though not a foolish calculation, almost brought another summer of disaster. The victories at Stalingrad and Kursk were designed by the military leaders, not by Stalin, who played a smaller role as the war went on and the General Staff matured into an effective command and planning unit. Stalin's role was to act as a spur, driving the whole war effort along. He harried and bullied his subordinates where he sensed failure or timidity. The threat of imprisonment or death hung over every mistake. Stalin placed particular worth on the willingness of his staff to present the true state of affairs, 'straightforward and unembellished'. Once it was realised that Stalin could actually be told the truth without risking life and limb, the prospect of operational miscalculation was much reduced. By 1944 Marshal Voronov, Chief of Artillery, recalled that the atmosphere at Supreme Headquarters had eased: 'Stalin was more balanced, far more even-tempered than before.'[52]

The man western leaders met at Teheran was very different from the rude and irritable despot, under 'intense strain', who had greeted the first western mission to Moscow in September 1941. He made a remarkable impression on General Brooke, the British Chief-of-Staff, whose splenetic views on American strategy were well known. During the course of the conference, at which Stalin did much of the talking for the Soviet military, Brooke formed the view that he had 'a military brain of the highest calibre'. He could not recall Stalin making a single strategic error. This was high praise indeed.[53] Commander of the American air forces General Arnold echoed Brooke's testimonial: 'brilliant of mind, quick of thought and repartee, ruthless, a great leader . . .'[54] The quality of any leader remains, of course, a matter of judgement. Stalin carried with him the disadvantages of dictatorship – the excessive centralisation, the pall of fear enveloping subordinates – but he brought a powerful will to bear on the Soviet war effort that motivated those around him and directed their energies. In the process he expected, and got, exceptional sacrifices from his besieged people. The 'personality cult' developed around him in the 1930s made this appeal possible in wartime. It is difficult to imagine that any other Soviet leader at the time could have wrung such efforts from the population. There is a sense in which the Stalin cult was necessary to the Soviet war effort. It provided a common focus of loyalty, and promoted a growing conviction about ultimate victory. That people suspended their disbelief, that they colluded with a myth later tarnished by revelations of the brutal nature of the wartime regime, should not blind us to the fact that Stalin's grip on the Soviet Union may have helped more than it hindered the pursuit of victory.

Franklin Delano Roosevelt was everything Stalin was not. He was born in 1882 into a wealthy New York family, on a 100-acre estate beside the Hudson river. He had every social advantage. Excessively pampered by his mother, Roosevelt was sent first to the exclusive school at Groton, Massachusetts, modelled on the English public school, then on to Harvard, and to law school. He was not outstanding academically, but the tall, distinguished, sociable patrician, a member of America's untitled aristocracy, needed little more than his name and background when he launched himself into politics in 1910. He had no particular scruples about which party to join; his cousin, Theodore, was the Republican President, but he adopted the Democrat ticket because they asked him to stand first. He soon became a political high-flyer. In 1913 he was appointed Assistant Secretary for the Navy. He was a solid administrator, and also an arch politician, utterly absorbed by the art of politics. A college friend remembered a man 'extremely ambitious to be popular and powerful'.[55] He had many Republican friends, whose social world he shared, but made his name fighting on issues for the common man. In 1920 he was chosen by James Cox, the Democrat candidate, to run as Vice-President. It was his first setback. He campaigned on support for the League of Nations, which America had not yet joined, but found the tide of opinion isolationist. In the Republican landslide he failed to win even his home state of New York. The following year, at the age of 39, he was struck by polio and paralysed from the waist down. He withdrew from politics to fight his affliction. Those who knew him well found him transformed by the struggle. The young politician had an arrogance, an intolerance of weakness, a hint of superficiality, that marred his energy and charm. During the seven years it took him to recover he became a more humble and more sympathetic personality. 'He was serious,' Roosevelt's Secretary for Labour Frances Perkins later wrote, 'not playing now.'[56]

In 1928 he returned to politics, winning a narrow victory as Governor of New York State. Four years later he was returned as the first Democrat President since Wilson, with overwhelming support. During the 1930s Roosevelt struggled to repair the damage inflicted on the American economy by the Great Slump, and to heal the social wounds of unemployment and poverty. He promised Americans a 'New Deal' on jobs and welfare, and used the powers of the state, underdeveloped by European standards, to secure it. The results were mixed. There were still nine million unemployed by 1940, when Roosevelt ran for an unprecedented third term. The New Deal legislation was dragged over the coals by the Supreme Court for the unconstitutional powers it gave to the President and the new state agencies he set up. He made a great many bitter political enemies in his attempt to strengthen Presidential rule. When the British ambassador Lord Halifax, recently arrived in Washington in 1940, met a group of Republican Congressmen he was astonished to be told that all present thought Roosevelt 'as dangerous a dictator as Hitler or Mussolini'.[57] When the Lend-Lease bill went through Congress, the isolationist Republican senator Arthur Vandenburg dubbed Roosevelt 'Ace Power Politician of the World'.[58]

Roosevelt was certainly an ambitious President, who disliked the obstruction of his policies. He devoted most of his energy to short-term political tactics, and was never choosy about the allies he found. He was obsessed with public opinion and his own popularity. He was an unsophisticated idealist, who once confessed that his political outlook could be summed up in two words: democrat and Christian. Though the idealism was genuine enough, friends and colleagues found his views on most issues ill-defined and pragmatic. Roosevelt's instinct for political survival created in him a distrust of ideological conviction. Charles Bohlen, who interpreted for him at Teheran, thought the President 'preferred to work by improvisation than by plan'. He disliked putting anything down on paper, and instead did much of his work in informal conversations, throwing round ideas, exploring options, testing the water. He could be disarming, flattering, cheerful, supportive, but was, by the general agreement of those around him, difficult to pin down. 'Not a tidy mind,' wrote an otherwise sympathetic British observer.[59]

Roosevelt the shrewd tactician and Roosevelt the idealist were difficult to reconcile. This was particularly so in time of war. Though his public stance in the 1930s against violence – 'I hate war' – helped to maintain domestic political support among a largely isolationist population, it was difficult for him to hide his hatred of fascism and his expectation that America at some point would become involved with keeping the peace abroad. The ambiguities in this position were sufficiently pronounced to make it almost impossible for the American public to decide just where their President stood on the issue of war, yet to make it just as difficult for Roosevelt to seize the initiative and side openly with the democracies in 1940 and 1941. When Japan attacked in December 1941 everything was simplified for people and President alike: isolationism was dead as a political force and Roosevelt could lead his people in war unfettered by hostile opinion. He brought to the role of war leader some admirably suitable qualities. His was a big personality, made larger by years of publicity and the calculated wooing of popular approval. He had unrivalled experience in politics, having spent eight years in the highest office in the land. When it came to a job of work he was not hostage to party prejudice but hired Republican and Democrat alike. He was adept at managing Congress, and at building bridges between the many constituencies – ethnic, political, religious – that made up American society.

The coming of war injected a lease of life into the Roosevelt administration. The President announced that 'Dr New Deal' was handing his practice over to 'Dr Win-the-War'. He insisted on calling himself Commander-in-Chief, and made it clear that he was not going to stand back, as Woodrow Wilson had done, from the day-to-day business of fighting the war. He saw himself as the ringmaster of the coalition: 'I am responsible for keeping the grand alliance together,' he told Marshall.[60] There was almost no established structure in American government for Roosevelt to play the role of military supremo, and the first frantic months of conflict were spent trying to devise one. The result was a chaos of appointments and committees. Eisenhower, recently promoted to the Pentagon, recorded his impression of life in the capital three weeks after Pearl Harbor: 'Tempers are short. There are a lot of amateur strategists on the job; and prima donnas everywhere . . .'[61] Gradually a central machinery was established around the military chiefs. A new Joint Chiefs-of-Staff Committee was established, dominated by the army chief, George Marshall, but chaired by the President's personal representative, Admiral William Leahy. He and Harry Hopkins were the only members of Roosevelt's entourage allowed to enter the Map Room in the White House, where the President kept all his most important correspondence under lock and key, and look at what had been written. Though Roosevelt kept a close interest in international politics, in reality he played a smaller part in the military deliberations than he had intended. He preferred the informal atmosphere of discreet, usually unrecorded, one-to-one interviews rather than large chiefs-of-staff committees. Even in the intimacy of a closed discussion he gave surprisingly few clear directives. He preferred to suggest and encourage rather than order; he relied on his chief appointments to read between the lines and act on their own responsibility.[62]

The flaws in Roosevelt's leadership came with him into war. His informal administrative habits − suggestions which should have been orders, spoken directives instead of a written brief − made it infuriatingly difficult, from all accounts, to know exactly what policy was. Roosevelt himself seems to have viewed this as a strength as much as a weakness. 'I am a juggler,' he told a group of businessmen in 1941, 'and I never let my right hand know what my left hand does . . . I am quite happy to mislead and to tell untruths if it will help to win the war.'[63] Stimson found this 'topsy-turvy, upside-down sytem' appalling for the conduct of government business, though on balance it does not appear to have inhibited the American war effort any more than excessive centralisation might have done, if anything rather less. Roosevelt cheerfully appointed plenipotentiaries who trampled roughshod over the established departments of state. The inner circle of unofficial advisers − Hopkins, Harriman, Leahy − allowed the President to bypass the normal channels, and excited jealousies among the permanent officials. Above all Roosevelt remained a trimmer, an arbiter, aware that the diffuse character both of American society and of its wide political class required a great deal of politicking to hold a wartime consensus together. Roosevelt with his bruising experience of the New Deal was better placed than most American politicians to keep Americans fighting in a common cause.

For all his inclination to compromise, on the big issues Roosevelt took a clear stand, backed by his closest advisers. He sustained aid for Britain against a good deal of popular criticism; he did the same for the Soviet Union. In 1942 he stuck to the strategy agreed with the British of defeating Hitler first. This was a choice vital to the outcome of the war, and it flew in the face not only of much popular opinion (isolationists were happy to fight Japan, but much less certain about fighting in Europe) but also of the Pacific-minded navy. All the while he stoked up popular enthusiasm inside and outside America with his idealistic vision of a new world order after the war based upon the principles of freedom and good-neighbourliness. These were ambitions honestly held but they had the added bonus of sustaining popular commitment to war after the initial thirst for revenge had been slaked.

The net effect of Roosevelt's leadership is difficult to judge. Like Stalin, as the war went on Roosevelt was able to leave much of the routine of war to the American and inter-Allied apparatus set up in 1942. He was a good judge of men, and appointed people who could do the job. Exuding confidence and optimism himself, he responded to these traits in others. When he made Marshall chief of the army in 1939 there were 33 generals more senior who might have hoped for the job. He liked Marshall because he told him the truth. What Roosevelt supplied was inspiration; he remained steadfast, unruffled even by defeats, supportive to all around him. He kept his anxieties to himself, just as he kept his disability from the public gaze. He had the strength to recognise his limitations, in itself a hallmark of intelligent leadership. At Teheran the critical Brooke observed that the President 'never made any great pretence at being a strategist'. But he impressed all who met him at the conference. Churchill's military aide, General Ismay, found him to be the perfect coalition chairman, 'wise, conciliatory, paternal'.[64] During 1944 his health deteriorated rapidly, worn down by a dozen years in office. He struggled on with his work, borne up by his confidence in victory, but on 12 April 1945, just weeks away from that outcome, he died. The country was stunned at the loss. In Moscow Molotov hurried to the American Embassy in the middle of the night, where, evidently deeply distressed, he spoke of Soviet respect for Roosevelt. Lord Halifax observed the effect in Washington: 'Such a gap did the withdrawal of Roosevelt's personality seem to leave that it was hard to imagine anybody filling it.'[65]

Much the same might have been said of Churchill had he not survived a heart attack and serious bouts of pneumonia during the war. But survive them he did. Though Churchill was eight years older than Roosevelt, five years older than Stalin, he outlived them both by a good margin – and this despite a regime of indulgence and little exercise for most of his later life. Churchill differed from Stalin and Roosevelt not only in temperament and background but also in the circumstances of his wartime leadership. He was the only one of all the major wartime leaders, on either side, who was appointed during the conflict as a *war* leader. In the second place he was a leader by the grace of Parliament, which put him there in May 1940 and had the right at any time to remove him. Roosevelt was President for four years; Stalin was a dictator; Churchill was a chief minister, responsible to the Commons. The constitutional position of such an appointment remained ambiguous; the limits of authority depended on how much could be achieved through sheer force of personality.

In Churchill's case personality carried him a long way. He was so much larger than life, it is difficult for the historian to judge the true dimensions of his leadership. Canonised as the saviour of western civilisation, vilified as the flawed commander who diminished the empire he led, Churchill defies neutrality. Throughout his long and chequered career he provoked bitter resentments and deep affections. He was born in 1874, the eldest son of a Tory peer, Lord Randolph Churchill, and his American wife, Jenny. Disliked and ignored by his irresponsible, spendthrift, philandering parents, Churchill grew up at the time of Europe's aristocratic *fin de siècle,* surrounded by a wayward family that staggered between debauchery and bankruptcy. He took the conventional upper-class trail, first public school at Harrow, then Sandhurst and a commission in the Hussars. He rode in one of the last British cavalry charges, at Omdurman in the Sudan in 1898. He found action irresistible. A year later he was a war correspondent in the Boer War, where he was captured. His dramatic escape made him into a popular hero. In 1900 he entered Parliament as a Conservative; four years later he changed sides in time to profit from the Liberal landslide of 1905. Twenty years later he switched back to his old party. In the interim he was in and out of high office. In 1910, at the age of only 36, he became Home Secretary; a year later he was moved to the Admiralty, where he masterminded the Dardanelles campaign in 1915. The calamitous failure of the enterprise cost Churchill his job, and almost ended his political career.

After two years in the political wilderness, serving on the Western Front, Churchill was recalled to serve as Minister of Munitions in Lloyd George's Cabinet, against strong Conservative resistance. Except for a brief spell between 1922 and 1924, he was in high office throughout the 1920s. He was never widely popular, either in the country at large or in Parliament. He was regarded as unprincipled and dangerous by the right; progressive opinion found him deeply reactionary. In 1929 he once again disappeared into the wilderness, and it was widely thought that this time it was for good. In the 1930s Conservative politics was dominated by Chamberlain, whose austerity, thoroughness and self-righteousness represented everything that Churchill was not. The two men despised each other. While Chamberlain wrestled with the crises of empire and of Europe, Churchill sat in virtual isolation on the back benches, berating the government ceaselessly for failing to hold the empire more firmly, and for rearming too slowly. Many found him to be a man out of touch with the modern age, a Victorian grandee whose values and habits were locked in the world before the war. If his career had ended in the 1930s he would be remembered largely as a political maverick, which was the reputation of his father.[66]

Churchill was saved by the war. In 1939 a reluctant Chamberlain brought him back to the Admiralty. His thirst for action was undiminished. His shrill bellicosity set him apart from the rest of the Cabinet. In April 1940 he pushed for invasion of Norway, for which British forces were manifestly unprepared. The failure of the campaign carried strong echoes of Gallipoli, but its chief victim was not Churchill but Chamberlain. Widespread popular disillusion-ment with the war effort inside and outside Parliament made Chamberlain's position untenable. Churchill was not the obvious successor. Anthony Eden, the Dominions Secretary, was more popular with the public; Lord Halifax, Foreign Secretary, was preferred by Chamberlain's party. But Churchill was impossible to ignore. He was a man, Lloyd George once remarked, who 'likes war'.[67] On 9 May, one day before Hitler attacked in the west, Chamberlain called Halifax and Churchill to Downing Street to discuss the succession. Churchill's version of events is melodramatic (and wrongly dated) – the long pregnant pauses in this, 'the most important' interview of his life, the silence finally broken by Halifax who ruled himself out of the running, the profound sense of responsibilities settling on his shoulders. In reality, Halifax had already decided that he did not have the stomach for war, and had told Chamberlain so. Churchill's appointment depended only partly on Halifax, and more crucially on the willingness of the Labour opposition to work with a new Conservative leader. On 10 May Chamberlain was told that Labour would work with Churchill, and he reluctantly informed the King, who preferred Halifax. Churchill's appointment as war leader was achieved against a great many odds, and at the behest of a Labour Party he disliked deeply. When Churchill entered the Commons the following day there was no more than a ripple of applause to greet him; the news was met in the Lords by complete silence.[68]

Some measure of the man who took office in 1940 can be found in his recollection of the triumph: 'At last I had the authority to give directions over the whole scene. I felt as if I were walking with destiny, and that all my past life had been but a preparation for this hour and for this trial . . .'[69] Churchill's outlook was incurably romantic and overblown. He treated life, the philosopher Isaiah Berlin once commented, 'as a great Renaissance pageant'.[70] His was not a subtle or devious personality. He saw things in black and white, right or wrong. He had a great respect for liberty and an intense dislike of tyranny, though he never defined either very deeply. He was steeped in history at the expense of the present, from which he seemed oddly alienated. He might have flourished in any age – a knight at arms, an Elizabethan captain, a cavalier, an

old-regime general like the ancestor he so admired, John, First
Duke of Marlborough. His great love was war. He was not
personally bloodthirsty, but took a boyish delight in military action.
He told French leaders on the eve of their capitulation that Britain
would 'fight on for ever and ever and ever'. That same summer,
when Halifax asked if he would consider transferring the government to Canada, Churchill disarmingly replied that if the Germans
invaded, 'I shall take a rifle (I'm not a bad shot with a rifle) and put
myself in the pillbox at the end of Downing Street and shoot till
I've no more ammunition.' Halifax had his doubts, but Churchill
said very much the same to Harriman two years later on a voyage
across the Atlantic on the *Queen Mary*, when he disclosed that his
lifeboat was to be fitted with a machine gun if they were torpedoed:
'I won't be captured. The finest way to die is in the excitement of
fighting the enemy.'[71]

No one ever doubted Churchill's bravery, nor the energy and
impetuosity that he brought to the war effort. But there was a
darker side to his character. He had a group of intimates with
whom he shared his extravagant, almost bohemian lifestyle, the
sybaritic habits of that *louche* nobility into which he was born, but
he was otherwise cut off from people. His wife regarded him as
selfish and egotistical, 'like Napoleon'.[72] He was a difficult man to
cross; he would pursue his point of view to the length of obsession.
He was temperamental and petulant, though, by all accounts, not
mean-spirited. Beneath the blustering exterior lurked a more
sensitive and insecure individual, prone to bouts of deep
depression, the 'Black Dog'. During the Teheran conference
Churchill's doctor found him unusually gloomy and desperate: 'we
are only specks of dust, that have settled in the night on the map of
the world . . .'[73]

Few people have doubted that in 1940 the hour found the man.
The defeat in France and the threat of invasion brought close the
collapse of Britain's war effort. A peace with Hitler was a
possibility, as Churchill told his ministerial colleagues at the time of
Dunkirk. But he recognised, as the French discovered, that peace
on Hitler's terms would be shortlived and one-sided. Had
Churchill made peace, German domination of the Continent
would have been assured. There would have been little prospect of

rousing the British public for a second war against Hitler's Germany if the peace turned sour. But Churchill did not think that agreement with Hitler was compatible with 'his own conscience or honour'. No one seriously challenged this view, and Churchill's personal defiance came to stand for that of the whole nation.[74] Churchill was the least likely figure to abandon the contest; he saw himself as a leader chosen to run the war, not to make the peace. The British decision to fight on in the summer of 1940 owed a good deal to the nature of Britain's new helmsman.

Churchill's main contributions in the first months of his premiership were not only his resolute will to fight on, but also his construction of a clear, centralised system to run the war. His memories of the First World War, of political crisis and military confusion, inclined him to fuse political and military responsibilities in his own hands. As Prime Minister and Minister of Defence Churchill, like Roosevelt, could oversee the whole war effort. He set up his own political and military secretariats to keep him in close touch with the detailed movement of events. The Chiefs-of-Staff Committee was quickly established as the main forum for the formulation of strategy, while a small War Cabinet discussed the wider political issues of war. Churchill introduced the novelty of communicating quickly by personal and often peremptory notes, sent directly to ministers and generals rather than left to wind their way through the usual bureaucratic channels. Churchill, again like Roosevelt, preferred informality to protocol. He disliked what he called the 'official grimace'. He saw it as his role to prod officials into action, to energise and invigorate, to expedite policy.[75]

Churchill's real interest was in strategy, and it is here that his leadership was at best a mixed blessing however generously it is assessed. His habit was to pursue every project that occurred to him if it seemed to promise swift or dramatic results. His instincts were all for the offensive, as if every operation were a cavalry charge. His long-suffering army Chief-of-Staff, General Brooke, found him 'erratic' and 'impulsive', working through intuition rather than analysis.[76] The one merit of this approach was that it kept subordinates on their toes, arguing against his more hare-brained schemes and thoroughly preparing their objections. If the case against was sensible, recalled one of his private secretaries, 'you had a fair hearing, and he was open to argument'. Churchill never overruled the Combined Chiefs-of-Staff if they were against him, though he would complain bitterly if he did not get his way.[77] He once ruefully observed that he did not have 'autocratic powers' like Stalin or Hitler, but Britain's war effort was almost certainly the better for it.

On two issues he was clear-sighted: his pursuit of American assistance and his support for air power. He had proposed the same strategy in the Great War: 'There are only two ways of winning the war', he announced in 1917, 'and they both begin with an A.'[78] In 1940 there were few other options. No major state could provide effective help for Britain except for America; the war could only be brought home to Germany through bombing. The pursuit of both in 1940 and 1941, though results at first were meagre, proved in the long run to be of inestimable value in the defeat of Germany. But on almost every other strategic issue Churchill's judgement was questionable. The Norwegian campaign was a disaster; Churchill was all for sending more aircraft to France in June 1940 to be frittered away in an unwinnable battle; and in the spring of 1941 he explored the possibility of laying on an 'air banquet', drawing together every available aircraft in Britain, even from the training schools, to launch a single all-out air attack against Germany.[79] It was Churchill who sent the *Prince of Wales* and *Repulse* to the Far East on the grounds that battleships could still fight their way past air power; Dakar, Greece, Crete – a positive gazetteer of poorly prepared, poorly supplied operations. Finally, Churchill's single-minded pursuit of the Mediterranean option, and his obsession with Turkey and the Balkans – again a hangover from the First World War – might well have inflicted serious damage on western strategy if he had won his way. Hemmed in by the Alps and the Balkan ranges, at the end of long supply lines, the western Allies would have inflicted much less damage on Hitler than they did in France, while the Soviet advance in the east would have been slowed up, as Stalin fully realised at Teheran. On these issues Churchill showed himself at his worst, angrily convinced of his own strategic insight, unable to concentrate on other issues, bullying and cajoling by turns. He remained certain at Teheran that he was right about the 'shining, gleaming opportunities in the

Mediterranean', but he could carry neither his allies nor many of his staff with him.[80] It was a conviction all the more difficult to understand in the light of Churchill's failure at Gallipoli, and the insignificant part the Balkans and Middle East played in the defeat of the Central Powers in 1918.

It is difficult not to conclude that Allied strategy succeeded despite Churchill, though his pugnacity and spirit remained a valuable symbol of the Allied will to win. Both his allies and his military staffs soon learned how to cope with their mercurial companion by diverting and ignoring his interventions. He was a poor administrator, and left much of the machinery of war, once it was established, to run itself. He grew out of touch with policy on the home front. In military affairs he met his match in Brooke, who managed to blunt his excesses. His notorious habit of interfering with front-line operations, and sacking generals and admirals he did not value, was curbed with the emergence of strong military personalities – Alexander in the Mediterranean, Montgomery in Europe – who ensured his bark remained much worse than his bite. By 1943 his influence on the war effort was much reduced; at Teheran he confided that he was 'appalled by his own impotence'.[81] Despite the postwar mythology, his popularity with the population was not as secure as he would have wished. In early 1942 opinion polls showed fewer than half of those asked in favour of his premiership. In July 1942 he was subjected to a Parliamentary vote of no confidence, though he survived it comfortably. However in 1945, two months after the victory in Europe, Churchill was heavily defeated in the General Election, 'immediately dismissed by the British electorate' he remarked in his memoirs, his bitterness scarcely concealed. Churchill had been the man for the hour, but for no longer. 'Don't you feel lonely without a war?' he asked his doctor a decade later. 'I do.'[82]

$$\star \quad \star \quad \star$$

There is no ideal war leader. For all their many strengths the coalition leaders had their share of flaws. What is striking, as the war effort went on, was the ability of the wartime apparatus to cope with those flaws. In each case the personal role of the leader moved from a deliberate prominence at the start of the conflict to a more subdued participation by its end. This was an inevitable product of the war, even for Stalin. No one man could hope to master every area of activity; delegation was an absolute necessity. The western war effort was run by large committees staffed by both Allies. These committees formed the apex of a pyramid of staffs and offices where the routine work of the war effort was conducted. The war was not so much led as administered.

What such systems needed were managers, and it is to the credit of all three Allied leaders that this was quickly recognised. Behind each leader there emerged a cohort of military managers and civilian officials who took on the real responsibility of running the war. In general these tasks were carried out by professionals, whose experience and qualities singled them out for office. The Allied wartime administration was on balance surprisingly free of political stoodges and dud appointments; incompetence at the highest level was difficult to conceal. Functional effectiveness rather than political loyalty governed promotion, and resulted in a valuable degree of stability and continuity at the higher levels of leadership of all three states.

Three examples may serve to demonstrate the force of this assertion. Each one illustrates the way in which Allied leadership came to be shared between commander-in-chief and professional specialist. The first is the case of Alan Brooke. A career soldier from Northern Ireland, Brooke made his reputation early in the war rescuing the defeated forces from north-east France. He was promoted on his return to head the Home Army, and set about the reorganisation and training of forces to repel invasion. In December 1941 he replaced Sir John Dill as Chief of the Imperial General Staff, the most senior military appointment. Dill was a clear-sighted and energetic chief, but he had not been able, on his own admission, to cope with the nervous strain of constantly arguing with Churchill. Brooke had no such qualms. He had a quick temper and a strong streak of stubbornness and was happy to argue issues out face to face with Churchill. He did so from a position of great strength. He was much more closely in touch with the war effort than Churchill; he had a quick mind and was able to grasp the strategic effort as a whole, to weigh means and ends precisely. His manner was terse to the point of rudeness; subordinates found him severe, demanding, and aloof. He worked hard and efficiently and expected nothing less from those around him. He was the very opposite of Churchill, sceptical where his chief was enthusiastic, consistently sensible rather than erratic, a thorough and unruffled administrator. He cultivated a mask of complete imperturbability. 'I considered it essential', he wrote after the war, 'never to disclose outwardly what one felt inwardly . . . It was of primary importance to maintain an outward appearance that radiated confidence.'[83] When in June 1942 he became Chairman of the Chiefs-of-Staff Committee, and the leading British representative among the inter-allied Combined Chiefs, he was able to influence the whole Allied military effort.

His brusque manner and and persona grated on his American opposite numbers, but his exceptional grasp of the complexities of global strategy, meticulous preparation and analytical power dominated the early Allied meetings as well. Churchill found him an uncongenial foil: 'When I thump the table and push my face towards him, what does he do? Thumps the table harder and glares back at me . . .'[84] But he recognised the virtues of the military manager and Brooke kept his place for the remaining years of the war, running every aspect of the conflict from the Cabinet War Room in Great George Street, a little way from St James's Park in central London. For the man regarded by the Secretary for War, James Grigg, as second only to Churchill in his contribution to victory, he features remarkably little in Churchill's history of the war, beyond the bland assertion that Brooke rendered 'services of the highest order'. But other colleagues with less to conceal testified to Brooke's exceptional qualities. Ismay, Churchill's go-between, observed the work of eight Chiefs of the Imperial General Staff at close quarters and considered Brooke 'the best of them all'. Though Eisenhower did not like Brooke's strange mannerisms when they first met, and thought him shrewd rather than wise, he ended by regarding him 'as a brilliant soldier'.[85] Brooke rather than Churchill was the architect of Britain's military revival from 1942. His one weakness, a product of his gruelling experiences in the retreat from France in 1940, was his excessive caution over direct cross-Channel assault, even though he knew it made strategic sense. He preferred the peripheral to the direct strategy in 1942 and 1943 on grounds of military realism, but he accepted the joint decision of the coalition to launch an invasion in 1944, and hoped to command it. Instead he remained at his existing post until February 1946 when he was succeeded by Montgomery.

The second example is less well known. During the course of 1942 in the Soviet Union the crises of retreat brought a rapid turnover of key military personnel. Zhukov and Novikov continued to enjoy Stalin's confidence, but it proved impossible to find a satisfactory Chief-of-Staff or Chief-of-Operations, positions of critical importance in the Soviet military structure. Between June and December there were no fewer than seven Operations Chiefs.[86] A settled Chief-of-Staff, Vasilevsky, was appointed in July, but he was compelled by common practice to spend long periods at the front directly coordinating operations. In December 1942 the remaining gap was filled by the appointment of General Alexei Antonov as Chief-of-Operations, and simultaneously deputy Chief-of-Staff, to act when the Chief was at the front. Antonov, 46 years of age and a distinguished staff officer, was Chief-of-Staff to the Trans-Caucasian front when the summons came from Moscow. He was an inspired choice. He made it clear that he was not going to follow his unfortunate predecessors by jumping to Stalin's tune. He spent a week in Moscow familiarising himself with the military situation before he visited Stalin. Instead of the brief sojourn everyone expected, Antonov stayed at this post until February 1945, when he was made full Chief-of-Staff. His deputy, General Shtemenko, regarded him as an officer of exceptional qualities, firm, even-tempered, clear-minded. Shtemenko never saw him lose his self-control in six years. Like Brooke he was impatient with less able men, intolerant of 'superficiality, haste, imperfections and formalism'.[87] He praised people seldom, planned his work with a meticulous care, spoke 'with brevity and clarity'. His calmness and breadth of vision restored morale at the army's head.

Stalin developed a great respect for Antonov, not least because he gave him straightforward accounts of the state of affairs at the front, however unpalatable. In addition, Antonov was not afraid to argue with Stalin, which was even more unusual. He quickly acquired a reputation for the skill with which he presented the General Staff case. Even Marshal Zhukov, never happy at sharing the limelight, allowed Antonov, 'a master at presenting material', to draw up the operational maps and schedules and to go over them with Stalin at the evening briefings.[88] Gradually the pattern of General Staff work altered. More and more of Stalin's directives were prepared by Antonov, and Stalin would sometimes sign them without even reading them. The balance between the generals and Stalin perceptibly changed. In the months leading up to the Battle of Kursk Antonov played a key part in the planning and preparation, and in allaying Stalin's fears that the 1943 campaign would be a repeat of 1941 and 1942. Antonov and Zhukov between them argued against Stalin's desire for a quick pre-emptive strike at the gathering German forces. Over the shape of Soviet strategy in the summer of 1943 the General Staff view prevailed. Antonov took a leading role in the planning of the drive into Poland, and the final assault on Berlin, but he was able to conduct much of his work from Moscow, by telephone, rather than suffer the time-wasting visits to the front forced on other senior commanders. After the war he retained Stalin's affections longer than Zhukov, but in 1948 he was suddenly demoted from the General Staff to command the Trans-Caucasian Military District. In 1954 he was reinstated and became from 1955 until his death in 1962 Chief-of-Staff of the Warsaw Pact.[89]

There has been little controversy about the merit of the last example of professional leadership, Roosevelt's army Chief-of-Staff, General George Marshall. He was the epitome of the modern military manager. Though he ran an army of more than eight million people, he himself had never experienced combat. He ran the American war effort from a desk in the Pentagon. Born in 1880, the son of a Pennsylvania businessman who wasted his fortune on a failed hotel, Marshall never wanted to be anything other than a soldier. His reputation as a serious, excessively industrious, and ascetic personality was formed at cadet school in Virginia. He saw himself as a man of action, but was always fated to miss it. He arrived in the Philippines in 1902 a week after the armed insurrection there was over; he was sent to France in the vanguard of United States forces in 1917, but ended up on staff and training assignments. When he was finally offered command of the invasion of France in 1944, Roosevelt rescinded the commission.

Marshall's real virtues were evident from the first. He was an outstanding organiser and manager, avid for responsibility. Before the Great War he collapsed twice from overwork. In an army light on managerial skills Marshall was indispensable. He organised much of the training programme for the American army raised from scratch in 1917–18, but like so many middle-aged American servicemen between the wars he remained stuck at a relatively junior rank with limited staff duties. He was rescued from obscurity by Roosevelt, in 1939, who wanted a Chief-of-Staff with experience of building up an army quickly. Marshall was the obvious man for the job.[90]

The new Chief was very different from Roosevelt, and although they came to respect each other deeply they never became close friends: in six years Marshall never once visited the President's estate at Hyde Park in New York state. His personality was outwardly calm and unassuming. He had a natural air of authority. He was reserved where his President was extrovert, sparse with words where his leader was garrulous, mechanically competent rather than informal and disorganised. Beneath the modest exterior, Marshall was known to be firm; he had a fierce temper which he tried to keep under control. He eschewed both sentiment and anger, once telling his wife: 'My brain must be kept clear.'[91] Colleagues found him aloof and unsociable, a strict taskmaster, a perfectionist who had not a great deal of tolerance for imperfections in others – in fact remarkably akin to Brooke and Antonov. His whole being was subsumed by work. He was famous for the disciplined regularity of his own life. He rose at 6.30 every day, rode half a dozen miles, arrived for work at 7.45, took lunch alone with his wife at home, and worked until five. No one, Marshall believed, had an original idea after five o'clock in the afternoon, and when he left the office, even during the war years, he cut himself off from the outside world until the following morning. After an evening ride, or canoeing on the Potomac river for an hour or so, he retired at nine. This routine was disturbed during the war only by the introduction of a general briefing meeting each morning. Outside office hours he answered the telephone only to the President or the Secretary for War, and invited no one to his home for fear that they would talk shop.[92]

His attitude to war reflected his personality. He described it as a managing director might define the operation of a giant company. In a speech to veterans in June 1940 Marshall reminded them that the old 'flag-waving days of warfare are gone'. The modern army, he continued, 'is composed of specialists, thoroughly trained in every aspect of military science, and, above all, organised into a perfect team'. In war, he believed, cold factual analysis was preferable to enthusiasm, common sense to sentiment.[93] He applied these technocratic views to the job of constructing America's new army and choosing its strategy. His experience of World War I convinced him that unity of command was essential. The development of an American equivalent to the British Chiefs-of-Staff was his inspiration. He dominated the proceedings of this new Joint Chiefs-of-Staff Committee, which reported directly to the President. During the war he assumed the role of senior adviser to Roosevelt. Unity between Allies he regarded as of paramount importance and he was a driving force in fusing together the British and American war efforts. He streamlined the top of the army, reducing from 61 to six the number of officials with direct access to his office, and he divided the organisation into three major elements, army, air forces and supply. A great deal of his energy was devoted to training and logistics. These were not just the preferences of a desk general. War for Marshall was a unity, from recruitment through to combat. He did not want to repeat his experience in France in 1917 when the United States 1st division arrived short of weapons and uniforms, with men who had not yet fired a gun.[94]

On the major issues it was Marshall rather than Roosevelt who perceived the necessity for a 'Europe first' priority, and battled with Fleet Admiral King and the navy to sustain it. It was Marshall who without reservation backed the cross-Channel attack when his President wavered. His approach was rooted in strategic rationality and good sense. He did not decide things lightly, but once plans were formulated he stuck to them. The decisions to fight in Europe and to attack the German main force in the west can both be regarded as central to Allied success. Marshall displayed great force

of character in carrying the plans through to fulfilment in the face
of considerable opposition. For Roosevelt he became 'the indispensable man'. By 1944 he drafted many of the President's military
papers, even replies to Churchill. But he remained a self-effacing
hero, the model of the citizen-soldier, the manager in uniform.
Churchill regarded him as 'a magnificent organiser', but he was not
a Churchillian soldier. There was nothing flamboyant, daring, or
even very courageous about Marshall. He was the personification
of a managerial culture absorbed by the business of war.[95]

★　　★　　★

Superficially Hitler's wartime leadership bore some resemblance to
that of his enemies. He too was Supreme Commander of the armed
forces, a post to which he appointed himself in February 1938. He
concentrated civilian and military power in his own hands. He
interfered ceaselessly in the conduct and planning of operations, as
Stalin and Churchill did. He was not happy with routine
administration, and preferred the private interview to the large
committee, like Roosevelt. But between Hitler and the three
Allied leaders there was one very great difference. Hitler took his
position as Supreme Commander literally. He alone devised
strategy; he decided on all major operational questions. The
delegation of responsibility in any meaningful sense of the term was
quite foreign to him. He had no sense of his own limitations.
Indeed as the war went on he became ever more convinced that he,
the humble veteran, knew more about the conduct of war than the
generals.

Assessing Hitler's achievement as a military leader has never been
easy. German generals after the war clamoured to demonstrate that
German defeat was the product of their commander's ineptitude.
Their testimony has never been regarded as entirely reliable. To set
against the later failures, there is the awkward evidence of German
successes up to the autumn of 1942, all of which were achieved
under Hitler's command as well. The temptation exists to argue
that he outdid the military leaders at their own game, that he
possessed a naive but intuitive grasp of military strategy. There were
certainly serious-minded soldiers in 1940 who suspended their

disbelief and hailed the Führer's genius. But it is a temptation that should be resisted. Hitler's credentials both as a strategist and as a commander were negligible. He had had no professional military education or staff training. Though Hitler regarded his experiences in the trenches as a harsh preparation for life, they hardly qualified him for supreme command. He had only a rudimentary familiarity with military affairs before 1938, and had a buff's eye view of military technology. His staff found him profoundly ignorant of the basic principles of command and of the art of war. In the view of General Walter Warlimont, deputy for operations at Supreme Headquarters, Hitler lacked any appreciation of 'the relative strengths of two sides, the factors of time and space'.[96] He was at best a half-hearted administrator. He brought to high command two principles of his own: pursue the offensive whatever the circumstances, and fight to the death rather than abandon ground. This was more Custer than Clausewitz.

On one thing Hitler's military critics agree: he exercised high command with exceptional willpower, the same quality he had shown in his earlier political struggle. Von Manstein, who struggled more than most with Hitler's incompetence, thought strength of will 'the decisive factor' in Hitler's military make-up.[97] Hitler certainly thought so. 'Genius', he told a general in December 1944, 'is a will-o'-the-wisp if it lacks a solid foundation of perseverance and fanatical tenacity. This is the most important thing in all human life . . .' On another occasion he observed: 'My task has been never to lose my nerve under any circumstances.' Hitler's was 'a will of iron'.[98] No doubt his fanatical self-belief, his refusal to compromise or take advice, his blinkered obsession with conquest, the heroic stances – 'never yield, never capitulate' – played some part in overcoming the hesitancy of his generals and in firing the war effort. Hitler took risks the generals would never have taken. In 1939 and 1940 the risks paid off. Against weak or disorganised enemies the fighting skills of German forces allowed them to prevail with relative ease. This demonstration of military competence had little to do with Hitler, though he took the credit. He did not attribute these successes to military professionalism but to his strength of will.

Hitler's concept of willpower needs to be treated with caution.

His messianic self-belief is not in doubt, but what he took for willpower might be regarded more properly as wilfulness. He was impervious to advice. He listened to but did not absorb the opinions of others. If critics persisted, wrote his operations chief after the war, 'he would break into short-tempered fits of enraged agitation.'[99] He brushed aside uncongenial facts so habitually that his staff began to filter out intelligence of the worst complexion. He displayed in all this not a shred of self-criticism. When things went wrong he blamed others. Sulky, vindictive, intolerant, irascible, Hitler's 'will' was the expression of poor powers of leadership, which he masked with a self-constructed myth of infallibility.

The weaknesses in Hitler's military capability were magnified many times over by the command structure he set up. When Hitler assumed the supreme command in 1938 he deliberately avoided the establishment of a command staff to run military affairs. He wanted to do this himself. He appointed a small group of officers to serve as 'administrative assistants', translating his decisions into orders, providing him with information, but neither formulating nor recommending strategy. Hitler had scant respect for the professional staff officer. He dominated the headquarters discussions and discouraged independent thought. Those around him were treated, according to Field Marshal von Richthofen, like 'highly-paid non-commissioned officers' – sweet revenge for the ex-corporal. Major strategic issues were discussed, if at all, with a close circle of party cronies. The military were left in ignorance of what to expect from their commander. As a consequence any kind of overall or long-term planning was out of the question. Hitler deliberately kept his plans to himself, as he felt befitted the guardian of the nation's destiny: 'My true intentions you will never know,' he told the army Chief-of-Staff, Franz Halder. 'Even those in my closest circle who feel quite sure they know my intentions will not know about them.'[100]

By avoiding any body equivalent to the British Chiefs-of-Staff, to ensure that he alone would determine strategy, Hitler also avoided establishing any unity of command for the three services. None of the other major areas of the war effort – production, logistics, manpower, intelligence – was coordinated by a single committee, or discussed by any kind of war cabinet. There was simply no forum in which the war effort could be viewed as a whole. Instead the three services and the civilian ministries competed with each other for Hitler's attention. The rivalry between them was never controlled. They cooperated loosely or not at all, a situation that helps to explain the poor economic effort and the failure to decide on priorities in the technological war. Hitler was, as he intended, the sole common denominator, the spider at the centre of the web. This suited his secretive nature, his intense dislike of committee work – the civilian cabinet ceased meeting in 1938 – and his distrust of military expertise. He was the first man 'since Charlemagne' to hold 'unlimited power', he told his commanders, and he 'would know how to use it in a struggle for Germany'.[101]

The Polish invasion was the only campaign in which the armed forces had any degree of independence in planning operations, though they had no influence whatsoever on the broader strategic issues as they did in 1914. After that Hitler dominated operational preparations as well. Senior officers risked a great deal in arguing with Hitler as the long list of demotions and sackings attests. They had to adapt themselves to the thinking and habits of a Supreme Commander whose outlook was ill-equipped for the demands of such an office. Hitler took his role seriously none the less. 'The Fuehrer always made the important decisions himself,' Wilhelm Keitel, Hitler's *chef de cabinet*, told interrogators in 1945. He would discuss issues, Keitel continued, only when he wanted to, but then often sleep on the decision for twenty-four hours. 'Then he would appear and say: "I have come to this decision and no more discussion will follow." '[102] The major activity of the day was the situation conference held around noon and lasting up to four hours. A second report would be made to Hitler later in the evening, sometimes as late as 1 a.m. Hitler often sat at the map table, but all the other officers stood. No stenographic notes or record of the discussions was taken, until a point in 1942 when Hitler, frustrated at what he saw as the deliberate flouting of his orders, commanded a regular record taken down for his personal use at every conference. These records were deliberately kept from the military leaders who continued to rely on their recollection of what Hitler had ordered.[103] Army leaders were usually in attendance but the air force and navy were often represented by junior liaison officers. Hitler's primary interest throughout remained the field army. Neither the navy nor the air force was fully integrated with the higher direction of the war. Much of the discussion at the conferences turned on small technical or operational issues, seldom on wider issues of strategy or operations. Senior officers found themselves compelled to discuss matters of relative triviality; Hitler's legendary memory allowed him to trump his officers time and again when they forgot technical details. Little of this activity was properly the purview of a supreme commander. Following the situation conferences came more meetings, and then long evenings after dinner, when Hitler took the chance to unwind with close party friends or invited guests. Here he indulged in long monologues in which issues of war featured surprisingly little.

At the end of 1941 Hitler added to his responsibilities direct command of the field army. Until then the army leaders had been able to use their own initiative within the operational guidelines laid down by the Supreme Command. With Hitler's assumption of control over the army, all surviving freedom of action was lost. From that moment Hitler neglected the wider issues of supreme command in order to concentrate on the front-line duel with Stalin, while the Supreme Headquarters staff were left to run the other theatres of war. Any pretence at central direction of the war disappeared. Hitler took his new duties as seriously as his old; deeply distrustful of the army leadership, and confident that he had mastered the military art, he rode roughshod over the army General Staff. Relations between the army leaders and Hitler, strained already in 1941, reached a new low. He regarded them as conservative, cautious and obstructive. General Franz Halder, the army Chief-of-Staff, despised his new commander: 'this so called "leadership" is nothing but a pathological reaction to the impressions of the moment . . .'[104] When Halder suggested in July 1942 that the German line withdraw at a point in the north to create a sounder defensive position, Hitler angrily denounced him as a coward. The tension between the two men poisoned effective staff work during the summer campaign in 1942. In September Halder was sacked. His replacement, Colonel Kurt Zeitzler, was chosen because he was sufficiently junior, and sufficiently pliant, to do what Hitler told him. In a speech to army staff Zeitzler said he expected one thing from them: absolute belief in the Führer and in his method of command.[105]

From that point onward Hitler denied his forces any initiative. It is no accident that German forces now experienced the long series of disasters that turned the tide of war. Hitler interfered with the smallest details of battle; regiments and air squadrons could be moved on the instructions of the Supreme Commander. The consequences were predictable. Instead of an overall strategy Hitler substituted a jumble of 'individual decisions and orders'. There was, Zeitzler later complained, 'no delegation of powers or coordinated action', no 'decisions on policy'.[106] Even allowing for professional jealousy, the generals' recollections of Hitler's leadership paint a uniform picture of a man quite out of his depth. During the course of the war, according to an admirer, General von Manteuffel, he developed 'a good grasp of how a single division moved and fought' but never understood the operation of an army.[107] As the army leaders struggled to ameliorate the misjudgements of their commander, conflicts became more common; Hitler sacked those who crossed him, or who refused to fight for every inch of ground. He trusted no one to take decisions for him. His concern with details was Napoleonic. He was a victim not just of his own military incompetence, but also of the overwhelming nature of his responsibilities. No one on the Allied side attempted to undertake a fraction of what Hitler took upon himself. The remarkable thing is that German forces continued to fight with so much determination that the war hung more in the balance than Hitler's flawed command merited.

Few leaders can have needed a military manager more than Hitler. Yet the nature of Hitler's personal command ruled out the emergence of any figure to compare with Marshall or Brooke. The chief of the Supreme Headquarters staff, Field Marshal Wilhelm Keitel, was appointed in 1938 for his weaknesses rather than his strengths. He was at best an office manager, channelling information to his chief, disseminating his orders. Keitel did not see it as his responsibility to recommend strategy, or to contradict his commander. He was known, without affection, by his nickname 'Lakeitel', a pun on the German word for lackey. The Operations

Chief at Hitler's Supreme Headquarters, Colonel Alfred Jodl, was generally judged to have had the aptitude to guide Hitler's operational choices, but he grasped early on that Hitler was not open to advice, and he gave it, when he did, without conviction. Everything Jodl said to Hitler, observed one General Staff officer, 'simply bounced off him with no visible effect'. Though Jodl understood Hitler's military deficiencies, he was loyal to Hitler the statesman. In his cell at Nuremberg after the war, awaiting execution as a war criminal, Jodl described Hitler as a true German hero, who bravely chose death and destruction rather than face humiliating surrender.[108]

There was no shortage of other candidates of high quality on the German side. Hitler could have made an effective deputy out of von Manstein or Guderian or von Rundstedt or Milch or many others; the list is a long one. It was simply not in Hitler's nature to share the responsibilities of 'leader'. For the Allies the war was a proving ground, a struggle for survival which by its very nature threw up commanders and organisers of high quality. Despite Hitler's Darwinian view of life, the opposite was generally true in Germany where the story of the war is littered with wasted expertise, men at the peak of their profession sacked, demoted, imprisoned, humbled. Not even Stalin, who could be as capricious as Hitler, dispensed with Soviet talent so freely, at least until the war was over. The men Hitler preferred to employ were recruited for their loyalty to him and the Nazi movement. The most notorious, and almost certainly the most damaging, was Hermann Göring, who survived as Commander-in-Chief of the German air force throughout the war. A veteran of the Great War and of the early Party struggles, Göring rose to become one of the most powerful figures in the movement. An astute and ambitious politician, his credentials for high military command were if anything more feeble than Hitler's, and only rather better than the unfortunate Udet's. He bears a large part of the responsibility for the failures of German air power, but he was too important a political figure to be replaced. In 1940 Hitler gave him the new title of *Reichsmarschall*, making him the highest ranking officer throughout the armed forces. He was the only other man allowed to sit at the daily briefings with Hitler, when he was around, which was seldom. The air force functioned as well as it did only because its commander, unlike Hitler, neglected the routine of command more as the war went on. In the last year of war senior air force officers risked a great deal by petitioning Hitler to sack Göring, but to no avail. Hitler placed great store on personal loyalties.[109]

Hitler chose to remain a lone commander. His isolation was more pronounced as the war went on. He seldom left the seclusion of his headquarters, or his retreat at the Berghof. He made almost no public appearances. He paid his own allies scant attention. Yet he retained a surprising degree of loyalty from his forces and from the wider German public. To those who saw him discussing the military crises, dominating the situation conferences, he exuded a powerful sense of optimism about the final outcome of the conflict. His personality continued to exert its compulsive, almost hypnotic effect on those around him. To those outside the Führer's orbit, the myth of the lone German saviour battling to stem defeat and rescue the nation, though seriously diminished by the accumulating evidence of catastrophe, was still able to mobilise fading energies in the German public. Only in the last weeks of the war did Hitler's will, and wilfulness, fade. Speer found a man who 'had let go of the controls'. He was a physical wreck; his limbs trembled, he shuffled rather than walked, his face was yellow and swollen. He moved between bouts of petulant, shrieking anger and listless disillusion-ment. His companions still gave the Hitler salute when he came into the room. He played out a charade of command, ordering non-existent aircraft and troops to imaginary battles. His last order was to appoint Admiral Dönitz his successor as chancellor in place of Göring, whose loyalty he finally doubted.[110]

*   *   *

When Hitler read the news of Roosevelt's death on 12 April 1945 he was overwhelmed with joy. That afternoon Speer was attending the farewell concert of the Berlin Philharmonic, at which they played Wagner's *Twilight of the Gods*. The piece was Speer's choice, an epitaph for the regime. When he returned from the concert he found Hitler in a state of agitated euphoria. 'The war isn't lost,' he told Speer. 'Roosevelt is dead!'[111] For weeks beforehand Hitler had

been repeating the story of Frederick the Great, whose imminent
defeat in the Seven Years' War was suddenly averted at the last
moment by the death of the Tsarina Elizabeth, and a switch in
Russian allegiance. For some months Nazi propaganda had been
making much of alleged divisions between the western Allies and
the Soviet Union. Hitler and Goebbels were gambling on
persuading the west to join Germany in a crusade against
Bolshevism. Goebbels hurried back to Berlin from a visit to the
front when he heard the news from America. He was convinced
that the new President, Truman, unlike the 'Jewish' Roosevelt,
feared communism more than Nazism. Hitler and he together
drafted an appeal to the soldiers of the east to hold the approaches
to Berlin at all costs. With the death 'of the greatest war criminal of
all times', German soldiers should expect 'a turning-point in the
war'.[112] It was all delusion. A day later all German armies in
the Ruhr surrendered, and Marshal Zhukov launched the assault
on the German capital that took the Red Army to Hitler's bunker
in a little over two weeks.

Hitler was right to detect serious strains in the coalition by 1945,
but he was wrong about their nature. There was no disagreement
about the defeat of Germany, only about the consequences of that
defeat. Stalin, it is true, nursed growing suspicions about his allies as
news filtered through of efforts by German soldiers and politicians
to reach a separate peace with the west. But there was no danger
here; there was nothing German leaders could offer that was
remotely attractive to Britain and America, least of all the prospect
of an anti-communist crusade. Marshall's 'impossible unity' was
maintained through to unconditional surrender on 8 May. The
willingness to fight in a common coalition for so long Marshall
regarded as the single greatest achievement of the war.[113]

Hitler, too, thought that Allied unity would not last. His own
relations with his closest allies were limited and secretive. He did
not share his strategy or his military thinking with anyone. Alliances
were made and unmade. When Ribbentrop was presented with a
casket by his staff containing all the alliances he had concluded as
Foreign Minister, it was found that almost every one had been
broken. When Hitler heard he is said to have laughed until the tears
streamed down his face. Hitler found it difficult to imagine what

cause had been sufficiently powerful to bind together such unlikely comrades-in-arms as capitalist west and communist east for so long. His own collaboration with the Soviet Union between 1939 and 1941 began to turn sour after less than a year; in under two years the two allies were at war. The only explanation Hitler could find for the survival of the coalition was a 'Jewish conspiracy' that married together the forces of Mammon and the forces of world revolution to bring about Germany's collapse.[114] It does not seem to have occurred to him that he himself was the reason.

The different attitudes to alliance reflected the deeper contrast in the way the two sides ran their war efforts. The Allied systems were centralised, unified and coordinated. They were run by a central staff with wide powers, but they all operated with a good degree of delegated responsibility. Each relied on leaders with solid professional and managerial skills who actually ran the war effort from day to day. Though each national leader could bring considerable influence to bear on the system, they waged what were essentially wars by committee, with the burden of duties spread over a wide administrative area. As the war went on military responsibilities devolved more to the professionals. No one could pretend that the system worked perfectly, but it was designed around checks and balances that reduced the element of arbitrary will or personal misjudgement, even in the case of the Soviet dictatorship.

Hitler's system was almost the exact opposite in every respect. Military affairs were dominated by the will of one man who considered himself to be a modern Charlemagne. There was no central staff to run the war effort. There was no unity of command, no formal structure to oversee military operations that united army, navy and air force around the same table. The idea of a committee war flew in the face of everything Hitler understood by the term leader. As the war went on Hitler delegated less rather than more, while his personal role in the war effort was further enlarged; so many responsibilities were there that no one, not even a leader more rational and more discriminating than Hitler, could have mastered them. Hitler dragged the military system and his own people behind him by the sheer force and momentum of his neurotic personality sustained by the willingness of the population to believe in him. If Hitler helps to explain why Germany lost the war, he is also the reason why German forces continued to fight long past any prospect of victory. At the end of the war the tank general Heinz Guderian was asked by his American captors to reflect on the lessons of command on the German side. Guderian had seen Hitler at close quarters. He concluded, predictably enough, that the essential feature in any war was unity of command under an agency that enjoyed complete authority 'over the whole military establishment', but one that was run not by the head of state but by a 'trained professional', a 'military man'.[115] Dictatorship on its own was not enough.

9

EVIL THINGS, EXCELLENT THINGS
The Moral Contest

'The chief reason why we are at a serious disadvantage compared with the Nazis over this business of "big ideas" is that the evil things for which they stand are novel and dynamic, whereas the excellent things for which we claim to be fighting may seem dull and uninspiring.

British Directorate of Army Education, Booklet i, November 1942

IN WAR THE gods are always on your side. Even in the Soviet Union, where God had been officially proscribed, religion was revived by the war. On the day of the German invasion Metropolitan Sergei, head of the Russian Orthodox Church, persecuted for years by the authorities, hounded by Emelian Yaroslavsky's League of the Godless, appealed to the Soviet faithful to do everything to help the regime: 'The Lord will grant us Victory!'[1] In the Soviet Union an estimated half of the population were still Orthodox Christians, forced to live a religious half-life under a thoroughly secular regime. The number of priests was reduced by the 1930s to a few thousand. The churches were destroyed or in disrepair. No Patriarch, supreme father of the Church, had been permitted since 1926.

With the coming of war everything changed. Stalin wanted national unity. Propaganda emphasised patriotism and tradition. In this the Church had a part to play. Stalin stamped out the crude anti-Christian activities of the Party zealots. Money was made available to restore churches; religious observance was openly encouraged. A commissariat was set up for Church affairs, popularly nicknamed *'Narkombog'*, People's Commissar for God. In 1943 Stalin finally approved the restoration of Church authority. Breakaway sects – particularly the so-called 'Living Church' of Father Vvedensky, previously supported by the Party as a counter to tradition – were closed down; church leaders in the German-occupied territories who collaborated with the invaders were deposed and excommunicated. The Patriarchate was restored in September 1943. Stalin, the ex-seminarian, permitted the reopening of seminaries, and the Church was legally allowed to own property. When the first Patriarch, Sergei, died in May 1944, he was succeeded by Metropolitan Alexei of Leningrad. Moscow hosted a magnificent coronation. Russian Orthodox leaders came from all over the world, even Metropolitan Benjamin Fedchikov of the Aleutians and North America, who ventured the hope that Moscow might now become a 'Third Rome' behind Constantinople and the Holy City itself.[2]

The faithful responded to the revival. By 1943 the churches of Moscow were so crowded at Eastertime that the congregations spilled out into the surrounding streets. Though Stalin did not go so far as to allow chaplains to accompany the troops, it was noticed that soldiers on leave began to use the churches in large numbers too. Priests incanted prayers for Stalin, who was treated as 'an anointed of the Lord'. The church gave 150 million roubles to the war effort collected from its congregations. Sergei presented the Red Army with a battalion of tanks paid out of church funds. It was named the Saint Dmitry Donskoi battalion, after a fourteenth-century Russian prince who routed the Tatars at Kulikovo. At the formal ceremony of presentation, the church's representative spoke of Russia's 'sacred hatred of the fascist robbers' and described Stalin as 'our common father'.[3]

Much of this was for foreign consumption. Stalin knew that anti-Soviet feeling in the United States was linked with the Soviet persecution of the churches. Soviet spokesmen were sent to the Allied powers to provide assurances that communism had turned a new leaf. Foreign churchmen were encouraged to visit Moscow. Stalin told the British ambassador that, in his own way, he 'also believed in God'. In Britain the Anglican Church sprang to support the Soviet alliance. From his pulpit in Chichester Cathedral, Bishop Bell called on his flock to 'thank God for the heroic resistance of our allies in Russia'. At the pageant for the Soviet Union in the Albert Hall on New Year's Day 1942 the Archbishop of Canterbury spoke of 'a beacon shining through the vast clouds of destiny', the burning light of Russia's 'indomitable faith'.[4]

It would be a simple matter to argue that Stalin's conversion turned the war into a Christian war, a latter-day crusade, right against wrong, the faithful confronting the heathen. Yet nothing illustrates more the ambiguities in any moral explanation of the conflict than the position of Christianity. Few American Christians took Soviet policy at face value. Roosevelt did believe in God, devoutly so. His faith carried him through the terrible years of illness. He was a lifelong Episcopalian, and his religious conviction was strengthened by his struggle with his disability, the successful outcome of which he attributed to Divine Providence.[5] The first official statement following the outbreak of the German-Soviet war, approved by Roosevelt and broadcast on 23 June, made no distinction between Nazi Germany and Soviet Russia on the question of 'freedom to worship God'. Both states denied this 'fundamental right'. The atheistic principles of communism were 'as intolerable and alien' as the doctrines of Nazism.[6] There existed a strong religious lobby in the United States, including another breakaway Orthodox sect, the Theophilites, that was never fully reconciled to the connection with the Soviet Union. Among the strongest opponents were American Catholics, many of them of Polish or Italian extraction. In 1937 Pope Pius XI issued an encyclical on 'Atheistic Communism', forbidding Catholics from collaborating with communism in any undertaking. The Catholic community was divided on the issue only because some saw Hitler as the more immediate evil and were prepared to fight Hitlerism first. When on 29 June 1941 the Pope broadcast to the faithful, some expected an endorsement of the German anti-communist crusade. The Pope instead abstained, calling on the faithful everywhere to place themselves in the hands of Providence.[7]

The paradox for Christians siding with communism was clear. At least two of the three enemy states were nominally Christian. Italy was the home of Roman Catholicism; Germany's population was one-third Catholic. Religion in both states lived in uneasy proximity with regimes that were strongly anti-clerical in outlook peddling new secular religions of their own. The same month that the Papacy condemned communism, a second encyclical was published, '*Mit Brennender Sorge*' ('With Burning Anxiety'), which condemned the Nazi persecution of the churches, Nazi racism and Mussolini's deification of the state. Though Hitler often invoked God or Providence when he spoke, he was a thoroughly lapsed Catholic. Hitler considered Christianity incompatible with the new National-Socialist age – it was 'merely whole-hearted Bolshevism, under a tinsel of metaphysics'. He deplored the survival of religious observance among German ministers and generals, 'little children who have learnt nothing else'. He regarded Christianity and communism as two sides of the same coin, sharing in St Paul a common Jewish ancestor.[8] Hitler took the German nation as his religion. This did not make him a pagan as was widely believed, although paganism was practised under the Third Reich. The German Faith Movement, under the banner of the golden sun-wheel, with the 'Song of the Goths' as their anthem, indulged in pagan festivals and invoked the gods of pre-Christian Germany. Heinrich Himmler's SS generated a pagan theology, a pagan liturgy, even a pagan *credo*.[9]

Many German Christians were alienated by Nazism. The German opposition counted thousands of priests of every denomination in its ranks; six thousand churchmen died in the concentration camps or in prison. However, most German Christians saw the defence of the nation as a Christian duty, and the Church authorities reached compromises with the regime rather than risk a more violent anti-clericalism. But Nazism and Christianity were fundamentally at odds, and the Party made little effort to disguise it. Roosevelt could quiet his religious critics by showing a German state even more hostile to the Christian message than Stalin's Russia. A few weeks before Congress finally approved Lend-Lease aid for the Soviet Union in November 1941, Roosevelt obtained a copy of a thirty-point programme for a German National Church drawn up by Hitler's chief of ideology, Alfred Rosenberg. In his Navy Day Address on 27 October Roosevelt told his audience that he had in his possession a secret document which showed a Nazi plan to 'abolish all existing religions', to supplant the Bible with *Mein Kampf*, and to replace the Cross with a sword and swastika.[10] Roosevelt was also armed with a personal assurance from the new Pope, Pius XII, that the encyclical condemning communism could be bent sufficiently to allow Catholics to support aid for the suffering Russian people. American Christians of all denominations could now rally to the cause with a clear conscience.

Religion was mobilised in every state to support the national war effort, though it was by no means the only, or even the prime, source of moral validation. The conscription of moral energies, like the mobilisation of technical and economic resources, was a necessary element in the war effort on both sides, particularly in this war when societies were fighting for their very existence, or thought they were. Like the religious wars of the sixteenth century, the Second World War was fought with a ferocity and desperation born of real fears and deep hatreds. It was not a war simply for balance-of-power politics, but also for ideals deeply held. Moral commitment to the cause was forged from a heady mix of outrage, vengeance, loathing and contempt, an intensity of feeling and a depth of anxiety not experienced since the days of French Revolutionary Europe or the Thirty Years' War. Populations on both sides sustained the struggle with some sense of the justness of their cause, with that 'sacred hatred'.

No account of Allied victory can afford to ignore the moral dimension in war. The conflict was never a simple one of right against wrong, as the confused battle-lines of Christianity demonstrate. Roosevelt was clear in his own mind that on the basic moral issues Stalin's dictatorship was of a piece with Hitler's. The first line of Churchill's declaration of support for the Russian people broadcast on the night of 22 June 1941 began with the words: 'The Nazi regime is indistinguishable from the worst features of communism.'[11] None of the Allies, the democracies included, fought an unblemished war by any standard. The real success of the Allies lay in their ability to win the moral high ground throughout the conflict by identifying their cause with progressive, post-Enlightenment values. Their enemies they tarred with the brush of reactionary, even barbarous, vices. The western democracies were self-righteous about their liberal virtues; but even the Soviet Union was regarded then as a system striving for social progress, a 'new civilisation' as the British socialists Sidney and Beatrice Webb called it. The Allied war effort was sustained by the powerful sense that the Allies were prosecuting just war.

The belief that their cause was on the side of progress in world history gave a genuine moral certainty to the Allies, which the Axis populations largely lacked. Popular commitment to war in the aggressor states was half-hearted and morally ambiguous. In the Allied communities, on the other hand, there was a powerful crusading rejection of the forces of fascist darkness. This helped to mask the deep doctrinal and political differences between the three major Allies, and encouraged the greatest of efforts, particularly from the Soviet people, in destroying their enemies. The moral forces at work on the Allied side kept people fighting in a common cause; but as the war went on Axis populations suffered a growing demoralisation, a collapse of consensus, and increasingly brutal regimentation of the home front. This contrast is an important part of any explanation for the eventual outcome of the war.

★　　★　　★

Hatred of Hitler and 'Hitlerism' was the moral cement of the Allied war effort. In the same speech in which Churchill bracketed communism and Nazism together, he gave overriding priority to the German threat: 'We have but one aim and one single, irrevocable purpose. We are resolved to destroy Hitler and every vestige of the Nazi regime.'[12] These sentiments were not just for public consumption. Churchill developed a deep personal loathing for Hitler; his defeat became an obsession, 'the destruction of Hitler' the driving force of his war effort.[13] When the Soviet ambassador to Washington, Maxim Litvinov, was asked to report to Moscow on the attitude of the western Allies to the war in 1943, he concluded that 'the struggle against Hitler' was the defining aim. Roosevelt, he found, hated Hitler, was a 'staunch anti-Nazi' and surrounded himself with advisers who wanted simply 'to eliminate Nazism'.[14]

Hitler exerted a special power in binding his enemies together. On the surface the reasons for this are clear. Hitler's unquenchable, unpredictable appetite for conquest was a threat to every other state and way of life. Yet the issue is more complex than this. Hatred of Hitler and Hitlerism long pre-dated the final evidence of the Holocaust and the catalogue of crimes exposed in 1945. Its roots go back to the 1930s, even before the outbreak of war, when Hitler was singled out from other dictators – Mussolini, Stalin – as the greater force for evil, and German aggression was feared far more than that of Japan and Italy, though both had actually waged particularly brutal wars in the 1930s, unlike Germany. Stalin's rule in the 1930s was atrocious by western standards, but Stalin was not demonised like Hitler until after the war. The English novelist George Orwell believed that most intellectuals preferred Stalin to Hitler, accepting 'dictatorial methods, secret police, systematic falsifications', as long as they felt 'it was on "our" side'. Even Orwell, whose own denunciations of Soviet oppression swam against the general current, thought Stalin's dictatorship 'a more hopeful phenomenon than Nazi Germany'.[15] When the American public was asked by pollsters in the summer of 1941 to choose between the Soviet Union and Nazi Germany, 4 per cent favoured Germany and over 70 per cent the Soviet Union.[16]

The threat that a Hitler-led Germany represented eclipsed every other menace faced by the Allies. It has been the general view, both then and since, that the war was Hitler's responsibility and the defeat of Hitler the Allies' primary purpose. As Churchill confessed, 'his life was much simplified' by seeing the war in personal terms. Unlike many of his countrymen, Churchill made a distinction between Hitler and Nazism on the one hand, and Germany on the other. He had an unlikely ally in Stalin, who wanted to smash German fascism, but not necessarily to eradicate a potentially communist Germany.[17] Roosevelt was much more inclined to see Hitler as typical of the German people as a whole. He displayed a deep prejudice against Germany, rooted in his childhood, when he was sent to school in Baden on one of many trips to German spa towns with his ailing father, who was strongly anti-German. The young man developed a strong dislike of German militarism, of German 'arrogance and provincialism'.[18] His later attitude to his childhood hosts was anything but Christian. At the time of Munich he told his Cabinet that all Germany's neighbours ought to combine to batter the German population from the air until their morale cracked. During the war his private comments on the German fate betrayed an unusual brutality. He favoured the punishment of Germany advocated by the Treasury Secretary, Henry Morgenthau, which was to turn Germany into an agricultural state, whose impoverished population would have to live from army soup kitchens. More than once Roosevelt suggested some form of extreme population control – castration was his recommendation, though it is difficult to take seriously – to prevent the reproduction of this militarised people.[19] Even the mild Eisenhower exhibited the most pronounced Germanophobia. At lunch with Lord Halifax in July 1944 he announced that he would recommend liquidating the German General Staff, the Gestapo and any Nazi officials above the rank of mayor. His treatment of German prisoners in American hands at the end of the war was brutal and neglectful.[20]

Among the wider Allied populations hatred of Hitler, Nazism and Germany was powerfully instilled. In 1942 the Soviet regime launched a formal campaign of hate against the aggressor. Hitler was shown in countless cartoons and posters as a scavenging animal – a hyena, a rat, a vulture – or a primitive, ape-like creature; his followers became vermin, 'foul hounds'.[21] Soviet journals preached a single-minded detestation of the enemy. 'May holy hatred', ran the *Pravda* editorial on 11 July, 'become our chief, our only feeling.' The Soviet poet, Konstantin Simonov, called on his readers to 'kill a German, kill a German, every time you see one!' Here are echoes of Churchill's casual slogan 'kill the Hun!'[22] Britain did not officially instigate a progamme of hatred, but one was widely evident. Lord Vansittart, Permanent Secretary of the Foreign Office in the 1930s, sold over half a million copies of *Black Record,* a diatribe against the German race, 'a breed which from the dawn of history has been predatory and bellicose', led by a fanatical anti-Christ.[23] In a series of lectures given at Oxford shortly after the outbreak of war, the historian R.C.K. Ensor told his audience that the Germans were not like other civilised peoples: they were aggressive, militaristic, prone to cruelty. 'Ask yourself whether a Dachau or a Buchenwald would be conceivable on English or French soil,' he continued. 'I do not think it would.'[24] Anti-German sentiment, rooted in hatred of Hitler and a crude racial stereotyping, was widespread well before the course of the war lent such views an evident validity. The war exacerbated them. 'Whether you like it or not,' ran an editorial in the *Daily Express,* shortly before Pearl Harbor, 'vengeance on Germany is becoming the prime war aim of all Europe.' NOT ENOUGH HATE was the headline above another editorial: 'You can't win a war like this without hating your enemy.'[25]

Hitler aroused emotions strong enough to hold together an

unholy alliance for the whole course of the conflict, strong enough to fire exceptional levels of popular hatred, strong enough to incite the peaceable Chamberlain to declare war in 1939 against all his instincts, and to rouse Roosevelt from neutrality to intervention. The strength of feeling owed something to the nature of the man himself. Hitler's populist roots, the demagoguery, the crude messianism singled him out from the generation of orthodox statesmen in the 1930s. His dishonesty in international dealings was regarded as quite distinct from the usual pursuit of flexible self-interest. Roosevelt considered Hitler's word worth no more than 'the bond of gangsters and outlaws'.[26] In the 1930s Hitler's world view was characterised outside the Reich as nihilistic, amoral, destructive. 'Hitler is the arch-destroyer,' ran a sermon preached at Canterbury Cathedral; 'wherever he goes he brings death.' The style of Hitler's leadership provoked abroad an apocalyptic language that curiously mirrored Hitler's own expectation of a final struggle, a reckoning of accounts, the new order versus the old. An English guest at a Nazi Party rally in Berlin recalled in his memoirs that he had observed what appeared to be flashes of blue lightning escaping from Hitler's body, and concluded that at certain moments Hitler was actually possessed of the devil.[27]

More significant for understanding the reaction to Hitler than the man himself is what he was taken to represent. Hitler was the personification of a German threat abroad that could be traced back to at least the 1890s. Popular anti-German feeling was composed of a number of strands. Fear of Prussian militarism and bellicosity was a well-established feature of it well before 1914; H. G. Wells's *The War in the Air,* published in 1908, when no bombers yet existed, painted a frightening picture of devastating attacks by 'German bomber fleets', which heralded 'the beginning of the end' of civilisation.[28] Blame for the Great War was laid at the door of German ambition by the victor powers in 1919. The continuities in German power-seeking were regarded as self-evident by much foreign opinion. 'The Germany of the Kaiser', wrote the American journalist Walter Lippmann, in 1944, 'was not nearly so evil a thing as Hitler's Germany. But it had the same fundamental design of conquest.'[29] In the 1920s, side by side with liberal sympathies for German mistreatment at Versailles lay the constant fear of German revival and rearmament. In the 1930s it was commonly assumed that Germany would violate Versailles and rearm in the air, posing the sort of annihilating threat first painted by Wells. The British Air Ministry expected the Germans to use bombing in a 'ruthless and indiscriminate' fashion because that was the German way.[30] Even before the Nazi revolution, Germany was regarded as the disruptive force in world affairs. Hitler played a part already written for him.

The representation of Hitler as anti-Christ, as a symptom of a more profound malaise in European life, exaggerated the German threat. Before 1914 a whole generation of European thinkers turned their back on simplistic notions of liberal progress. There emerged a prominent cultural pessimism, fears of decline and decay for western civilisation, premonitions of war and social chaos, a yearning for spiritual renewal. Both extreme right and extreme left rejected the solid virtues of bourgeois Europe – scientific rationalism, the pursuit of peace and prosperity, the values of Christian respectability. They turned instead to the cult of violence and overthrow, exalting the strong at the expense of the weak; life was seen as Darwin saw it, a struggle for existence, the survival of the fittest. The sense of impending doom, of the end of an age, was fuelled by the horrors of the Great War and the revolutionary upheavals at its end. Violence and social chaos were established facts. The liberal order was visibly in decay. The moral certainty of the pre-war age was destroyed. 'What are the roots that clutch, what branches grow/Out of this stony rubbish' asked the poet T.S. Eliot in *The Waste Land*, written in 1922.

The belief that the European order was bankrupt, that the pursuit of progress was a dead end, was widely held in post-1918 Europe. The economic crisis in 1929, which was so severe that the very future of capitalism seemed in doubt, was further proof of the fragile nature of the existing system. The idea that some kind of violent, purifying transformation was necessary to renew European civilisation, or the white race, was not confined just to the radical political fringe. A New Order based on harsh authority, eschewing sentiment and compromise, was a more serious prospect in the 1930s than the survival of traditional liberalism. Hitler reflected, but did not cause, the profound moral crisis of the inter-war years. He embraced the idea of a New Order; he accepted violence and struggle as the elements of human existence; he gloried in his harsh post-Christian morality. Hitler appeared as nothing less than the harbinger of a new Dark Age. 'A Nazi victory', wrote the American philosopher, Melvin Rader, in 1939, 'would not only destroy the freedom of Europe; it would jeopardize every noble ideal of human culture, every high concept of human morality, every fine achievement of hard-won democracy.' Rader thought the war a turning-point in world history, upon whose outcome depended 'the fate of mankind'.[31] These were large claims for a conflict ostensibly over the fate of Danzig. Hitler was taken to stand for all those forces worldwide making for disintegration and violence. An English bishop wrote in response to the Pope's 'Five Peace Points' in May 1940, that there was a choice 'between the Christian religion and nihilism, the destruction of humanity that Hitlerism brings'.[32] As a focus of Allied hatred Hitler was the object not only of the accumulation of half a century of anti-German sentiment, but also of the deeper fear that civilisation was now confronted by barbarism, order by chaos, good by evil.

* * *

The firm conviction that what they were fighting was a wicked thing greatly simplified the Allied war effort. The willingness to fight the war to the end was nourished by the image of 'just war'. Moreover the fact that the Allied powers were all the victims of aggression also simplified the task of constructing the wartime consensus. Where the aggressor powers had to find ways of justifying aggression to the home populations, the Allied peoples fought wars of self-defence, in which victory was pursued for its own sake. Defence was less morally ambiguous than attack. Roosevelt did as much as he could during 1941 to avoid having to declare war first because of the strength of pacifist and isolationist opinion. The Japanese assault on Pearl Harbor, and the German declaration of war four days later, solidified American public opinion behind a patriotic war of defence. The Soviet war effort was presented by Stalin not so much as a clash of ideologies, but as a sacred mission to save Mother Russia. "The Red Army's strength', Stalin announced in his Order of the Day, to Soviet forces in February

1942, 'lies above all in the fact that it is not waging a predatory, imperialist war, but a patriotic war, a war of liberation, a just war.'[33]

The Soviet situation differed from that of Britain and the United States because the Soviet Union was actually invaded and occupied. Britain and the United States fought their wars of self-defence on foreign soil; the Soviet people fought to regain lost territory. The German conquest of the western Soviet Union also precipitated a crisis in Soviet society that had no parallel in the west. At times in 1941 and 1942 the situation resembled the military disasters and economic chaos that sapped the Russian war effort in the First World War and brought the collapse of the Tsarist system. Sustaining Soviet morale was essential to the revival of Soviet material power during the critical middle years of the war. The sacrifices both expected of and endured by the Soviet people required an exceptional level of moral mobilisation. Indeed, how that resolve was summoned up remains one of the central questions of the war.

The problem in any assessment of the Soviet will to war is to distinguish genuine commitment from the fear of punishment that hung over every Soviet head. Behind the front operated an estimated three-quarters of a million NKVD security troops, whose task was to root out defeatism, to hunt down saboteurs and fifth columnists, and to shoot soldiers who deserted or fled before the enemy. Stalin's order 227, published in July 1942, that retreat meant death, was not literally applied to every soldier who retreated, but was certainly used on occasion as an encouragement to others. Order 270, in which Stalin declared that all Soviet soldiers who fell into enemy hands were 'traitors to the Motherland' faced the ordinary Soviet soldier with a grim choice.[34] Yet it is difficult to reconcile this image of a terrorised soldiery with all the evidence of spontaneous commitment and self-sacrifice. A great many Soviet people approved of the harshness because it matched the expectations of the time. One soldier recalled his reaction to hearing the order 'Not a step back': 'Not the letter, but the spirit and content of the order made possible the moral, psychological and spiritual breakthrough in the hearts and minds of all to whom it was read . . .'[35] A story from the wartime journals of a British war correspondent reveals the inadequacy of any explanation based entirely on Stalinist terror. At a railway station in Moscow at the height of the November fighting in Stalingrad he observed an elderly Siberian soldier waiting on the platform for his train to the front. Suddenly over the loudspeaker system came a low but distinct voice. The soldier gave a start and listened with deep attention. Then he whispered to himself the name 'Stalin' and solemnly performed the sign of the cross.[36]

The Soviet leadership was well aware that a whole population cannot be made to fight and work at the point of a gun. The war was presented in Soviet propaganda not as a war for communism but as a war for the Russian motherland. The very name chosen for the conflict, the Great Patriotic War, emphasised the continuities between the war with Hitler and the earlier wars against Napoleon, or the Teutonic knights. The rediscovery of Russia's past history allowed the current Soviet conflict to be identified with an almost mythic contest between German and Slav that stretched back seven centuries. The term Soviet was used less and less frequently, to be replaced by a new vocabulary of nationalist endeavour. The communist hymn, the 'Internationale', was suppressed, to be replaced with a new Russian national anthem.[37] The regime presented the war as a people's war, sustained by the daily heroism of ordinary Russians. Above their struggles stood the Russian Father, Stalin. The symbol of the great leader, whose wisdom and resolution would hold the war effort together, was a necessary symbol, widely accepted by the Soviet public. The young Petro Grigorenko, a later Soviet dissident, remembered after the war that everyone he mixed with agreed that Stalin had produced the turnabout in the fortunes of war. He ended the war still believing that 'without Stalin's genius' victory might not have come at all.[38] Soviet soldiers were said to charge into battle shouting 'For Motherland and Stalin!', in that order.

The entire Soviet war effort was suffused with a grim atmosphere of sacrifice and death. The people were immersed in a popular culture that exalted death in battle and preached violence towards any enemy of the war effort. Soviet newspapers were full of stories of extraordinary heroism for others to emulate. 'Every soldier must be ready to die the death of a hero,' ran a *Pravda* editorial in July 1942, and there is enough evidence from the German side to corroborate the tales of suicide attacks on tanks and Soviet soldiers refusing to surrender until cut down to the last man.[39] The Russian attitude to death in battle was not a casual one, as is sometimes supposed, but was rooted in a social psychology that had long endorsed self-sacrifice and leadership by example. There was little formal code of honour like the rules governing Japanese military behaviour, but there was a strong sense of a deeper social and ethical commitment beyond loyalty to Stalin or the Party. Some of that commitment can be explained by the image in the mind of the Soviet people of the enemy they faced. German forces were presented, with good reason, as utterly amoral and bestial. Atrocity stories crowded every news page. Soviet propaganda created a dehumanised image of the enemy, not unlike the American image of the Japanese as apes or sub-humans, or German depiction of the Jews as vermin. The public language used against the German enemy was violent in the extreme: the conflict was described as 'a war of extermination', aimed at 'annihilating to the last man all Germans'. Soviet soldiers told to expect mutilation and death if they were caught routinely meted out harsh treatment to German soldiers. Russians were in a very real sense absorbed by the immanence of death. They sought death in battle, they inflicted death, they confronted death. Which other people could have suffered the loss of more than twenty million of their number and continued to fight?[40]

For Britain and the United States death was the exception rather than the rule. American civilians were not subject to mainland attack at all, while British civilian casualties ran to sixty thousand, a small fraction of civilian losses in the other warring powers. It was a priority in both states to keep casualties to a minimum. About 3 per cent of those mobilised for military service lost their lives. The American censors deliberately played down the theme of death in the first two years of conflict. *Life* magazine did not show a dead American until September 1943. The Information Manual produced for Hollywood by the newly created Office of War Information instructed film directors to limit scenes of death or injury: 'In crowds unostentatiously show a few wounded men.' Between May and November 1942 only five out of sixty-one war films showed deaths in combat.[41] American authorities remained anxious that the reality of war might dent morale. Pictures showing weeping were banned. Only in 1944, and partly in response to popular demand, did the American media show the war in its truer colours. American society was less prepared for the traumas of battlefield violence than the people of Europe and Japan. The first American forces to fight in North Africa suffered a 25 per cent casualty rate from psychological disorders. A report by army psychologists on one American division found that during heavy combat a quarter of the soldiers soiled themselves, and another quarter vomited.[42]

Unlike the Soviet Union, the two western Allies were democracies, whose populations were used to a high standard of living and amenity. They could not be regimented to the extent of Soviet citizens, or their armies terrorised to fight. Throughout the whole war only forty British soldiers were executed.[43] Those who suffered what was euphemistically called 'lack of moral fibre' were demoted or redeployed, but not thrown into Gulags or shot. In the American forces a good deal of emphasis was placed on mental health. One million conscripts were rejected on neuropsychiatric grounds. Almost one million soldiers were admitted for treatment of psychiatric disorder during the war, and almost half a million of these were subsequently released permanently from fighting. Combat zones with a high death rate – long-range bombing, for example – were staffed entirely by volunteers.[44] On the home front it was necessary to establish some kind of moral consensus. It was not difficult to paint the enemy in the blackest terms. But it was necessary to paint over the war effort of Britain and America with a thick moral gloss as well, to make it clear that the democracies were, on balance, fighting a liberal war. Since their major ally, Soviet Russia, was anything but democratic the presentation of the wider Allied cause had to be delicately handled.

The moral commitment of the home populations was easy to mobilise when there was a direct and violent threat. The Battle of Britain and the Blitz repaired Britain's dented morale after the defeat in France. They provided the core myths of invincibility and steadfastness for the remainder of the war. The period of bombing saw the creation of a war effort more united than it had been in the first months of war.[45] In the United States the attack on Pearl

Harbor had a galvanising effect on domestic opinion. Hatred towards the Japanese, drawing on decades of anti-Asian racism, was immediate and general. The Japanese were regarded as sub-human, racially and physically inferior, fanatical and heathen. Public loathing for Japanese forces was almost certainly stronger than anti-German feeling, until the unveiling of the horror of the camps in 1945, after the German surrender. Even the mild General Marshall talked publicly of the 'treacherous barbarians' of the East, and had no compunction about contemplating the fire-bombing of Japanese cities.[46] In opinion polls during the war a tenth or more of those polled favoured the physical extermination of the Japanese race. What is startling is not so much the savage response as the fact that the question was put at all. A comparable poll on the German enemy excluded the extermination choice.[47] Despite the priority given to the European theatre, vengeance against the Japanese was a core element in popular attitudes to the war, fuelled by a diet of atrocity stories which helped to dull resistance to the indiscriminate destruction of Japanese cities and the eventual use of the atomic bomb.

Both the Blitz and Pearl Harbor came at the onset of hostilities. After that the direct threat of invasion receded. British and American forces were only intermittently in heavy combat. The battlefields were for the most part remote from the home populations. For the first years of war there were long periods of reverses or stale inactivity which made it more difficult to maintain popular enthusiasm. Though the news was carefully filtered, it was impossible to disguise the reality of the war, or to stifle criticism. Even when the tide of war had turned in the Allies' favour, moral commitment needed sustenance. During 1944 and 1945 when victory appeared certain the willingness to accept sacrifices for a cause already secure began to weaken. In Britain and the United States war production was eased off during 1944, and larger numbers of consumer goods began to reappear in the shops. The partial relaxation of the domestic war effort coincided with the period of the highest western casualties of the war, a juxtaposition that was hard to reconcile for those still doing the fighting. The Allies by then had little choice but to fight to the finish. Since most of the war had been spent presenting the enemy as an abomination, there was no question of negotiation or compromise. Roosevelt's call for unconditional surrender, though made almost casually at the end of the Casablanca conference, was the logical outcome of the Allied view of the enemy. The moral chasm between them had been made too wide to bridge.

British and American propaganda was chiefly concerned to reinforce the positive moral stance from which it was argued victory would spring. This was expressed in the conventional language of freedom against tyranny, barbarism crushed by civilisation. 'We wanted to make the world safe for democracy – and protect the Four Freedoms,' wrote the American General Wedemeyer after the war.[48] When the film producer Frank Capra was recruited by General Marshall to produce a series of documentaries on *Why We Fight,* to help educate American opinion, Capra took as his working theme the 'enormity' of the enemy's cause, 'and the justness of ours'. The western war was presented as a decent war, as the 'good war'.[49] This was not difficult to do in the light of Japanese and German atrocity, and the vicious and illiberal regimes that ran the Axis war effort. It was more difficult when explaining morally complex issues such as western policy on bombing enemy civilians, or the fact that the west was fighting for democracy and freedom at the side of the Soviet Union.

The ethical issues raised by bombing were never clearly in focus during the war years. What criticism there was came from the belief that the western states should maintain the values of liberal decency in the way they conducted the war. The authorities were concerned enough with avoiding such criticism to highlight the military nature of the targets they attacked, even when those targets were whole industrial cities. The attacks on the enemy war economy stretched the term 'military target' out of all recognition, but everything possible was done to ensure that the two western states were not seen to be engaged in indiscriminate terror-bombing, of the kind Germany was alleged to have carried out against Warsaw, Rotterdam and Coventry, or Japan against Nanking. The American emphasis on the precision bombing tactic, though its results were known to be exaggerated, was publicly promoted to create the illusion of good bombing against bad. These efforts reflected serious public concern. Londoners polled during the Blitz were almost evenly divided on the question of whether to give the German people an equal measure of terror. Public concern did little in the end, however, to inhibit the use of the bomber in inflicting almost a million civilian deaths in the name of democracy.

The most striking moral paradox of the war years was the willingness of ostensibly liberal states to engage in the deliberate killing of hundreds of thousands of enemy civilians from the air. Serious consideration was at times even given to the use of chemical and biological weapons. There was little evidence of moral scruple in discussions about the use of atomic weapons. This paradox can be explained partly by the deliberate choice made by the western democracies to save the lives of their own populations by resorting to technological solutions rather than strategies with high manpower losses. The atomic bomb was the supreme expression of the reliance on technology to inflict insupportable damage, while reducing the democracies' losses to virtually nil. The use of technology produced a distance between those who planned and executed attacks and the victims themselves. A western ground army would never have run amok in Hamburg, murdering forty thousand people. Bombing permitted a kind of moral detachment, evident in the language surrounding it. The attack on city suburbs was called 'de-housing', as if the buildings could in some way be separated from the families inside. The fire-bombing in Japan was justified on the grounds that the residential districts were scattered with small-scale industry, which had to be burnt out, along with indistinguishable but innocent households. A second, and less charitable, explanation is simply that the western states did react with a crude vengeance. They adopted a strategy of lynch-law against those states that violated the world order, and did not accept that there was a moral case to answer when dealing with outlaws. The sense of outraged decency provoked by bombing has grown with the passage of time. During the war the awkward moral position was side-stepped or ignored.[50]

Few even moderately informed Britons or Americans could have been unaware of the political complexion of their Soviet ally, although it was true that information from the Soviet Union was difficult to obtain. During the period when Hitler and Stalin were in temporary alliance the Soviet and Nazi systems were treated by much western opinion as varieties of the same warped totalitarianism. When German aggression brought the Soviet Union into the anti-Hitler camp it was impossible to sustain the image of a 'decent war' in combination with a regime run by a one-man, one-party dictatorship, full of concentration camps and secret policemen. The moral coalition worked only to the extent that the west was able to suppress or at least lighten their ally's dark image. This was done deliberately. Capra found that the first seven authors assigned to write the scripts for *Why We Fight* contributed what he regarded as 'Communist Propaganda'. Roosevelt promoted Soviet sympathisers and played down public criticism. Soviet missions in Washington and London were permitted to circulate newsletters and books presenting the Soviet view. In Britain *Soviet War News* sold over fifty thousand of each edition.[51] The British Ministry of Information published during the war a special manual providing journalists with 'Arguments to Counter the Ideological Fear of Bolshevism'. The first suggestion was that the 'Red Terror' should be portrayed as a figment of the Nazi imagination, a mirror-image of German behaviour; the publication went on to recommend that British propaganda should then construct 'a positive picture of Russia' – the patriotism, the contributions of Soviet scientists and artists to knowledge and culture, the increased encouragement of savings and possession of personal property, and the improved Soviet attitude to religion. This last point was to be used 'only in addressing people already sympathetic to the USSR on other grounds'.[52] These guidelines proved too much for George Orwell, who gave up his weekly war commentaries for the BBC in 1943 to write his powerful satire on Soviet life, *Animal Farm*. The Ministry of Information banned its publication until after the war. Then Orwell began work on *Nineteen Eighty-four,* depicting a world in which history is falsified, truth becomes untruth.[53]

To strengthen their positive stance both western Allies played on the promise of a new world after the war in which all states, including the Soviet Union, would play a part in securing international peace and cooperation. The war was presented as a crusade for the defeat of tyranny and aggression, and not a new wave of conquest and empire-building. The declaration of the so-called United Nations in January 1942, to which 49 countries had subscribed by 1945, was a public expression of world opinion against the Axis New Orders, a reassertion of public morality in dealings at home and abroad. The Soviet Union found itself pledging to preserve 'life, liberty, independence and religious freedom' and to observe 'human rights and justice'. Since the document was also signed by Poland, Yugoslavia, Haiti, Bolivia, Abyssinia, China, Cuba and Persia, the concept of human rights and justice was interpreted loosely at best. What mattered at the time was not so much the moral credentials of the signatories, but the sense conveyed in the document that those who signed it walked on the side of the angels against 'savage and brutal forces seeking to subjugate the world . . .'[54]

The moral ambiguities in the Allied coalition were never strong enough to undermine the image of a righteous cause. The wartime consensus was built around simple shared aims. There were no deep conflicts about the main war aims, or about the necessity of fighting the war to a victorious conclusion. With language of liberation, freedom and reconstruction, the Allies developed a positive moral outlook which worked in countless ways to keep people fighting and labouring with the promise of better to come. This helps to explain why from early on in the conflict, well before victory was even a remote prospect, the certainty of a just cause promoted confidence in Allied victory. 'I am absolutely convinced', wrote the German novelist Thomas Mann, from his exile in the United States in October 1941, 'that Hitler's game is up and that he will be destroyed – no matter how many detours and how much unnecessary effort is required to complete the work . . .'[55] Churchill's private secretary wrote as early as February 1941: 'I am confident that we have won. We shall see much serious damage and undergo many trials and dangers . . . the ultimate issue cannot be in doubt.'[56] Stalin in November that year argued that the 'moral degradation' of the German invader made their final rout 'inevitable'. When a few weeks later General Wedemeyer observed the British Chiefs-of-Staff in Washington discussing future strategy with their opposite numbers he was struck by how little affected they seemed by the desperate situation facing the Allies: 'There was nothing in their demeanour to reveal concern or doubt about final victory.'[57]

*   *   *

On the Axis side the war was fought with much less moral certainty or popular commitment. At the outset of the conflict there was no clear consensus in favour of war, and a great deal of evident misgiving. When the American journalist William Shirer heard the news of the British declaration of war on Germany on 3 September 1939, he was standing among a crowd of 250 Berliners. He watched them listen carefully to the announcement from the street loudspeakers. 'When it was finished', he wrote in his diary, 'there was not a murmur. They just stood there . . . Stunned.' All that day he observed only 'astonishment and depression' on the faces all around him.[58] On the day that Italy entered the war, 10 June 1940, Mussolini addressed a dispirited crowd in front of the Palazzo Venezia. The news aroused little enthusiasm. In his diary for that day the Italian Foreign Minister, Count Ciano, who had opposed running the risk of war with the west, expressed nothing but profound regret at the decision to fight: 'I am sad, very sad. May God help Italy!'[59] In Japan, which had been at war with China for ten years, the news of war with the US was greeted, according to one witness, the journalist Kazuo Kawai, with 'indifference and shock'.[60]

Aggressive war was not a popular choice in any of the three Axis populations. It was the aim of narrow sectional interests. In Germany and Italy war was declared because of the ambitions of two dictators, who carried their country into conflict against the strong advice of political colleagues and military leaders. In Japan war was promoted by the military elites, who persuaded the Emperor, against a chorus of civilian protest, that war was an unavoidable necessity. All bar two of the elder statesmen, or *jushin*, summoned to meet the Emperor late in November 1941, counselled peace and negotiation. Not even the military were confident that they could win outright a war against the United States; the risk of war was taken in the hope that Japanese forces could make the cost of reconquering lost territories too high for America, and as a result arrive at a negotiated settlement favourable to Japan. Once war broke out there was a patriotic response in all three states, though nothing quite like the drive for retribution and justice, the unity of purpose, that animated the populations of their enemy. Enthusiasm for war was least evident in Italy; in July 1943

Mussolini was overthrown by the army, and peace was agreed with the Allies before the final defeat of Italian forces. In Germany and Japan morale never collapsed completely, but the willingness to fight over the last two years of war was only maintained with large doses of propaganda and terror.

Popular morale in Japan was already subdued when war broke out after the draining years of warfare in China. The early victories brought about a sudden revival of enthusiasm, but it was short-lived, as conditions on the home front rapidly deteriorated. The defeat at Midway, though represented at home as a great victory, brought home to a great many officers that the war could not be won.[61] The ordinary soldier and civilian knew little in general about the true course of the war. Censorship stifled all forms of communication and was enforced with a brutal thoroughness. Official propaganda turned every defeat into victory. In 1943 the army invented a new verb, *tenshin*, to march elsewhere, in order to avoid having to say retreat.[62] Little effort was made to tell people what the war was actually for. Instead it became an opportunity to rally round the Emperor. Loyalty to the sacred ruler and to *'yamato damashii* – Japan's divine racial spirit – was summoned up to encourage a spirit of sacrifice and endeavour. When American pollsters explored Japanese morale after the war they found that almost half those asked believed that Japan's spiritual values represented the country's greatest source of strength.[63] On the eve of war Admiral Ugaki, who became Commander-in-Chief of the 5th air fleet by 1945, wrote in his diary that the coming conflict was 'sacred', a war in which the highest honour was 'to die as martyrs for our empire' and whose simple purpose was to display 'single minded loyalty to His Majesty'.[64]

Religion mattered a good deal more in the Japanese war effort than in that of any other combatant power. The Emperor had a divine status in the eyes of the population. To die in battle for the Emperor was to die a holy death. The ashes of each dead soldier were solemnly returned to his family in a brief religious ritual. Even in educated Japanese society the motto 'We are guarded by the gods above' was taken with great seriousness. Propaganda made a great deal of Japan's 2,600 years without a defeat. Divine prov-idence was assumed to be the reason, The Japanese population expected the gods to intervene quite literally in the course of the war. A doctor from Hiroshima explained to American interrogators after the war what this meant in practice: 'There is a big difference between the way they think about Christ in Europe and America, and the way the Japanese think about *kamisama* [gods] . . . This is one big place where Japan falls below America. It is all right to believe in the gods, but it is pure foolishness to think that the gods will help you out of holes like this.'[65] The high levels of sacrifice displayed by Japanese troops were sustained by religious belief. Each officer cadet had to learn by heart in the first three days of training the 27,000 sacred words of the Emperor on the duties of a soldier. The suicide charge on the battlefield, the refusal to surrender, the stark fear of dishonour were instilled in all Japanese. In the Pacific war almost half of all the soldiers committed by Japan to the island battles were killed, a proportion of loss that eclipsed even the Soviet figures.[66]

The popular emphasis on spiritual armament was supplemented by a good deal of coercion. The civilian workers were allowed no independent organisation but all belonged to the single Industry Patriotic Society. Factory police *(kempei)* were posted in every war plant, listening out for dissent or complaint, punishing offences on the spot in front of the workforce, or removing offenders into a gruelling police custody. Everywhere police *agents provocateurs* worked to eradicate defeatism by deliberately prompting comments on the futility of war, or the brutality or misgovernment on the home front, and punishing those foolish enough to loosen their tongues. Military police attacked and on occasion killed those suspected of pacifist sentiment. The military tapped the telephones of their civilian colleagues, and harassed and threatened ministers and officials who were less than wholeheartedly committed to the war.[67] By 1943 a negative outlook was more and more common. The civilian population, restricted to slim food rations, poorly informed about the state of the war, brow-beaten by police and military both in the workplace and through the thousands of 'neighbourhood associations' established throughout Japanese society to keep it patriotic, became progressively disillusioned with the war effort, and with the military leadership. After Midway it was widely recognised by Japan's rulers that Japanese victory could come only in the wake of German success. In February 1943 the officials of the Japanese Foreign Office met to discuss Japanese options after Stalingrad. They guessed, rightly, that Germany's new offensive in 1943 would fail; seeing little chance of a German victory, they proposed that Japan should 'reorient her policy' towards a peace settlement with America. At the end of March 1943 Emperor Hirohito himself expressed the wish that the war be ended without delay. The military leadership was unmoved. But by the spring of 1944 even they could see the writing on the wall. A commission set up under Rear-Admiral Takagi reported in February 1944 that Japan could not possibly win the war and should seek a compromise peace.[68]

From 1943 onwards Japan's war effort kept going at the insistence of the military hardliners for whom surrender, even a negotiated surrender, was anathema. Ordinary Japanese citizens were aware of the wide gulf that had opened between the crude propaganda of victory and the reality of the war, particularly after the onset of bombing. 'The government kept telling us that we would defeat the United States,' complained one Japanese after the war, 'but as my house was burned down and I had no food, clothing or shelter, I didn't know how I could go on.'[69] In 1944 it was impossible to disguise the reality of the war. The Japanese loss of Saipan, where only a thousand Japanese survived out of the 32,000 on the island, was reported by the General Staff nine days later. 'We were staggered at this dreadful news,' wrote a trainee pilot. 'It was obvious that Japan had no hope at all of regaining supremacy on the sea or in the air.' Bombing brought the reality home. One-third of the urban population lost their homes and possessions; over eight million were evacuated; two-fifths of the industrial workforce was absent from work for more than two weeks during 1945.[70] Little attempt was made to persuade the population that there was a moral purpose in continued belligerency beyond the fear of dishonour. Waverers were victimised and terrorised, but behind the scenes Japanese politicians tried to find a way of ending the war that would satisfy the Allies and the die-hard militarists at the same time. The Japanese war effort was riddled with moral ambiguity. Behind a façade of national unity and confidence in victory, both leaders and led understood that the war was effectively lost. The American postwar study of Japanese morale found that by 1945 68 per cent were convinced the war was lost. Only 28 per cent were willing to continue fighting and embrace death rather than dishonour.[71]

The German case was every bit as ambiguous. The German population was little prepared for the conflict when it came. The early victories provided a moral boost, as they did for the Japanese. But in Germany, too, all information was carefully controlled and information distorted to mask the reality of war. Again there was no clear indication from the authorities about the nature of war aims, or the moral purpose behind the conflict. The war against France and Britain was presented as a revival of the 1914 war against Germany's restrictive 'encirclement' by other European powers. Victory in 1940 was hailed as an end to the hated Versailles system, and revenge for the humiliation of 1918. To justify the war against the Soviet Union, the regime returned to the 1930s propaganda of Bolshevik menace. The population found the sudden escalation of the war difficult to accept. A security service report showed the initial reaction as one of 'bewilderment' leavened with 'sober confidence'. Even Hitler's propaganda supremo Joseph Goebbels found it difficult to switch on hatred for the new enemy with real plausibility. The party line was to stress 'the treachery of Bolshevist leaders'. As the campaign dragged on into the winter of 1941 there was every evidence that popular confidence in the war effort was ebbing, not least, as Goebbels told his own staff, because the official propaganda made no mention of reverses and was simply not believed.[72]

For Hitler himself the war with the Soviet Union was the one he had been waiting for throughout his political career. Though there were practical reasons for the campaign – to deny Britain an ally on the Continent, and to forestall any Soviet moves in eastern Europe – Hitler's main aim, as he told Goebbels, was finally to eliminate 'the Bolshevik poison' from Europe. Hitler made no pretence that he was fighting a virtuous war, despite the strong moral antipathy to communism evident throughout Europe. Three days before Barbarossa was launched Hitler had a revealing discussion with Goebbels, later recorded in the minister's diary. 'Right or wrong', Goebbels noted, 'we must win . . . And once we have won, who is going to question our methods? In any case, we have so much to answer for already that we must win . . .'[73] Hitler gave deliberate instructions that the campaign should be waged with unremitting brutality. The harsh treatment of populations in the east began with the invasion of Poland in 1939, when the murder of Polish intellectuals and leaders, and Polish Jews, had been sanctioned by the regime. The legitimisation of savagery developed its own momentum. Army discipline deteriorated during the Polish campaign. Efforts by regular soldiers to prevent the brutalities were swept aside by Himmler's army of officials and policemen. Atrocity was permitted in the name of the 'higher' law of racial survival.

Hitler revelled in his rejection of conventional morality. He poured scorn on what he termed the 'beatific liberalism' of the west.[74] The war with the Soviet Union was not only a war of ideologies, but also a struggle for survival, a conflict of nature. Before the campaign the old laws of war were torn up. The army was issued with the 'Commissar Order' which permitted the murder of any Communist Party functionary found with the Red Army. In June 1941 the German army was freed by the regime from any restrictions under the 1899 Hague rules on the conduct of land warfare. The security services under Himmler were prepared to follow the armies into the Soviet Union specifically to murder anyone defined as an enemy of Germandom. Himmler asked his forces to behave with merciless violence against the races of the east who know 'with animal instinct why they are fighting'. In July 1941 Himmler's instructions to the security forces were to act against any populations defined as potentially anti-German or racially inferior by shooting indiscriminately all males, deporting the women and children, seizing food and valuables and burning the villages to the ground. Hitler sanctioned even the murder of women and children if it served his principle of preserving at all costs the lives of German soldiers.[75]

At every level, from the chief of state down to individual army units in the field, the war was fought as a racial conflict of the most savage kind in which any methods, criminal or otherwise, were sanctioned. The armed forces were given a general amnesty before the campaign for anyone guilty of murder or looting. German commanders accepted the criminalisation of warfare because of the special nature of the enemy they believed they faced. In June 1941 the armed forces' 'Information for the Troops' backed up the order to murder commissars: 'anyone who has once looked into the face of a Red Commissar knows what Bolsheviks are . . . It would be insulting animals if you described those mostly Jewish features as animal-like.' Not surprisingly the ordinary soldiers soaked up the constant dehumanisation and demonisation of the enemy. 'Hardly ever do you see the face of a person who seems rational and intelligent,' ran one letter from the front; 'these sons of the steppe, poisoned and drunk with a destructive potion, these incited sub-humans . . .' ran another; 'We have seen the true face of Bolshevism . . . communist scoundrels, Jews and criminals.'[76]

The result of the indoctrination of the troops with the image of a bestial enemy was to give licence to a wave of barbarisation in the east that could not be controlled by those soldiers or officials with more scruples or humanity. Many were outraged by it, but the war soon developed its own savage codes on both sides which nothing could reverse. Millions of Soviet prisoners died in the first winter of the campaign from studied neglect. Thousands more soldiers were simply shot down as they found themselves behind the lines of their rapidly advancing enemy. Within weeks of arrival at the front-line troops or police with no previous experience of violence or crime became infected with the barbarous virus. Group solidarity explained some of it. One unit of ordinary police from Hamburg assigned to the routine shooting of Jews in the east produced a small number of dissenters. When interrogated after the war none of them expressed moral revulsion at what they had been asked to do, but they did reveal a deep sense of shame for letting down their fellows, who had had to do their dirty work for them.[77] The lead was taken from the top. Hitler's utter lack of conscience, his view of war as a lawless state of nature, his moral detachment which always succeeded in making the victim seem the perpetrator, set the tone throughout the whole war effort. The war was never presented as a moral crusade. 'We shall not place too much emphasis on fighting "for Christianity",' Goebbels wrote in his diary on the second day of the campaign. 'That would, after all, be just too hypocritical.'[78]

The law of the jungle might have assisted German fighting spirit in the Soviet campaign, but its moral effects were otherwise entirely negative. The criminalisation of warfare produced a growing indiscipline and demoralisation among German forces themselves. The German armed forces condemned to death 22,000 of their own men and executed between 15 and 20,000, equivalent to more than a whole army division.[79] A further 23,000 were sentenced to long prison terms, and another 404,000 to shorter periods in prison or penal battalions. As a proportion of total mobilised manpower, these figures were higher than they were for the Red Army, 3.3 per cent against an estimated 1.25 per cent. Desertion or refusal to obey orders increased as the war went on, and the law of the jungle seeped into the military structure itself. The struggle for survival had a remorseless logic. The regime imposed ever more draconian terror on its own forces to keep them fighting until the very end of the war when Hitler, amidst the dying embers of his Reich, ordered any saboteur or deserter shot on the spot.[80]

The effect of Germany's conduct of war in the east on the rest of world opinion was bleak. The Allies were able to stoke up the fires of moral indignation almost effortlessly with the string of well-attested atrocities laid at Germany's door. Though German allies and sympathisers – Italy, Spain, Hungary, Romania, Slovakia – sent troops to help fight the Bolshevik threat, their treatment at German hands was arrogant and discriminatory. Germany was feared and hated by most of Europe, and everything it did in the Soviet Union reinforced this image, even among those non-Russian nationalities who had at first welcomed the German armies as liberators from Russian-dominated communism. In the occupied territories German apparatchiks became a byword for criminality and violence – their rule was harsh in the extreme, their economic policies a mixture of looting and exploitation. This was not true of all German officials, and neither was barbarism practised by every German soldier, but the dominant image abroad was dictated by those who did thrive on crime and vice. The Spanish ambassador to London told the Japanese ambassador in Madrid in December 1942 that the United Nations were 'utterly certain of whipping the Axis'. The Germans, he thought, ran their war effort mechanically, inflexibly and 'when it comes to diplomacy' he considered 'their heads are as hard as lead'. 'There is not a single country', he continued, 'which in its heart is following the Germans. France, Belgium, Holland, all hate the Germans . . .' Neither Hitler nor his entourage were worried by the state of world opinion in 1942, but their outlook unquestionably gave the moral field to the Allies. 'If we win,' Goebbels remarked in his diary, 'we shall have right on our side.'[81]

At home the moral bankruptcy of the eastern campaign stimulated the conscientious rejection of the regime by sections of German society, and brought more officers, horrified by what they had experienced in the east, into the German opposition. Many were drawn from the upper reaches of German society, recruited from field-marshals and generals, diplomats and senior officials. They were united by their detestation of Hitler. Twice, in 1938 and in 1939, the leading opponents had considered a coup d'etat, but had lost their nerve at the last moment. The war against the Soviet Union they regarded as a disaster: 'a frightful, senseless and unfathomable war', wrote one of their number in a diary.[82] The resistance worked at the very heart of the German war effort. It included the head of German counter-intelligence, Admiral Wilhelm Canaris; there were a circle of opponents in the Foreign Ministry; in the Air Ministry there operated until 1942 the largest communist spy-ring in Germany, the Red Orchestra; even General Halder, the first army Chief-of-Staff under Hitler's Supreme Command, was counted among the military opponents of the regime. The roll-call of prominent Germans from all walks of life who opposed Hitler's war and the immorality of the regime revealed the extent to which the war effort lacked any broad base of public support. From resistance circles came a long run of peace feelers to the west, searching for some way of both ending the war and destroying Hitler with Allied cooperation. Neither proved possible. The west distrusted the motives of many of the conservative Germans who approached them; by 1943 the three Allies were committed to unconditional surrender, which the resistance could not deliver. The destruction of Hitler was frustrated time and again by the qualms of conscience or political caution of his domestic enemies, and also by exceptional bad luck. There were no fewer than 42 failed attempts on Hitler's life.[83]

Support for the war effort among the German population as a

whole began to decline steadily from 1941. The failure to clinch defeat of the Soviet Union in 1941, when Hitler had confidently claimed victory in October, reflected poorly on the credibility of Hitler's promises. In a postwar survey of German wartime morale it was found that 38 per cent of the respondents thought the war was already lost by January 1942. In April 1942 Hitler gave one of his last major speeches. It was not a good performance. His demands for greater sacrifice from a home front already short of food and fuel, his hints of another winter campaign in the Soviet Union, struck the wrong note. Goebbels thought it sounded like the 'cry of a drowning man', and insisted that German propaganda at home should be couched in more realistic, even pessimistic, tones to prepare people for a harsh struggle.[84] By the autumn deep pessimism needed little fabrication. Through the winter of 1942, during the long slugging match at Stalingrad, the residual confidence of the German people in the prospect of victory slowly evaporated. News of the Soviet counter-attack seeped out in Berlin through the official wall of silence. The heavy lossess could not be disguised. An air of deep gloom descended on the German population. Off the record Goebbels told a group of foreign correspondents in December that 'only a hair's breadth separates us from the abyss'.[85]

The defeat at Stalingrad brought about a real moral crisis in the German war effort. The mood in the capital was one of sorrow and exasperation. Rumours abounded that Germany might sue for an armistice. One neutral observer noticed in the mood of 'desperation' and 'anxiety' an open willingness to blame Hitler for what had gone wrong.[86] The regime made a virtue of necessity and used the crisis as an opportunity to shift the moral ground of the conflict from uncritical confidence in victory to a sombre defence of the homeland against the barbaric Bolshevik threat. Goebbels was the inspiration behind the change. Hitler disliked the idea of a war of defence, since it smacked of weakness, but he accepted Goebbels's suggestion that moral justification should now be based on the idea of a life or death struggle between European civilisation, shielded by Germany, and Asiatic barbarism. Goebbels thought the message should be uncompromising so that even non-Nazis could see that 'everybody will have his throat cut if we are defeated'. The propaganda was never quite so blunt, but Goebbels relayed to the nation on 30 January 1943, in a speech in the Berlin Sportpalast, the crux of Hitler's new realism: 'In this war there will be neither victors nor vanquished, but only survivors and annihilated.'[87] The issue was no longer a triumphant war of imperial conquest, but the survival of the German people. No one could fail to see that the new language suggested a nation under siege. The Party took up the theme of Europe's final defence against the east, inverting the reality of the war entirely, but in the process succeeding to some extent in rallying forces more willing to defend the fatherland than to conquer living-space. Confidence in eventual victory, on the other hand, continued to decline. The postwar survey of morale found that by January 1944 77 per cent of the sample regarded the war as lost. By early 1944, the diplomat Ulrich von Hassell wrote in his diary that 'anxiety and horror hold sway'. Hitler, he thought, had caused nothing but 'spiritual confusion and moral deterioration'. A secret police report from March 1944 observed that morale had reached the lowest point since the war began.[88]

It was against this background of moral decline that the German resistance took the decision that the only way to end the war was to kill Hitler. This was not a simple choice. For the soldiers of the opposition it meant a betrayal of their oath of loyalty, and mutiny against their Supreme Commander. For all the assassins, plotting Hitler's death was high treason. During the autumn of 1942 the resisters considered carefully the justification for so radical a step. They appealed to a higher morality. They searched for historical precedents. They yielded to the argument that their duty to the fatherland was greater than their duty to any one individual, especially one whose orders were manifestly criminal. 'If ever in history an assassination was justifiable', wrote one of the few survivors of the plot, 'this was one.' The soldiers who were willing to break their oath of allegiance did so in the belief that they acted in accord with 'the highest standards of ethics, morality and patriotism'.[89]

The real problem was not squaring their consciences but doing the deed. Men who moved whole armies around on a battlefield found the job of murdering one man almost impossibly complex.

They discussed shooting him at the daily conference, but thought the risk of failure too high. Baron Georg von Boeselager offered to storm Hitler's headquarters with his whole regiment but was never posted near enough to carry out the threat. Finally it was decided that a time bomb was the answer. This brought new difficulties. German explosives were not well suited to assassination. The fuses available made a hissing noise. The plotters found that the best material was British plastic explosive, triggered by British-made timing devices. The army had a quantity of both, picked up from supplies parachuted by British planes to agents in Europe. The bomb was disguised as two bottles of Cointreau, a present for someone at Hitler's headquarters. On 13 March 1943 General Henning von Treschkow and Fabian von Schlabrendorff arrived at von Kluge's headquarters in Smolensk for a meeting with Hitler. The Führer arrived surrounded by bodyguards. At lunch, prepared by his personal cook, Hitler's physician tasted every course first. At the airport, following the conference, the parcel of liqueur was handed to one of Hitler's staff, the fuse primed. Hitler's aircraft, with his armour-plated cabin and seat fitted with a parachute, had been divided into compartments to reduce the effects of bomb blast; it flew off to East Prussia surrounded by a shield of fighter planes. The bomb failed to detonate. Before it had even been discovered, the conspirators recovered the package the following day on the pretext that the wrong gift had been sent.[90]

Another fifteen months passed before a second serious attempt. This time the plotters enjoyed an even wider circle of sympathisers among senior military officers; the assassination was planned as part of a general coup d'etat which would, it was hoped, precipitate an end to the war. The assassination was to be masterminded by an outstanding young staff officer, Colonel Count Claus Schenk von Stauffenberg. From a devout Catholic background, Stauffenberg was an unconventional soldier. In his youth he was brought under the spell of the German symbolist poet Stefan George, whose spiritual rejection of the modern age and its soulless materialism, his mystical appeal to the eternal values of German culture, linked the young soldier with traditions of chivalric behaviour and patriotic duty. He came to see the killing of Hitler as an act of spiritual redemption, St George slaying the dragon.[91]

He was none the less not an ideal assassin. He had been badly wounded in North Africa, where he lost an eye, his right hand and two fingers of his left hand. He refused to abandon his military calling, and a few months after his injuries he returned to active duty in Germany as Chief-of-Staff for the General Army Office in Berlin, organising recruitment and training. While he helped to coordinate the plans for a coup, the resistance attempted again and again to kill Hitler. A second consignment of the British explosive, shipped to East Prussia, mysteriously exploded in its hiding-place. Then it was planned to use a display of new uniforms as the opportunity for the assassination. A soldier modelling the new kit, primed with explosives, was to throw himself on the Führer and detonate the bomb, killing them both. Though the event was scheduled three times, Hitler never had time to inspect the new outfits. Finally in desperation it was agreed that Hitler should be lured to army headquarters in Russia, and there, like Caesar in the Senate, be done to death by the circle of men around him. But Hitler never went.

The conspirators became increasingly disillusioned. Some argued that they might just as well wait for the Allies to finish the job for them. Finally von Stauffenberg undertook to do the deed himself. Twice in July 1944 he arrived at conferences carrying in his briefcase a time bomb which he had learned to prime with his remaining three fingers. At the first Himmler was not present and Stauffenberg preferred to wait until he could kill him too. At the second meeting Hitler left before the bomb could be detonated. Finally on 20 July at Hitler's East Prussian headquarters at Rastenberg Stauffenberg succeeded in smuggling his bomb through the three security zones surrounding the buildings, and into the conference room itself. Instead of the meeting being held in the usual heavy concrete bunker, where the blast would have been deadly, it was held that day in a small wooden building. The bomb was primed, and Stauffenberg slipped it under the large oak map table, inches from Hitler. He left the conference to answer a fictitious telephone call, too soon to notice another officer, having bumped his legs on the briefcase, push it farther under the table, behind the large oak support. When the bomb exploded the hut was torn apart. Stauffenberg, crouched in hiding outside, saw bodies hurled into the air. Satisfied that Hitler must be dead he bluffed his way through the security net, and flew to Berlin to seize power from the Party. The bomb blast had not done the job. The thin wooden walls released the impact, throwing most of those present out of the room. Four people were killed, but Hitler was shielded by the thick oak table. He emerged shaken and grazed, his trousers torn to shreds, consumed by a livid rage against the traitors. Within hours of arriving in Berlin Stauffenberg was arrested by fellow officers and executed in the courtyard of the War Ministry.[92]

The failure of the assassination attempt on 20 July had dire consequences for the opposition to the war. Hitler took the opportunity to exact a savage revenge. Thousands of senior soldiers, ministers, and officials were rounded up and subjected to torture, imprisonment and, for many, a grisly death. Confessions were extracted with all the apparatus of a mediaeval dungeon – thumbscrews, the rack, the iron maiden. The heart was torn out of the resistance. The Party and all those who followed willingly in its wake reaffirmed their fanatical loyalty to their leader; even those hostile to the regime, and conscious of Germany's imminent defeat, were forced into line. Himmler's SS were unleashed in a wave of lawlessness across Germany. Any hint of defeatism or demoralisation was punished with a brutal indifference. Between July 1944 and the end of the war, the full apparatus of terror and barbarism practised in the east was turned on the German people. Caught between a remorseless enemy and a vicious despotism the population continued to fight and work in conditions of increasing desperation. All but a fraction accepted that the war was lost. Postwar findings suggested that by the end of 1944 almost three-quarters of Germans wanted to give up the war at once. The proportion willing to fight to the very end, 29 per cent, may not all have been Nazis, but the figure was close to Hitler's share of the vote in the last free elections, in November 1932.[93]

In both Germany and Japan confidence in victory ebbed away long before the final battles. In both states important sections of the ruling class were hostile to the war. Popular enthusiasm was muted. A hard core of committed supporters kept the war effort going despite the widespread demoralisation on the home front. Soldiers and civilians alike were subjected to ever harsher discipline to keep them fighting and working in conditions of terrible hardship. Yet both states continued their defence right up to the point where further resistance was no longer possible. As the end drew nearer both peoples were consumed with a dread at what might happen to them after their defeat. In Japan the official line made it clear that the enemy was bent on annihilation and oppression. When Japanese were asked after the war what they expected the Americans to do to them, more than two-thirds believed they would suffer starvation, enslavement or annihilation, and only 4 per cent expected to be treated humanely. One munitions worker explained that at school young Japanese were told that the Americans would torture and murder them: 'I thought it would be better to be dead than captured.'[94] In Germany the war effort in 1944 and 1945 owed a good deal to popular dread, fuelled by the propaganda machine and rumours from the east, of what would happen when the Soviets arrived. A mirror image of German atrocity was projected on to the advancing enemy. Most anti-Nazis were united in the view that they did not want to be ruled by Stalin any more than by Hitler.

The effect of being caught between terror on the home front and a terrible enemy was to produce a pronounced fatalism among both peoples. As the end neared there developed a do-or-die mentality among the troops. Japanese society prepared for the last stand, or *kessen*, when thousands of their number were to perish in suicidal attacks on the invading enemy. The military wanted to fight with everything to hand, to die with honour. Faithful to his view that the war was a sacred conflict, Admiral Ugaki refused to accept Japan's surrender. On the day it was announced, he arrived at an air squadron headquarters, commandeered an aircraft, and set off to crash into the American fleet at Okinawa. He left a suicide note behind: 'I am going to ram into the arrogant American ships, displaying the real spirit of a Japanese warrior.'[95] A young German student later recalled the odd air of calmness in the young men she knew preparing to leave for the final battles of the war: 'There was nothing morbid about the way they accepted their fate, although none of them wanted to die for one man's insanity. When they left, they knew that soon . . . they would be killed, and that the war was already lost for Germany.' They were sustained by the view that their death would in some sense atone for the cruelty and folly all around them. Not one of them survived.[96] While German soldiers came to terms with the reality of death and defeat, their masters prepared to abandon them. Thousands of Nazi functionaries began to flee the threatened areas of Germany. When the war ended thousands more committed suicide. Among their number was Heinrich Himmler, the man whose empire of genocide and terror had turned Germany into a moral desert. He disguised himself in a police sergeant's uniform and took the papers of another sergeant shot for 'defeatism'; his familiar moustache was shaved off and he wore a black patch over one eye. Two weeks after the end of the war he was caught at a British checkpoint, though no one at first recognised who he was. A few days later he declared his identity. In the middle of a strip search by British interrogators he bit on a phial of cyanide and could not be revived.[97]

*   *   *

The Allies' moral coalition lived on after the war. The remaining Nazi Party leaders and military chiefs, together with a host of lesser officials, soldiers and businessmen, were taken into custody to await trial as war criminals. The decision to indict the leaders of the Nazi state was taken late in the war. Up to the last months the predominant view was in favour of summary execution by military firing squad, an idea proposed by the British. To their surprise Stalin strongly opposed the suggestion, on the grounds that the Allies would be accused of being afraid to give their enemies a fair trial. Roosevelt did not reject the idea of treating German leaders harshly, even of finishing the job swiftly in 'kangaroo courts'. But his successor as President, Harold Truman, was horrified by the suggestion that a liberal state should be engaged in lawless killing. In May 1945 he insisted that war criminals should be brought before an international tribunal, to answer for their crimes at the bar of world opinion.

This was easier said than done. There were arguments about who was, and who was not, a war criminal. There was widespread unease at the absence of any precedent in international law – bar the exile of Napoleon to St Helena – for formally imposing the victors' justice on the vanquished. There was the vexed question of what precisely the leading war criminals could be accused of. The decision to indict them for 'Crimes against Peace' and 'Crimes against Humanity' was regarded in some quarters as a mockery as long as Soviet judges sat on the bench, while its legal propriety was clearly questionable. The trials finally opened at Nuremberg, spiritual home of the Nazi movement, on 20 November 1945. The opening statement by the American justice, Robert H. Jackson, indicated the wider moral purpose of the trials, which was nothing less than to set on record, for all the world to see, the contrast between 'imperilled civilisation' and the evil cause she had fought: 'Against their opponents . . . the Nazis directed such a campaign of arrogance, brutality and annihilation as the world has not witnessed since the pre-Christian era . . .'[98]

The tribunal took nine months to demonstrate the justness of the Allied cause. The Soviet judges treated the occasion like a Stalinist show trial, bullying and hectoring the defendants. At a dinner in honour of the Soviet deputy Foreign Minister, Andrei Vishinsky, who had been Stalin's chief prosecutor during the Moscow show trials of 1936–8, proposed a toast to the defendants: 'May their paths lead straight from the courthouse to the grave!'[99] The trials gave more opportunity than their instigators could ever have intended for the chief war criminals to argue their side of the case. The prisoners reacted to their moral indictment in different ways. Some showed no remorse. Göring stood by everything he had done and tried to force the other prisoners to do the same. Speer on the other hand admitted his guilt and that of all those beguiled by the system, a confession that probably saved his life. In fact most of them, when brought unavoidably face to face with what they or their companions had done, were shocked or stunned by the realisation.

Early on in the trial, on 29 November, the defendants were shown a film taken by American forces of the liberated concentration camps. A psychologist was posted at either end of the dock to note the reaction of the prisoners. Even allowing for calculated expressions of remorse, the reactions are worth recalling: 'looks pale and sits aghast . . . has head bowed, doesn't look . . . covers his eyes, looks as if he is in agony . . .blinks eyes, trying to stifle tears . . . Goering looks sad . . . Doenitz has head buried in his hands . . . Keitel now hanging head . . .' The film pricked all but the coldest conscience. When the psychologists visited the cells that same evening, many of the prisoners were still in shock, most were horrified and shamed by what they had witnessed. Hans Frank, the Nazi ruler of wartime Poland, burst into a sobbing rage when asked about the film: 'Don't let anybody tell you that they had no idea! Everybody sensed that there was something horribly wrong with this system. To think that we lived like kings and believed in that beast . . .'[100] From within the wretched remnants of Hitler's elite there surfaced, in varying degrees, a recognition of the immoral character of the regime they had served.

The history of the Nuremberg Tribunal exemplifies the moral contrast between the two sides (and the awkward morality behind a victorious coalition of democratic and communist powers). The indictments were an extension of the Allied conviction that they had fought a just war against aggression and barbarism. The justness was demonstrated by the fact that the victors had not, as Justice Jackson put it, exacted immediate vengeance while 'flushed with victory and stung with injury', but had submitted their case to the due process of law. The same procedure was adopted when Japan's leaders were brought before a second International Tribunal, at which a catalogue of appalling atrocities against civilians and soldiers was paraded in horrifying detail. The revelations in both trials confirmed the picture created during the war to sustain the Allied war effort, of primitive savages in the east, and devious barbarians in Europe. This image had both simplified and strengthened the Allied cause. During the war, hatred of Hitlerism papered over the deep cracks in the Allies' own coalition of interests and ideologies, and it continued to do so, if falteringly, during the trials. It was a hatred that had sustained the most significant moral effort of the war, the mobilisation of the Soviet will to win. Whatever the rights and wrongs of the Allied cause, the belief that they fought on the side of righteousness equipped them with powerful moral armament.

There were many on the Axis side who would have agreed. War was not widely welcomed, nor were its purposes understood. Popular propaganda was distrusted. A hard core of enthusiasts saw the war as a way to impose a brazen 'new morality', rooted in racism, violence and enslavement. But many more continued to fight only through fear, or struggled, like the German resistance, to reassert a conventional morality. As the war deteriorated for the Axis states, the instruments of terror were turned on their own people and soldiers. They fought from sheer survival instinct, but the underlying moral dilemma of fighting an aggressive war in which brutalisation and atrocity had become routine was inescapable. The repeated efforts to murder Hitler revealed a system divided against itself, just as the wave of suicides at the war's end surely revealed uncomfortable consciences. Historians are loth to pronounce on moral issues, even where the balance of right and wrong seems clear-cut. But can there be any doubt that populations will fight with less effect in the service of an evil cause?

*'There has been no instance yet in the history
of wars of the enemy jumping into the abyss
of himself. To win a war one must lead the
enemy to the abyss and push him in to it.'*

Joseph Stalin, Order of the Day,
23 February 1944

As ALLIED ARMIES closed in for the kill in the spring of 1945, and German leaders urged their battered forces to stand and die like heroes, Hitler took time to reflect on why he had lost the war. His remarks were faithfully recorded by his indispensable secretary, Martin Bormann, who followed his leader with pad and pencil so that posterity would be denied none of Hitler's prophetic wisdom. There they sat as Germany crumbled around them, the thick-set, boorish stenographer, a dull sounding-board for his master, and Hitler, isolated, physically broken, consumed with hatred and self-pity, but clear-headed enough to look back over his years as warlord to see where he went wrong.

The start of his troubles Hitler traced back to the Munich crisis of 1938. He regretted his failure to keep his nerve and conquer Czechoslovakia in defiance of Britain and France. He was convinced that had he done so the west would have backed away, German domination of the Continent would have become fact, and the great war to the east could have been postponed until Germany was thoroughly prepared. He regretted his friendship with Mussolini: 'anything would have been better than having [Italians] as comrades in arms . . .'[1] Italy drew Hitler into the Mediterranean and the Balkans, when the Soviet Union was the priority. Looking back, Hitler realised that he should have attacked Stalin in May 1941, and won an extra five weeks of dry weather. Better still, he should not have fought a two-front war against Britain and the Soviet Union. He was forced to attack the Soviet Union because Britain's 'stupid chiefs' refused to make a sensible peace: perhaps, he reflected, he should have struck south, seized Gibraltar and swept into the Near East to smash British resistance.

But then, standing in the wings was Stalin, just waiting for the moment to strike.[2]

The remarkable thing about Hitler's reflections was how little blame attached to him. At every move it was other people, other forces, that compelled him to act. 'I, perhaps better than anyone else, can well imagine the torments suffered by Napoleon,' Hitler told Bormann towards the end of the dictated testament, 'longing, as he was, for the triumph of peace and yet compelled to continue waging war, without ceasing . . .'[3] Munich was the fault of Neville Chamberlain who 'really intended to wage ruthless war against us'; Hitler was let down by Mussolini, frustrated by Stalin, served by a German elite composed of feeble 'petty bourgeois reactionaries'. Above all German defeat was the work of the Jews, a refrain that echoes through Bormann's jotted notes. The war, Hitler believed, was 'typically . . . and exclusively Jewish'. It was sustained by the 'most powerful bastion' of world Jewry, the United States, whose President, 'the elect of the Jews', worked tirelessly, in Hitler's view, to keep war against Germany going. 'If we should lose this war' – and this in February 1945 – 'it will mean that we have been defeated by the Jews.'[4] Hitler was quite unable to grasp the extent of his own responsibility. Germany was a plaything for fate, doomed by the forces of world history to fight on 'until our last drop of blood has been shed'. Hitler thought the suffering would be redemptive, purifying, good for Germany. Out of the ashes of defeat, a new Reich would arise.

No one doubts that the war was ultimately Hitler's responsibility, or that Hitler made mistakes on a grand scale. In most postwar explanations of the outcome Hitler's failings stand at the head of the list. The story is a familiar one. German victories early in the war were the result of short, opportunistic campaigns against enemies who were weaker and isolated. In 1941 Hitler made the mistake of invading the Soviet Union in the belief that the tactics of 'lightning war' would bring victory in four months. In December 1941 Germany found herself at war with a combination of the three largest industrial economies outside Continental Europe, a war that Germany, allied to economically weak states, could never hope to win. Hitler's belief that a German super-power could tear up the political structure of Europe and western Asia and replace it with a Party-led authoritarian empire was always irrational and deluded.

Much of this argument comes with hindsight. The idea that the whole imperial enterprise was flawed from the outset is a postwar rationalisation. Moreover, Eastern Europe *was* dominated for forty years after the war by an authoritarian super-power, run by single-party dictatorships which denied civil rights and smothered society with secret policemen and a thick blanket of ideological conformism. The Soviet bloc lacked the wanton destructiveness and deadly racism that a Nazi empire would have displayed, but there was nothing deluded or irrational about the new system. The dominance of communism and the Red Army was achieved as a direct consequence of the power they had built up in the military defeat of Hitler's empire.

The assumption that German defeat was a result of fighting a 'two-front war' is also questionable. There is no necessary link between military defeat and fighting a two-front war. The United States fought a war on three fronts – five if the bombing offensive and the Battle of the Atlantic can be defined as fronts in their own right. All of those fronts competed with each other for resources of manpower, shipping and weaponry, and all bar the Atlantic were thousands of miles from the security of the home country, situated at the end of long and vulnerable shipping lanes. The Soviet Union was the only major combatant power to fight a one-front war, although for much of the critical central period of the war Germany too fought on one main front until the western Allies threw the full weight of their forces into France in the summer of 1944. For much of the First World War Germany survived a two-front war until the Russian war effort collapsed in 1917, but paradoxically she was defeated in a one-front war in 1918.

Clearly the fact of a two-front war is not an explanation for defeat as such. But is it any sounder to argue that Germany was over-whelmed by the economic size of the coalition assembled against her from 1942? This has always been a popular view. In 1946 the economist Raymond Goldsmith claimed that Gross Domestic Product won the war: the Allies simply had more of it than the Axis. Even during the war such a view was not uncommon. When Maxim Litvinov, deputy to the Commissar of Foreign Affairs, heard the list of American and British supplies read out at a meeting in Moscow in September 1941, he broke all the rules of Soviet negotiation, leapt out of his chair and shouted: 'Now we shall win the war!'[5] This was before American belligerency turned the United States into a fighting power rather than just so much inviting GDP.

The drawbacks in this argument have already been laid out, but they are worth a curtain call. Economic size as such does not explain the outcome of wars. China had on paper a large economic product in the 1930s, but it did not help to make China a significant warring state. If the explanation covers only the product of industrial powers then there remains the awkward evidence that Germany had greater industrial capacity than Britain in 1940, and access by 1941 to a good deal more than Britain and the Soviet Union together, and yet was unable to bring either power to defeat. And had Germany prevailed in Europe before 1942, could the United States really have used its larger GDP to reconquer the Old World? The balance of economic product explains everything and nothing. Political will, technical modernity, a popular willingness to accept sacrifice, the simple constraints of geography, these are just some of the many variables that affect the mobilisation of economic resources. The line between material resources and victory on the battlefield is anything but a straight one. The history of war is littered with examples of smaller, materially disadvantaged states defeating a larger, richer enemy. General Eisenhower, listening to politicians in Washington in the spring of 1942 talk glibly about the economic defeat of the Axis, observed in his diary that 'not one man in twenty in the government realises what a grisly, dirty, tough business we are in. They think we can buy victory.'[6]

There was no other way for the Allies to dislodge the Axis states from their conquests in 1942 than to defeat them on the battlefield. As Stalin put it, they were not going to jump into the abyss without being pushed. Some way might have been found of ending the war by negotiation, but this would certainly have meant making concessions to Axis imperialism. Fighting, and fighting better, was the only way to expel Germany and Italy from the European New Order, or to drive Japan from her new sphere of influence in Asia. Fighting power owed something in the long run to the large surplus of weapons available to the Allies, though in the critical battles of

1942 and 1943 that surplus was not as large as it became in 1944 and 1945 when Axis defeat was much more certain. In the first years of war the chief Allied states did not fight well, or were ill prepared for conflict. They were at a distinct disadvantage against Germany and Japan whose fighting skills prompted their rulers to risk war with industrially rich powers in the first place. Neither the Japanese nor German leaders rated Allied fighting power very highly, and they thought even less of it after their early successes. The Japanese military in the southern zone became over-confident. Rear-Admiral Takata remembered after the war the views he had heard: 'They said the Americans would never come, that they would not fight in the jungle, that they were not the kind of people who could stand warfare . . .'[7] Hitler formed the same dismissive view of the enemy. The first reports on American troops in North Africa suggested that they were simply 'rowdies' who would 'take to their heels very quickly'. Hitler thought America could never become 'the Rome of the future' with such poor spiritual stock.[8]

To win the war the Allies had to learn to fight more effectively, just as in the early 1800s the Coalition partners learned to tame Napoleon. They had to be prepared to fight together, and to continue to fight until the end of the conflict. Issues of morale and politics intimately affected the fighting power of the Allies, as did the strictly military elements of command, training, equipment and tactics. In Marshall's view the will to collaborate was the nub of the matter: 'In my opinion,' he told an audience at Yale in February 1944, 'the triumph over Germany in the coming months depends more on a complete accord between the British and American forces than it does on any other single factor, air power, ground power or naval power . . .'[9] Marshall's audience, who would have known nothing of the arguments between the Allies, must have puzzled at his cryptic remarks on the harmful effects of past 'discord', but unity of purpose and plan was not something to be taken for granted. It was always under strain: in the bombing offensive (where British and American air forces fought different campaigns, by day and by night); in the arguments about the route of re-entry to Continental Europe; in the tension between Soviet demands for a Second Front and western hesitation. It is no coincidence that Germany was defeated during the nine-month period when all three Allies, assisted by the exiled forces of the conquered European lands, put the main weight of their military effort together for the first time.

The Allies would have been the first to admit that fighting power was their real weakness. After the terrible defeats of the summer and autumn of 1941 the Soviet General Staff began a comprehensive review of what had gone wrong and right in Soviet practice. By March 1943 five volumes of critical analysis were published, covering everything from the use of tanks and aircraft to the laying of smoke-screens and the use of gunboats on Russia's rivers.[10] The fruits of Allied reflection have already been discussed, and need only a brief summary here. The reforms covered both the organisation of forces and their equipment and operational skills. The purpose was to achieve improvement in the qualitative performances of all Allied forces and technology, without which quantitative supremacy would have availed little. With better training and improved weapons the morale of Allied troops was also raised appreciably. Mistakes were still made, but the gap between the two sides narrowed in every sphere of combat.

The Allies did not depend on simple numbers for victory but on the quality of their technology and the fighting effectiveness of their forces. Axis forces did little to alter the basic pattern of their military organisation and operational practice, or to reform and modernise the way they made war. They were not under the same urgent pressure as their enemies and they responded more slowly to the sudden swing in the balance of fighting power evident in 1943. There is a deeper contrast here. In Germany and Japan much greater value was placed on operations and on combat than on organisation and supply. Here were societies where military endeavour ranked as the highest social duty, where military elites dominated the waging of war. The best military brains were at the battlefront, not in the rear. It is inconceivable that a Marshall or an Eisenhower, with no combat experience between them, could have won supreme command in either the German or Japanese war effort. The German army was notorious for its stubborn inability to release from conscription men whose scientific or managerial skills were in desperate need on the civilian front. Staff officers were obliged to spend time fighting at the front, which explains why even the armed forces were starved of large numbers of experienced planners and organisers. One-quarter of all air force staff officers were killed or captured at the front.[11]

In both Germany and Japan less emphasis was placed upon the non-combat areas of war: procurement, logistics, military services. In the Pacific War there were eighteen American personnel for every one serviceman at the front. The ratio in the Japanese forces was one to one. The postwar bombing survey of Japan observed the marked failure of Japanese air forces to provide 'adequate maintenance, logistic support, communications and control, or airfields and bases . . .' Young Japanese men did not want to be maintenance engineers; they preferred to fly.[12] In the German army in Europe there were roughly two combatants for every non-combatant; but the American army had a ratio almost exactly the reverse, one fighter for every two service personnel. Some measure of the emphasis Marshall placed upon military services can be seen in his decision to divide the army into three separate components, ground forces, air forces and services, each with equal representation on the main staff committees.[13] The American back-up for its combat troops was formidable. One German divisional commander in Normandy reported back the visible effects of the American supply system:

> I cannot understand these Americans. Each night we know that we have cut them to pieces, inflicted heavy casualties, mowed down their transport. But − in the morning, we are suddenly faced with fresh battalions, with complete replacements of men, machines, food, tools and weapons. This happens day after day . . .[14]

Stalin paid the same close attention to servicing Red forces. The Red Army, like the American, had a large and influential service sector under the Main Directorate of the Red Army Rear, whose chief sat on the Defence Committee on equal terms with the combat commanders. Its director, General Khrulev, remembered after the war that his job was regarded by Stalin as 'operational work, organically linked to the combat operations of the troops . . .'[15]

In the 1940s both Germany and Japan were highly militarised

societies. The forces controlled the selection and development of weapons; they bullied and harassed manufacturers to produce what they wanted; they jealously guarded military prerogatives, even to the extent of long and damaging feuds between the separate services. They resented and rejected interference and direction from mere civilians. Allied forces were by contrast supported by a large civilian apparatus; many of the senior commanders were civilians in uniform, and very conscious of it. The United States was so unprepared for war in 1941 that there was no alternative but to bring in men with virtually no military experience at all. Industry was freer to concentrate on large-scale production; scientists were freer to develop and experiment; managerial skills were rewarded as well as combat skills. No doubt Axis militarism brought its own rewards, but it made it harder to build a strong coalition of civilian and military expertise, or to capitalise on the early victories once the Allies were able to mobilise their economic, intellectual and organisational strengths for the purpose of waging war. The Allies, both dictatorship and democracies were committed to the rational exploitation of modernity. The Axis states sought to press modernity into the service of irrational or reactionary causes.[16]

For all this, the margin of victory or defeat was often so slender that general theories look out of place. Battles are not pre-ordained. If they were, no one would bother to fight them. The decisive engagement at Midway Island was won because ten American bombs out of the hundreds dropped fell on the right target. The victory in the Atlantic came with the introduction of a small number of long-range aircraft to cover the notorious Atlantic Gap. The bombing offensive, almost brought to a halt in the winter of 1943–4, was saved by the addition of long-range fuel tanks to escort-fighters, a tiny expense in the overall cost of the bombing campaign. The Battle of Stalingrad depended on the desperate, almost incomprehensible courage of a few thousand men who held up the German 6th army long enough to spring a decisive trap. The invasion of France hung on the ability to keep the enemy guessing, against all conceivable odds, the centre of operational gravity, and then on the weather. It is hardly surprising that Churchill thought at the end of the war that Providence had brought the Allies through.[17]

The Second World War was not of course won in a day, or in a month of good luck. But the narrowness of the margins returns us to the question of winning battles. For all the solid achievements of military reform and of military production, of science and intelligence, which made the Allied forces so much more effective by 1943–4 than they had been only two years before, the military contest still had to be won. Eisenhower's 'grisly, dirty, tough business' had to be endured to the bitter end. Of all the conflicts that made up the war, the one that mattered most to the Allies was the struggle with Germany. On their own Italy and Japan might have made regional gains. The most likely scenario is that without the success of German arms to shield them neither state would have risked war at all. Germany's success in overturning the old European balance of power in less than twelve months opened up possibilities that neither of her allies could have created alone, and even then they hesitated to follow in the German wake. But the German situation was different. The quantity and quality of its armed forces were in another league altogether. The chief instrument of the whole Axis war effort was the German army, more than two hundred divisions strong by 1941, supported by the most modern and effective tactical air force. It was this military strength that had to be defeated to secure Allied victory. The United States devoted only 15 per cent of its war effort to the war with Japan. The other 85 per cent was expended in the defeat of Germany.[18]

If the defeat of the German army was the central strategic task, the main theatre for it was the conflict on the eastern front. The German army was first weakened there, and then driven back, before the main weight of Allied ground and air forces was brought to bear in 1944. Over four hundred German and Soviet divisions fought along a front of more than 1,000 miles. Soviet forces destroyed or disabled an estimated 607 Axis divisions between 1941 and 1945. The scale and geographical extent of the eastern front dwarfed all earlier warfare. Losses on both sides far exceeded losses anywhere else in the military contest. The war in the east was fought with a ferocity almost unknown on the western fronts. The battles at Stalingrad and Kursk, which broke the back of the German army, drew from the soldiers of both sides the last ounces of physical and moral energy. Both sides knew the costs of losing – neither victors nor vanquished, Hitler announced in January 1943, only 'survivors and annihilated'. The other main fronts involved for most of the war much smaller forces. The German army fielded only 20–30 divisions at most in the Italian theatre, but succeeded in preventing Allied victory there for two years. The war in France in 1944, where Germany could have employed over fifty divisions, mostly understrength and some indifferently armed, was fought in its decisive phase between fifteen Allied divisions and fifteen German. Throughout the war German forces kept more tanks, guns and tactical aircraft in the east than on the other fronts.

The German air force, unlike the army, owed its defeat to the western war effort. The revival of Soviet air power in 1942 did a good deal to blunt German tactical air forces, but the main axis of defeat lay across the Reich itself, and was the product of the bombing offensive. The German air force was compelled to adjust to a strategy it had not anticipated, as the bulk of German fighter aircraft had to be moved back to the Reich, behind a rampart of radar, anti-aircraft guns and searchlights, to fulfil the unaccustomed role of defending German industry. By September 1944 80 per cent of the fighter force was based in Germany on anti-bombing missions. The whole air effort was thrown off-balance as priority was given to the production and maintenance of fighters and the proportion of bombers and dive-bombers in the German air force as a whole fell to less than a quarter from well over a half. In the summer of 1944 there were four times as many serviceable fighters as bombers.[19] The advent of the long-range fighter in the spring of that year was fatal for the Germans. The Allies were able to bring the fight all the way to the defensive fighters, which were neither able to attack the bombers – their stated mission – nor to fight off enemy escorts effectively. The German air force was drained of planes and pilots; Allied bombers hemmed in aircraft production and heavily damaged the plants producing aviation fuel. The switch to an anti-bombing strategy, and the destruction wreaked in the battles over the Reich, fatally weakened German air power at the front-line, which had relied on large numbers of medium-bombers. Allied air forces enjoyed a superiority of seventy to one in the invasion of France. German ground forces were compelled to fight the last two years of war with limited or non-existent air support. When the last Luftwaffe chief-of-staff, General Karl Koller, sat down at the end of the conflict to address the question 'Why We Lost the War', he reduced it to a single formula: 'What was decisive in itself was the loss of air supremacy.'[20]

In Keller's view, partisan though it no doubt was, 'Everything depends on air supremacy, everything else must take second place.' There is certainly a strong connecting thread of air power trailing through all the major zones of combat. In the war at sea naval power was not entirely superseded by aircraft, but the major naval engagements of the Pacific War were all determined by aircraft firing bombs or torpedoes. In the Atlantic, escort carriers and long-range anti-submarine aircraft brought the U-boat threat to a halt. In 1943 aircraft claimed 149 out of the total of 237 German submarines sunk.[21] Patrolling aircraft kept German naval forces at bay throughout the invasion of France. Allied aircraft took a heavy toll of the merchant shipping in the Mediterranean and in the Pacific. In the invasion battles in 1944 air power was regarded by both sides as the critical factor. 'There is no way by which, in the face of the enemy air force's complete command, we can find a strategy which will counterbalance its positively annihilating effect without giving up the field of battle . . .': so wrote Field Marshal Kluge to Hitler on 21 July, a few days before the Cobra operation broke open the German front.[22] On the eastern front Soviet air power from the summer of 1943 was able to do what German aircraft had done in Poland, France, Yugoslavia, Greece and Ukraine. Underlying everything was the bombing offensive, whose far-reaching effects on German economic potential and on the German home front was sufficient to limit the expansion of German military might to a point where the Allied ground forces could fight on more than equal terms.

Air power did not win the war on its own, but it proved to be the critical weakness on the Axis side and the greatest single advantage enjoyed by the Allies. Koller, in his reflections on defeat, did not simply attribute the failure in the air to German ineptitude, but recognised also that Germany's enemies drew the right conclusions from the early years of war 'and with an iron tenacity built up a superior air force which alone could lead to victory'.

Koller was right that the strategy was a self-conscious one. Perhaps the most important decision was taken in the summer of 1941 in Washington, when General Marshall instructed the air force to draft its contribution to the Victory Programme – in effect to plan America's air war. A small group of airmen led by Colonel Hal George was given the almost impossible task of drafting the plan in nine days at the height of a sticky Washington summer. They sat up until past midnight each night, and twice the team worked all through the night. They knew that Roosevelt had given the green light to produce as much as possible. Each of them worked out the figures for a particular element of air power – heavy-bombing, very-long-range bombing, tactical air support, training, and so on. When they came together at the Washington Munitions Building on a Sunday lunchtime to add all the figures up, they were astonished at the quantities needed to defeat the Axis. The plan was collated and approved, and the work was begun on establishing massive American air power even before America entered the war. The plan was leaked to the press some days before Pearl Harbor. In its essentials it was sent to Hitler by German agents in America. His staff drew up a draft directive – No. 39 – to hold the line in Russia, while they planned a course of action to keep American aircraft and men away from Europe. Hitler refused to accept the recommendation, and stuck to the contest with the Red Army.[23]

That decision locked Germany into a strategy in which the defeat of Soviet ground forces was the principal mission. The failure to achieve this ambition exposed German forces not only to a revamped and angry Red Army, but also to the eventual realisation of that massive western air power anticipated in the Victory Programme. Both together defeated German forces, leaving Germany's weaker allies to be destroyed in turn. When the Allies devoted their full attention to Japan in the summer of 1945, the Red Army swept Japanese forces in China aside in a matter of days, while American air forces transformed Japanese cities into open crematoria. The Red Army went on to become the backbone of the Soviet super-power, while American air power became the central pillar of the new military giant in the west. Both developments were the fruit of German aggression. To deny Germany's bid for world power, the Soviet Union and the United States had to become world powers themselves. The war was won in 1945 not from German weaknesses but from Allied strengths.

Even then, complete victory was not attained just through the reform of Allied fighting power, and its effective exploitation on the field of battle. The will to win, to continue through periods of intense crisis, stalemate or defeat, to keep the prospect of victory in sight and to mobilise the psychological and moral energies of a people under threat, proved to be inseparable from the ability to fight better. There is no doubt that at times in the war moral confidence was badly dented; in each Allied state enthusiasm for war had to be actively maintained. Individuals reacted to the demands of war in so many different ways that easy classification is defied. But on balance the commitment of both leaders and led to prosecuting the war to the finish, despite high levels of sacrifice and, in the Soviet case, exceptional losses, proved to be a positive element in the Allied cause. This had little to do with Gross Domestic Product, even if most soldiers had known what it meant. To those in battle the facts of aggregate economic strength were meaningless. People fought for many reasons, from fear, from hatred, from some sense of moral or racial superiority, from loyalty or patriotism. They did not fight to prove the statistics right, but from an effort of will.

There have been ample opportunities since 1945 to show that material superiority in war is not enough if the will to fight is lacking. In Algeria, Vietnam and Afghanistan the balance of economic and military strength lay overwhelmingly on the side of France, the United States, and the Soviet Union, but the will to win was slowly eroded. Troops became demoralised and brutalised. Even a political solution was abandoned. In all three cases the greater power withdrew. The Second World War was an altogether different conflict, but the will to win was every bit as important – indeed it was more so. The contest was popularly perceived to be about issues of life and death for whole communities rather than for their fighting forces alone. They were issues, wrote one American observer in 1939, 'worth dying for'. If, he continued, 'the will-to-destruction triumphs, our resolution to preserve civilisation must become more implacable . . . our courage must mount'.[24]

Words like 'will' and 'courage' are difficult for historians to use as instruments of cold analysis. They cannot be quantified; they are elusive of definition; they are the products of a moral language that is regarded sceptically today, even tainted by its association with fascist rhetoric. Yet in war these are the qualities, this the language, generated by conflict. German and Japanese leaders believed that the spiritual strength of their soldiers and workers would in some indefinable way compensate for their technical inferiority. When asked after the war why Japan lost, one senior naval officer replied that the Japanese 'were short on spirit, the military spirit was weak . . .'[25] and put this explanation ahead of any material cause. Within Germany, belief that spiritual strength or willpower was worth more than generous supplies of weapons was not confined to Hitler by any means, though it was certainly a central element in the way he looked at the world. The irony was that Hitler's ambition to impose his will on others did perhaps more than anything to ensure that his enemies' will to win burned brighter still. The Allies were united by nothing so much as a fundamental desire to smash Hitlerism and Japanese militarism and to use any weapon to achieve it. This primal drive for victory at all costs nourished Allied fighting power and assuaged the thirst for vengeance. They fought not only because the sum of their resources added up to victory, but because they wanted to win and were certain that their cause was just.

The Allies won the Second World War because they turned their economic strength into effective fighting power, and turned the moral energies of their people into an effective will to win. The mobilisation of national resources in this broad sense never worked perfectly by any means, but worked well enough to prevail. Materially rich, but divided, demoralised, and poorly led, the Allied coalition would have lost the war, however exaggerated Axis ambitions, however flawed their moral outlook. The war made exceptional demands on the Allied peoples. Half a century later the level of cruelty, destruction and sacrifice that it engendered is hard to comprehend, let alone recapture. Sixty years of relative security and prosperity have opened up a gulf between the present age and the age of crisis and violence that propelled the world into war. Though from today's perspective Allied victory might seem somehow inevitable, the conflict was poised on a knife-edge in the middle years of the war. This period must surely rank as the most significant turning-point in the history of the modern world.

EPILOGUE

'. . . it is not so bad to be defeated in this war . . .'

Shigeru Yoshida, August 1945

THE WORLD ORDER established after the Second World War was quite different from the fragile structure built in 1919. This time the system was dominated by those states with the power to maintain it, the United States and the Soviet Union. Britain and France, the key actors in 1919, found their postwar international position fatally weakened by their inability to stop Germany in 1940. Without allies there would have been no way that Britain could secure her empire, let alone defeat her enemies, once the French army was out of the contest. After 1945 Britain and France became powers of the second rank. Their evident weakness during the war encouraged nationalist struggles in both the British and French empires, and within a generation the empires were mostly gone. Of the western Allies, Britain lost most from the war – the old balance of power, the empire and a dominant role in the world's economy.

Nevertheless, British democracy survived. In 1945 the sense that the forces of light had triumphed against the forces of darkness was overwhelming, even more so as the grim catalogue of German and Japanese crimes was fully exposed for the first time to the public gaze. The end of the war was welcomed as a break with a decadent, disordered past, as an end to unemployment and slumps, as an end to crude geopolitics and racism. The stale atmosphere of the pre-war age, the morbid contemplation of decline, was blown aside by a widespread hope for a new beginning. The Axis New Orders, and the ideological baggage they carried behind them, were confined to history's scrap heap. In the west the ideals of democracy and international collaboration had the field of opinion to themselves.

The hopes for a progressive postwar order in the west rested almost entirely upon the United States. They alone possessed the economic strength and military power to prevent the return to economic stagnation and international instability, however unpalatable other western governments and peoples found this fact to be. In 1945 the willingness of American statesmen to assume the responsibility for maintaining a new international order could not be taken for granted. There were isolationists in the United States who wanted their country to do what it did in 1919, to withdraw from the political and military responsibility of remaking the post-war world. Gradually it dawned on American leaders that this time the Old World was neither able nor willing to maintain the peace, any more than it could win the war. Shortly after taking office in 1945, President Harry Truman recognised the transformation of America's position: 'we have emerged from this war the most powerful nation in the world – the most powerful nation, perhaps, in all history.' A different world system, with American power at its heart, was unavoidable. 'The whole world structure and order that we had inherited from the nineteenth century', wrote the American politician Dean Acheson, 'was gone.'[1]

The United States succeeded Britain as the major global power. From being a military stripling in 1940, America by 1945 had twelve million men in the armed forces, over seventy thousand naval vessels and almost 73,000 aircraft. America also possessed the atomic bomb, and, despite the spread of this technology world-wide, has retained a lead in nuclear weaponry down to the present day. Finally, the American economy emerged in 1945 strengthened by the war, capable of outproducing all the other great powers together. It was an economy committed to the capitalist values of liberal trade and market competition. American politicians and businessmen were determined that the world economy should not slip back into the bad habits of the inter-war years, of trade-blocs and tariffs, and exerted American political power to build a more open market. Generous funds were made available for recon-struction in the war-torn areas of the world outside the Soviet bloc. The siege-mentality of the pre-war era gave way to international collaboration through the new instruments of world-market regulation, the IMF and GATT. America's economic priorities gave the world economy a kick-start after 1945 which made possible the long economic boom. Prosperity dulled political antagonism and the thirst for conquest. Battle shifted to the boardroom, and has stayed there.

The American succession to world leadership in 1945 was the most significant change. The Soviet Union also emerged from the war as a major world power, but Russia had been a leading player in the international system for more than two centuries, and the Soviet state already possessed the world's largest military forces before 1939. The war threw Soviet power more clearly into focus. In the absence of any powerful neighbour, Stalin's state became the dominant political force throughout eastern Europe and Asia. It was a battered giant. The task of economic reconstruction and the search for military security absorbed Soviet energies for twenty years or more. But the war did secure the survival of communism. Victory in 1945 gave the Party a new lease of life. The political veterans of the war dominated Soviet life through to the 1980s. More than that, the war was used to present the Soviet Union as a force for world progress. Schoolchildren chanted: 'By defeating Hitler's Germany, the Soviet Nation saved mankind from annihilation . . . and preserved world civilisation.' The triumph of the Red Army was used to underpin the illusion that communism was Khrushchev's 'wave of the future', the system to which all societies were historically moving.[2]

Victory in the Great Patriotic War helped to establish the core myths of the postwar Soviet state. When memories were fading in the 1960s, the Soviet leader, Leonid Brezhnev, resurrected the war experience in order to bolster Party power. In 1965 Victory Day was made an official public holiday, and a medal was struck for all participants of the war. Associated state propaganda played up themes from the history of the war acceptable to Party leaders. Any discussion of the early defeats, or suggestions of Stalinist incompetence, were stamped on by the censors. The publication of wartime memoirs was halted for fear that the true dimensions of Stalin's leadership, or of the cost of the war, would escape into the open.[3] It is no coincidence that the torrent of uncensored publications on the war in the 1980s contributed to the unravelling of the myths of communism, and ultimately of the whole Soviet system.

The confident belief in a communist future engendered by victory found expression in the political success of communism throughout much of the defunct Japanese New Order, in China, North Korea and Vietnam. Communism filled a power vacuum left by the collapse of the old colonial empires, Japan included, and by the weakness and corruption of the Chinese regime under Chiang Kai-Shek. Most of the area fought over during the war was run by communist regimes by the mid-1950s, bent on modernising their states along Soviet lines. In all these countries high levels of state-directed economic growth transformed backward rural societies. No communist people became prosperous in the western sense, and neither did the war bring democracy any nearer for them. But communism presented itself as an ideological counterpart to fascism and militarism which had been laid low by Soviet arms in 1945.

The consolidation of communist regimes in most of Asia and eastern Europe did not fit with the American ideal of the free world and open economy proclaimed in 1945. One of the first casualties of the peace was the moral consensus that bound the Allies together in the common defeat of Hitlerism. Once that enemy was removed, both western Allies were able to return to a relationship with their totalitarian companion with which they were more morally at ease. The Cold War began where it had been left off in 1941, with profound western distrust of Soviet motives, and an ideological divide every bit as deep as that between liberalism and Nazism. Only two years after the end of the war the American Air Policy Commission reported to Truman that the essential 'incompatibility of East and West' called for the build-up of a 'devastating' force of bombers and missiles equipped with nuclear weapons capable of operating at a range of 5,000 miles.[4] American strategists moved effortlessly from one Manichaean world to the next.

The transition from World War II to Cold War made possible the most remarkable consequence of all: the integration of the three Axis states with the western, anti-communist world. This did not happen immediately, and it was not simply imposed on the vanquished willy-nilly. For the three or four years following the war, conditions for the defeated populations remained bleak; food was scarce, industrial production a fraction of what it was pre-war, much of the urban area had been flattened by bombing, and it was far from clear what the intentions of the occupying armies were. Although there was a sizeable fraction of the population prepared to embrace communism, even in the Soviet-occupied eastern zone of Germany, where communism had been experienced in the raw, in all three defeated powers there had always existed a significant majority attracted to western values and western economics. What disappeared in 1945 was any belief that the path to recovery lay in a revival of the violent imperialism and economic *dirigisme* of the 1930s. America in 1945 was self-confident and rich. The window to the west, closed in the depression years of the 1930s by American isolationism and economic crisis, was now invitingly open. A future Japanese Foreign Minister, Saburo Okita, recalled how in the depths of a wretched defeat people thought, ' "It's miserable now, but in time Japan will get back on its feet again, not through military power, but by new technology and economic power." '[5] The new Japanese constitution included Article 9, renouncing war 'forever' as a means of settling disputes and Japanese soldiers have not seen action since. When a western German state was set up in 1949, its constitution prevented German soldiers from ever serving outside the frontiers; the German forces permitted by the Allies by the mid-1950s were for defence only.

The Cold War hastened the realignment of the Axis states with their former enemy. There was more than a touch of irony in German and Italian populations once again finding themselves confronting the Soviet colossus they had fought in the war, but this time at the side of Britain and the United States. But by the early 1950s, during the Korean War, the first serious fighting between 'east' and 'west' since the war, America needed to establish a firm military alliance in Europe to contain the Soviet Union, for which German and Italian participation was essential. She also needed the industrial potential of the two states to contribute to fuelling the Korean War boom. During the 1950s the remaining restrictions on industrial development imposed on Axis economies in 1945 were lifted, and their populations rushed for the economic growth they had been starved of for a generation. As one German observed in 1960: 'The American outlook is based on enthusiasm for a better standard of living . . . the fellow who can figure out how soon he too can own a motor-bike or a Volkswagen won't dream any longer of the day when there will be a *Gauleiter's* job [Nazi provincial leader] opening up in Central Asia . . .' The scramble for economic success had other causes too. There was no prospect for any Axis state to improve its lot by reviving pre-war territorial ambitions. Militarism and racism were thoroughly discredited, and the elites turned to more conventional paths to wealth and success.

This still begs the question of how societies that failed so comprehensively at the task of waging war have succeeded so dramatically in the economic contest. No doubt the concentration of national energies on the achievement of material well-being did owe something to the deliberate choice made in the immediate postwar years to reject the failed policies of the 1930s and the war. Defeat in war shifted the values and status-systems of Japanese and German societies to an entirely new footing. Rational, civilian activities have been adopted because they have brought results: victory for the United States and Britain in 1945, economic miracles for Germany and Japan (and Italy too) in the fifty years that followed. In the long term the country that contributed the most to the defeat of the Axis, the Soviet Union, is the one that has lost most. Soviet militarism survived. There was no economic miracle. The manifest inability of the communist system to provide what the liberal west could offer, despite the smothering propaganda of socialist progress, eventually locked the Soviet system into a political dead-end. Its collapse after 1989 was not in any sense a certainty, but the choices made there after 1945, like Axis choices in the 1930s, made a circle impossible to square. The cost of maintaining the super-power status and the arms race finally outran the ability of the regime to persuade the people that the communist new order was worth that cost. The collapse of the Soviet Union was not caused directly by victory in 1945. The distance of time is too great for that. But victory proved a poisoned chalice. The Soviet people did not win freedom or prosperity, but their sacrifices have made it possible for all the other warring states to enjoy them both.

Weapons Production of the Major Powers
1939-45

	1939	1940	1941	1942	1943	1944	1945*
Aircraft							
Britain	7,940	15,049	20,094	23,672	26,263	26,461	12,070
USA	5,856	12,804	26,277	47,826	85,998	96,318	49,761
USSR	10,382	10,565	15,735	25,436	34,900	40,300	20,900
Germany	8,295	10,247	11,776	15,409	24,807	39,807	7,540
Japan	4,467	4,768	5,088	8,861	16,693	28,180	11,066
Major Vessels†							
Britain	57	148	236	239	224	188	64
USA	-	-	544	1,854	2,654	2,247	1,513
USSR	-	33	62	19	13	23	11
Germany (U-boats only)	15	40	196	244	270	189	0
Japan	21	30	49	68	122	248	51

(table continues overpage)

	1939	1940	1941	1942	1943	1944	1945*
Tanks‡							
Britain	969	1,399	4,841	8,611	7,476	5,000	2,100
USA	–	c.400	4,052	24,997	17,565	17,565	11,968
USSR	2,950	2,794	6,590	24,446	24,089	28,963	15,400
Germany	c.1,300	2,200	5,200	9,200	17,300	22,100	4,400
Japan	c.200	1,023	1,024	1,191	790	401	142
Artillery Pieces§							
Britain	1,400	1,900	5,300	6,600	12,200	12,400	–
USA	–	c.1,800	29,615	72,658	67,544	33,558	19,699
USSR	17,348	15,300	42,300	127,000	130,300	122,400	31,000
Germany	c.2,000	5,000	7,000	12,000	27,000	41,000	–

Dashes indicate reliable figures unavailable.

* figures for Britain, USA and Japan for January–August; for the USSR all year for aircraft, January–March for artillery; for Germany January–April

† excluding landing-craft and smaller auxiliary vessels

‡ includes self-propelled guns for Germany and the USSR

§ medium and heavy calibre only for Germany, USA and Britain; all artillery pieces for the USSR. Soviet heavy artillery production in 1942 was 49,100, in 1943 48,400 and in 1944 56,100.

1 Unpredictable Victory
EXPLAINING WORLD WAR II

1 R.A.C, Parker, *Struggle for Survival: The history of the Second World War* (Oxford, 1989), p. 86.

2 G.T. Eggleston, *Roosevelt, Churchill and World War II Opposition* (Old Greenwich, Conn., 1979), p. 127.

3 B.B. Berle, T.B. Jacobs (eds), *Navigating the Rapids 1918–1971: From the papers of Adolph A. Berle* (New York, 1973), pp. 374–5, diary entry for 31 July 1941.

4 M. Ferro, *The Great War 1914–1918* (London, 1973), p. 129.

5 Imperial War Museum, Speer Collection, Box 368, Report 67, p. 14.

6 R.L. DiNardo, A. Bay, 'Horse-Drawn Transport in the German Army', *Journal of Contemporary History* 23 (1988), pp. 130-9.

7 G. Weinberg, *World in the Balance* (New England University Press, 1981), p. 7.

8 W. Maser (ed.), *Hitler's Letters and Notes* (New York, 1974), pp. 52–6, 94. On the continuity of Hitler's outlook on the world, compare Maser (ed.), *Hitler's Letters*, esp. pp. 212–49 (speeches from 1920), and pp. 279–83 (synopsis for the 'monumental history of mankind') with the ideas expressed in 1936 on the coming war reproduced in W. Treue (ed.), 'Der Denkschrift Hitlers über die Aufgaben eines Vierjahrsplan', *Vierteljahreshefte für Zeitgeschichte* 3 (1954). On the formation of Hitler's world outlook, see G. Stoakes, *Hitler and the Quest for World Dominion* (Leamington Spa, 1986).

9 The 'waiter' quotation is in M. Kater, 'Hitler in a Social Context', *Central European History* 14 (1981), p. 247. There are good accounts of Hitler's 'split personality' in A. François-Poncet, *The Fateful Years: Memoirs of a French Ambassador in Berlin 1931–1938* (London, 1949), pp. 236–8, 289–92 and W. Schellenberg, *The Schellenberg Memoirs* (London, 1956), pp. 110–12.

10 H. Ickes, *The Secret Diary of Harold L. Ickes* (3 vols, London, 1955), III, p. 37.

11 F.C. Jones, *Japan's New Order in East Asia* (Oxford, 1954), p. 469.

12 W. Boelcke (ed.), *The Secret Conferences of Dr Goebbels, October 1939 to March 1943* (London, 1967), p. 184.

13 J. Toland, *Adolf Hitler* (New York, 1976), p. 685; H. K. Smith, *Last Train from Berlin* (London, 1942), pp. 60–4.

14 *Akten zur Deutschen auswärtigen Politik 1918–1945*, series D, vol. XIII, pp. 839–40, discussion between Ribbentrop and ambassador Oshima, 23 August 1941, appendix 4.

15 J. Colville, *The Fringes of Power: 10 Downing Street diaries 1939–1955* (London, 1985), p. 347, entry for 26 January 1941.

16 Colville, *Fringes of Power*, p. 382, diary entry for 2 May 1941.

17 M. Toscano, *The Origins of the Pact of Steel* (Baltimore, 1967), p. 378.

18 A. Salter, *Slave of the Lamp: A public servant's notebook* (London, 1967), pp. 151–2.

19 W. Averell Harriman with E. Abel, *Special Envoy to Churchill and Stalin 1941–1946* (London, 1976), p. 67.

20 E. O'Ballance, *The Red Army* (London, 1964), p. 164.

21 J. Barber, 'The Moscow Crisis of October 1941' in J. Cooper, M. Perrie and E. A. Rees (eds), *Soviet History 1917–1953: Essays in Honour of R. W. Davies* (London, 1995), pp. 201-18; M. M. Gorinov 'Muscovites' Moods, 22 June 1941 to May 1942' in R. Thurston, B. Bonwetsch (eds), *The People's War: Responses to World War II in the Soviet Union*, pp. 122-5.

22 O. Bartov, *Hitler's Army: Soldiers, Nazis and war in the Third Reich* (Oxford, 1991), p. 96. In addition to the deaths some 420,000 were imprisoned. The figure for British forces was only forty deaths.

23 Runciman Papers, Newcastle University Library, letter from Arthur Murray to Walter Runciman, 5 September 1939.

24 F. von Papen, *Memoirs* (London, 1952), p. 453.

25 B. Bond (ed.), *Chief of Staff: The diaries of Lieutenant-General Sir Henry Pownall: Vol. I, 1933–1940* (London, 1972), p. 221, entry for 29 August 1939.

2 Little Ships and Lonely Aircraft
THE BATTLE FOR THE SEAS

1 H.V. Morton, *Atlantic Meeting* (London, 1943), pp. 23–8, 35–7; D. Dilks (ed.), *The Diaries of Sir Alexander Cadogan* (London, 1971), pp. 395–6.

2 W. Averell Harriman with E. Abel, *Special Envoy to Churchill and Stalin 1941–1946* (London, 1976), p. 75.

3   Morton, *Atlantic Meeting*, pp. 48–9; T. A. Wilson, *The First Summit: Roosevelt and Churchill at Placentia Bay* (London, 1969), pp. 75–6.

4   E. Roosevelt (ed.), *The Roosevelt Letters: Vol. III, 1928–1945* (London, 1952), p. 364, Roosevelt to Churchill, 4 May 1941; W. Kimball (ed.), *Churchill and Roosevelt: the complete correspondence: Vol. I, Alliance Emerging* (Princeton, 1984), p. 103, Churchill to Roosevelt, 7 December 1940.

5   Harriman, *Special Envoy*, p. 75.

6   Dilks (ed.), *Diaries of Sir Alexander Cadogan*, p. 402; Morton, *Atlantic Meeting*, pp. 132–5.

7   W.S. Churchill, *The Second World War* (6 vols, London, 1948–54), III, p. 551.

8   See generally P. Kennedy, *The Rise and Fall of British Naval Mastery* (London, 1976); C. Barnett, *Engage the Enemy More Closely: The Royal Navy in the Second World War* (London, 1991).

9   F. Ruge, *Der Seekrieg: The German Navy's story 1939–1945* (US Naval Institute, Annapolis, 1957), p. 46.

10 C.S. Thomas, *The German Navy in the Nazi Era* (London, 1990), p. 187.

11 H. Trevor-Roper (ed.), *Hitler's War Directives* (London, 1964), p. 64, Directive no. 9, 'Instructions for warfare against the economy of the enemy'.

12 M.A. Bragadin, *The Italian Navy in World War II* (US Naval Institute, Annapolis, 1957), pp. 365–6.

13 K. Poolman, *Focke-Wulf Condor: Scourge of the Atlantic* (London, 1978); E. van der Porten, *The German Navy in World War II* (London, 1969), pp. 174–8.

14 S.W. Roskill, *The Navy at War 1939–1945* (London, 1960), pp. 110–11, 127–37.

15 Ruge, *Der Seekrieg*, pp. 43–4; S. W. Roskill, *The War at Sea 1939–1945* (3 vols, London, 1961), III (part ii), p. 479, appendix ZZ; on German decrypts see D. Kahn, *Hitler's Spies: German military intelligence in World War II* (London, 1978), pp. 218–19. The German naval decrypting office, the B-Dienst, lost access to the British codes in August 1940, but regained it after only seven weeks, and could read most of what was needed down to the middle of 1943.

16 J. Rohwer, *The Critical Convoy Battles of March 1943* (London, 1977), pp. 15–18; K. Dönitz, *Memoirs: Ten years and twenty days* (London, 1959), pp. 118–23, 127–30 on the organisation of submarine warfare in 1941.

17 J. Colville, *The Fringes of Power: 10 Downing Street diaries 1939–1955* (London, 1985), p. 358, entry for 26 February 1941; On the import crisis, C.B.A. Behrens, *Merchant Shipping and the Demands of War* (London,

1955), pp. 190–7; M. Olson, *The Economics of Wartime Shortage* (Durham, North Carolina, 1963), pp. 125–8.

18 Colville, *Fringes of Power*, p. 372, diary entry for 9 April 1941; Churchill, *Second World War*, III, pp. 107–9.

19 P. Beesly, *Very Special Intelligence: The story of the Admiralty's Operational Intelligence Centre 1939–1945* (London, 1977), pp. 88–95; J. Winton, *Ultra at Sea* (London, 1988), pp. 96–101.

20 Roskill, *War at Sea*, III (part ii), p. 479; Behrens, *Merchant Shipping*, p. 178. The loss rate of crewmen on torpedoed ships was 53.9 per cent in 1941.

21 Harriman, *Special Envoy*, pp. 111–12; J.G. Winant, *A Letter from Grosvenor Square* (London, 1977), pp. 198–200.

22 Lord Ismay, *Memoirs* (London, 1960), p. 241.

23 Cited in Barnett, *Engage the Enemy More Closely*, p. 440.

24 United States Strategic Bombing Survey (USSBS), Pacific Theatre, Report 73, 'The Campaigns of the Pacific War' (Washington, 1946), pp. 38–9. Allied losses totalled 34 naval vessels from December 1941 to March 1942, including 1 battleship, 1 battlecruiser, 5 cruisers and 13 destroyers.

25 M. Ugaki, *Fading Victory: the Diary of Admiral Matome Ugaki 1941–1945* (Pittsburgh, 1991), p. 65, entry for 1 January 1942; USSBS, Pacific Theatre, Report 72, 'Interrogation of Japanese Officials', vol. II (Washington, 1946), pp. 318–20, interrogation of Admiral Toyoda (Chief of Naval Staff, 1945).

26 S. E. Morison, *Coral Sea, Midway and Submarine Actions May 1942–August 1942*, History of US Naval Operations, vol. IV (New York, 1950), pp. 4–6; M. Fuchida, M. Okumiya, *Midway: The battle that doomed Japan* (US Naval Institute, Annapolis, 1955), pp. 68–77.

27 J. Dower, *War Without Mercy: Race and power in the Pacific War* (New York, 1986), pp. 20–41 for an excellent account of this transition.

28 M. Matloff, E. Snell, *Strategic Planning for Coalition Warfare* (2 vols, Washington, 1953–9), I, pp. 60–2, 97–119; R. M. Leighton, R. W. Coakley, *Global Logistics and Strategy* (2 vols, Washington, 1955–68), I, p. 716.

29 Morison, *Coral Sea*, pp. 11–13; P. S. Dull, *A Battle History of the Imperial Japanese Navy 1941–1945* (Cambridge, 1978), pp. 115–22.

30 Morison, *Coral Sea*, p. 60.

31 Ugaki, *Fading Victory*, p. 122, diary entry for 3 May 1942; E. P. Hoyt, *Japan's War: The great Pacific conflict* (London, 1986), pp. 282–3.

32 Fuchida, Okumiya, *Midway*, pp. 92–100; R. Lewin, *The Other Ultra* (London, 1982), p. 109.

33 Dower, *War Without Mercy*, p. 36; Morison, *Coral Sea*, p. 81.

34 E.J. King, W.M, Whitehill, *Fleet Admiral King: A naval record* (New York,

1952), pp. 147–9, 243; on US naval aviation, see N. Polmar, *Aircraft Carriers* (London, 1969), pp. 40–51.

35 Lewin, *Other Ultra*, pp. 85–106.

36 Morison, *Coral Sea*, pp. 81–2.

37 Details from Ugaki, *Fading Victory*, pp. 130–8; Fuchida, Okumiya, *Midway*, pp. 94–7.

38 Morison, *Coral Sea*, pp. 97–104; Fuchida, Okumiya, *Midway*, pp. 170–80.

39 J. S. Thach, 'A Beautiful Silver Waterfall', in E.T. Wooldridge (ed.), *Carrier Warfare in the Pacific: an oral history collection* (Washington, 1993), p. 58.

40 Fuchida, Okumiya, *Midway*, pp. 180–91; Ugaki, *Fading Victory*, pp. 151–3; D. van der Vat, *The Pacific Campaign* (London, 1992), pp. 192–3.

41 USSBS, Report 72, 'Interrogation of Japanese Officials', vol. I, p. 266, interrogation of Admiral Takata (Naval General Staff), and vol. II, p. 331, interrogation of Admiral Mitsumasa Yonai.

42 USSBS, Report 72, vol. I, p. 262, interrogation of Admiral Takata, and Report 73, 'The Campaigns of the Pacific War', p. 60.

43 USSBS, Report 72, vol. II, item 86, 'Naval aircraft strength and wastage'; USSBS, Pacific Theatre, Report 46, 'Japanese Naval Shipbuilding' (Washington, November 1946), p. 2; Polmar, *Aircraft Carriers*, p. 753. From 1942 onwards there developed a remarkable disparity in the losses of Japanese and American aircraft, 25,744 Japanese planes lost for 2,421 American.

44 Van der Vat, *Pacific Campaign*, pp. 383–4.

45 Leighton, Coakley, *Global Logistics and Strategy*, I, p. 638.

46 M. A. Stoler, *The Politics of the Second Front: American military planning and diplomacy in coalition warfare, 1941–1943* (Westport, Conn., 1977), pp. 34–6.

47 H. Payton-Smith, *Oil: A study in wartime policy and administration* (London, 1971), pp. 103, 322. Some 60 per cent of oil imports into Britain came from the eastern United States ports, and 40 per cent from the Caribbean.

48 M. Howard, *The Mediterranean Strategy in the Second World War* (London, 1968), pp. 30–40; Matloff, Snell, *Strategic Planning*, I, pp. 322–7.

49 E. Raeder, *Mein Leben* (2 vols, Tübingen, 1957), II, p. 277; *Fuehrer Conferences on Naval Affairs 1939–1945* (London, 1948, reissued 1990), p. 273, conference of 13 April 1942, and p. 285, conference of 15 June 1942.

50 *Fuehrer Conferences on Naval Affairs*, pp. 281–2, conference of 13–14 May 1942.

51 Dönitz, *Memoirs*, p. 235.

52 P. Padfield, *Dönitz: The last Führer* (London, 1984), chs 2–4.

53 Dönitz, *Memoirs*, p. 479, appendix I.

54   Ruge, *Der Seekrieg*, pp. 252–5; Beesly, *Very Special Intelligence*, pp. 102–10;
     M. L. Hadley, *U-Boats against Canada* (London, 1990), pp. 52–81; M.
     Milner, 'Anglo-American Naval Co-operation in the Second World
     War', in J. Hallendorf, R. S. Jordan (eds), *Maritime Strategy and the Balance
     of Power* (London, 1989), pp. 252–4.

55   D. Kahn, *Seizing the Enigma: The race to break the German U-boat codes*
     (Boston, 1991), pp. 214–17; Winton, *Ultra at Sea*, pp. 103–7.

56   Churchill, *Second World War*, II, p. 529; on oil, see J. Terraine, *Business in
     Great Waters: The U-boat wars 1916–1945* (London, 1989), pp. 514–15.

57   M. Milner, 'The Battle of the Atlantic', *Journal of Strategic Studies* 13
     (1990), pp. 46, 54–5; B. B. Schofield, 'The Defeat of the U-Boats during
     World War II', *Journal of Contemporary History* 16 (1981), pp. 120–4.

58   Rohwer, *Critical Convoy Battles*, p. 36; Beesly, *Very Special Intelligence*,
     p. 182.

59   Dönitz remark in *Fuehrer Conferences on Naval Affairs*, p. 294, conference
     of 28 September 1942; J. Buckley, 'Air Power and the Battle of the
     Atlantic', *Journal of Contemporary History* 28 (1993), pp. 146–50.

60   Details from B. Johnson, *The Secret War* (London, 1978), pp. 250–55;
     Buckley, 'Air Power', pp. 147–8.

61   D. Howse, *Radar at Sea: The Royal Navy in World War 2* (London, 1993),
     pp. 99–109, 132–143, 149; Johnson, *Secret War*, pp. 231–5; G. Hartcup,
     The Challenge of War: Scientific and engineering contributions to World War II
     (Newton Abbot, 1970), pp. 91–2.

62   Johnson, *Secret War*, pp. 237–8.

63   Dönitz, *Memoirs*, pp. 253, 315.

64   Bragadin, *Italian Navy*, pp. 365–6; Polmar, *Aircraft Carriers*, p. 124; W.
     Adair, 'The War in the Mediterranean', in Viscount Cunningham of
     Hyndhope, *A Sailor's Odyssey* (London, 1951), pp. 673–4.

65   Bragadin, *Italian Navy*, pp. 238–49; Roskill, *War at Sea*, II, pp. 342–5.

66   On Torch convoys, see Beesly, *Very Special Intelligence*, pp. 148–51.

67   W. S. Chalmers, *Max Horton and the Western Approaches* (London, 1954),
     pp. 150–62.

68   Ibid., pp. 163–74.

69   Dönitz, *Memoirs*, pp. 321–2.

70   Beesly, *Very Special Intelligence*, pp. 154–6; Winton, *Ultra at Sea*, pp. 108–9;
     F. H. Hinsley et al., *British Intelligence in the Second World War: Vol. II*
     (London, 1981), pp. 548–53.

71   Details in Rohwer, *Critical Convoy Battles*; Roskill, *Navy at War*, pp.
     272–5; Dönitz, *Memoirs*, pp. 328–30.

72   Roskill, *Navy at War*, p. 224; Dönitz, *Memoirs*, pp. 330–1; Terraine,
     Business in Great Waters, p. 569.

73 Chalmers, *Max Horton*, pp. 186–94; Roskill, *War at Sea*, II, pp. 363–4, 366–7.

74 Howse, *Radar at Sea*, pp. 143–9, 132–42.

75 Roskill, *Navy at War*, p. 276; Dönitz, *Memoirs*, pp. 339–40; S. E. Morison, *The Atlantic Battle Won, History of US Naval Operations*, vol. X (New York, 1956), pp. 76–80.

76 *Fuehrer Conferences on Naval Affairs*, pp. 331–2, conference of 31 May 1943. Dönitz reported that loss rates of operational boats were running at 30 per cent a month.

77 Chalmers, *Max Horton*, p. 203.

78 On air power, see Buckley, 'Air Power', p. 155.

79 Churchill, *Second World War*, II, p. 524.

80 Behrens, *Merchant Shipping*, p. 178.

81 Chalmers, *Max Horton*, p. 106.

82 Bragadin, *Italian Navy*, pp. 365–6; USSBS, Pacific Theatre, Report 48, 'Japanese Merchant Shipbuilding' (Washington, January 1947), pp. 18–20.

83 Roskill, *War at Sea*, III (part ii), pp. 439–42, appendix T, 'Nominal List of British Commonwealth Major Warship Losses'.

84 Ibid., appendix ZZ, table II, 'Annual Allied Merchant Ship Losses'.

85 Kennedy, *Rise and Fall of British Naval Mastery*, p. 300; F. C. Lane, *Ships for Victory: A history of shipbuilding under the US Maritime Commission in World War II* (Baltimore, 1951), pp. 5–7; S. E. Morison, *History of United States Naval Operations: Vol. XV, Supplement* (Washington, 1962), 'Ships of the United States Navy 1940–45'. In August 1945 the United States still had in commission 100 aircraft carriers, 23 battleships, 74 cruisers, 475 destroyers, 402 destroyer escorts, 253 submarines and 345 minelayers/minesweepers. Some of the smaller vessels had been converted by 1945 to auxiliary roles of one kind or another.

86 D. M. McKale, *Hitler: The survival myth* (New York, 1981), pp. 137–8; Morison, *Atlantic Battle*, pp. 360–1.

87 Morison, *Atlantic Battle*, p. 361.

3 Deep War

STALINGRAD AND KURSK

1 H.C. Cassidy, *Moscow Dateline 1941–1943* (London, 1944), pp. 220–1; A. Seaton, *Stalin as Warlord* (London, 1976), p. 39.

2 I. Deutscher, *Stalin: A political biography* (London, 1966), p. 476.

3 H. Trevor-Roper (ed.), *Hitler's War Directives 1939–1945* (London, 1964),

pp. 178–83, Directive no. 41, 'Our aim is to wipe out the entire defence potential remaining to the Soviets'.

4   C. Andrew, O. Gordievsky, *KGB: The inside story* (London, 1990), pp. 224–5; D. M. Glantz, *The Role of Intelligence in Soviet Military Strategy in World War II* (Novato, California, 1990), pp. 49–51.

5   W. Warlimont, *Inside Hitler's Headquarters* (London, 1964), pp. 246–7; D. Irving (ed.), *Adolf Hitler: The medical diaries* (London, 1983), pp. 98–100.

6   Trevor-Roper (ed.), *Hitler's War Directives*, pp. 193–7, Directive no. 45, 'Operation Brunswick'.

7   Warlimont, *Inside Hitler's Headquarters*, p. 248; M. Cooper, *The German Army 1933–1945* (London, 1978), pp. 416–20.

8   A. Speer, *Inside the Third Reich* (London, 1970), pp. 237–8.

9   A. Werth, *Russia at War 1941–1945* (London, 1964), pp. 409–13; J. Barber, M. Harrison, *The Soviet Home Front 1941–1945* (London, 1991), pp. 31–2, 72.

10 Cassidy, *Moscow Dateline*, p. 243.

11 *Great Patriotic War of the Soviet Union 1941–1945: A general outline* (Moscow, 1970), p. 117; G. K. Zhukov, *Marshal Zhukov's Greatest Battles* (New York, 1960), pp. 154–15,127.

12 Zhukov, *Greatest Battles*, pp. 132–3; G. K. Zhukov, *Reminiscences and Reflections* (Moscow, 1985), pp. 86–7.

13 O.P. Chaney, *Zhukov* (Norman, Oklahoma, 1972), chs 1–4.

14 Zhukov, *Reminiscences*, pp. 96–7, 119; Zhukov, *Greatest Battles*, pp. 139–44.

15 On the role of artillery, engineer and supply services see L. Rotundo (ed.), *Battle for Stalingrad: The 1943 Soviet General Staff study* (Washington, 1989), chs 9–12.

16 V. I. Chuikov, *The Beginning of the Road: The story of the battle for Stalingrad* (London, 1963), pp. 14–27; Werth, *Russia at War*, pp. 559–60.

17 W. Goerlitz, *Paulus and Stalingrad* (London, 1963), pp. 4–6, 24–7, 47–8; M. Middlebrook, 'Paulus' in C. Barnett (ed.), *Hitler's Generals* (London, 1989), pp. 361–5.

18 Chuikov, *Beginning of the Road*, p. 78.

19 Werth, *Russia at War*, pp. 454–5; Chuikov, *Beginning of the Road*, pp. 94–9.

20 Cited in Chuikov, *Beginning of the Road*, p. 132.

21 Chuikov, *Beginning of the Road*, p. 191.

22 S. J. Zaloga, J. Grandsen, *Soviet Tanks and Combat Vehicles of World War Two* (London, 1984), pp. 152–4.

23 Von Hardesty, *Red Phoenix: The rise of Soviet air power* (London, 1982), pp. 97–104; *The Soviet Air Force in World War II* (London, 1974, from the

Russian original), pp. 114–34. J. S. Hayward *Stopped at Stalingrad: The Luftwaffe and Hitler's defeat in the East 1942–1943* (Lawrence, Kans., 1998), pp. 322–3.

24 Warlimont, *Inside Hitler's Headquarters*, pp. 272–3.

25 Chuikov, *Beginning of the Road*, pp. 217–18.

26 *Great Patriotic War*, pp. 161–2; D. Kahn, *Hitler's Spies: German Military Intelligence in World War II* (London, 1978), pp. 435–9.

27 *Great Patriotic War*, p. 161; Zhukov, *Greatest Battles*, pp. 157–9.

28 Zhukov, *Reminiscences*, pp. 124–5.

29 J. Erickson, *The Road to Berlin: Stalin's war with Germany: Vol. II* (London, 1983), pp. 1–6; 'Stalingrad, a brief survey in retrospect by Field Marshal Paulus', reproduced in Goerlitz, *Paulus and Stalingrad*, pp. 283–5.

30 Warlimont, *Inside Hitler's Headquarters*, p. 277.

31 J. Fischer, 'Über den Entschluss zur Luftversorgung Stalingrads. Ein Beitrag zur militärischen Führung im Dritten Reich', *Militärgeschichtliche Mittellungen* 6 (1969); R. Suchenwirth, *Command and Leadership in the German Air Force*, USAF Historical Study no. 179 (New York, 1969), pp. 77–80.

32 W. Murray, *Luftwaffe* (London, 1985), pp. 141–4.

33 Details in D. Glantz, *From the Don to the Dnepr* (London, 1991), pp. 41–73; Erickson, *Road to Berlin*, pp. 15–31; E. von Manstein, *Verlorene Siege* (Bonn, 1955), pp. 359–80.

34 Chuikov, *Beginning of the Road*, p. 254. On the conditions facing German soldiers, see the account in J. Wieder, *Stalingrad und die Verantwortung des Soldaten* (Munich, 1962), esp. pp. 42–6.

35 Warlimont, *Inside Hitler's Headquarters*, p. 284.

36 Erickson, *Road to Berlin*, p. 114.

37 Werth, *Russia at War*, pp. 537–8.

38 Warlimont, *Inside Hitler's Headquarters*, p. 284. On the intelligence errors, see Erickson, *Road to Berlin*, pp. 47–9.

39 Chuikov, *Beginning of the Road*, pp. 258–9.

40 Werth, *Russia at War*, pp. 540–1; Middlebrook, 'Paulus', pp. 371–2.

41 Werth, *Russia at War*, p. 543.

42 Zhukov, *Greatest Battles*, p. 192. The figures for tanks and aircraft are almost certainly too high. The Luftwaffe lost almost 500 transport aircraft to the airlift, against a Soviet estimate of 676. Total German losses on the eastern front from September to January were 1,646 aircraft, while large numbers were rendered unserviceable. See Von Hardesty, *Red Phoenix*, p. 110, and Murray, *Luftwaffe*, pp. 107, 138, 142–4.

43 Werth, *Russia at War*, pp. 560–3.

44 Cassidy, *Moscow Dateline*, p. 222; A. Calder, *The People's War: Britain, 1939–1945* (London, 1969), pp. 401–2.

45 W.S. Churchill, *The Second World War* (6 vols, London, 1948–54), IV, p. 661.

46 M. Muggeridge (ed.), *Ciano's Diary, 1939–1943* (London, 1947), p. 555.

47 F. Gilbert (ed.), *Hitler Directs His War: The secret record of his daily military conferences* (New York, 1950), pp. 19–22.

48 Speer, *Inside the Third Reich*, pp. 250–1: 'Our enemies rightly regarded this disaster at Stalingrad as a turning point in the war. But at Hitler's headquarters the only reaction was a temporary numbness followed by a rush of feverish staff work . . .'

49 W. Keitel, *The Memoirs of Field Marshal Keitel* (London, 1965), p. 182.

50 K. Rokossovsky, *A Soldier's Duty* (Moscow, 1970), pp. 175–7.

51 Glantz, *Role of Intelligence*, pp. 80–2.

52 Cooper, *German Army*, p. 452.

53 P. Carell, *Hitler's War on Russia* (2 vols, London, 1970), doc. 1, 'Operation Order No. 6 (Citadel) of 15.4.1943', pp. 564–8; von Manstein, *Verlorene Siege*, pp. 473–95: H. Guderian, *Erinnerungen eines Soldaten* (Heidelberg, 1951), pp. 280–2.

54 S.M. Shtemenko, *The Soviet General Staff at War* (Moscow, 1970), pp. 152–61; Glantz, *Role of Intelligence*, pp. 84–5, 93–7. On improvements in Soviet intelligence gathering for Kursk, see T. P. Mulligan, 'Spies, ciphers and "Zitadelle": Intelligence and the Battle of Kursk 1943', *Journal of Contemporary History* 22 (1987), pp. 246–50. Soviet forces conducted 105 reconnaissance-in-force operations, 2,600 night raids and 1,500 ambushes to obtain detailed information from captives.

55 Andrew, Gordievsky, *KGB*, pp. 246–9; Zhukov, *Greatest Battles*, pp. 207–9, 214–22.

56 Rokossovsky, *Soldier's Duty*, pp. 186–90.

57 Zaloga, Grandsen, *Soviet Tanks*, pp. 131–7.

58 P. Rotmistrov, 'Tanks against Tanks' in *Main Front: Soviet leaders look back on World War II* (London, 1987), pp. 109–10; Zaloga, Grandsen, *Soviet Tanks*, p. 166.

59 Von Hardesty, *Red Phoenix*, pp. 152–8; on Soviet use of radio at Kursk, see D.R. Beachley, 'Soviet Radio-Electronic Combat in World War II', *Military Review* 61 (1981), pp. 66–72.

60 Zhukov, *Reminiscences*, p. 166, 179.

61 Mulligan, 'Spies, ciphers', pp. 238–9; Glantz, *Role of Intelligence*, pp. 100–3.

62 Glantz, *Role of Intelligence*, pp. 103–4; Erickson, *Road to Berlin*, pp. 130–1. On the Voronezh front a smaller artillery barrage had already been released at 10.30 p.m. on the 4th, following exploratory attacks by

German aircraft and tanks from Belgorod earlier in the day.

63 Zhukov, *Reminiscences*, p. 183.

64 Rokossovsky, *Soldier's Duty*, pp. 199–202.

65 C. Sydnor, *Soldiers of Destruction: The SS Death's Head Division, 1933–1945* (Princeton, 1977), pp. 283–8; Erickson, *Road to Berlin*, pp. 137–40.

66 Rotmistrov, 'Tanks against Tanks', pp. 109–10.

67 Ibid., pp. 112–13.

68 Sydnor, *Soldiers of Destruction*, pp. 288–90; Rotmistrov, 'Tanks against Tanks', pp. 114–18; Erickson, *Road to Berlin*, pp. 144–7.

69 Sydnor, *Soldiers of Destruction*, pp. 290-1; F. W. von Mellenthin, *Panzer-Schlachten* (Neckargemünd, 1963), pp. 163–5; 'our tank forces were almost bled white'.

70 Guderian, *Erinnerungen*, p. 283; Zaloga, Grandsen, *Soviet Tanks*, p. 166.

71 Zhukov, *Greatest Battles*, pp. 245–6.

72 Shtemenko, *Soviet General Staff*, pp. 177–84.

73 Ibid., pp. 190–2, 193–4.

74 Churchill, *Second World War*, V, p. 321; Shtemenko, *Soviet General Staff*, p. 197.

75 Zhukov, *Reminiscences*, p. 226.

76 Von Manstein, *Verlorene Siege*, p. 508; von Mellenthin, *Panzer-Schlachten*, p. 257; K. Zeitzler, 'Stalingrad' in W. Richardson, S. Freiden (eds), *The Fatal Decisions* (London, 1956), pp. 118–19.

77 Warlimont, *Inside Hitler's Headquarters*, p. 304.

78 Soviet losses in G.F. Krivosheev (ed.) *Soviet Casualties and Combat Losses in the Twentieth Century* (London, 1997), p. 85-91; see too B. Sokolov 'The Cost of War: Human Losses for the USSR and Germany 1939–1945', *Journal of Slavic Military Studies* 9 (1996), pp. 152–93.

79 Zhukov, *Greatest Battles*, pp. 256–7; Zhukov, *Reminiscences*, pp. 195–226. See too the discussion in J. Erickson, 'New Thinking about the Eastern Front in World War II', *Journal of Military History* 56 (1992), pp. 284–92; J. F. Gebhardt, 'World War II: The Soviet side', *Military Review* 72 (December 1992), pp. 91–3.

80 I. Ehrenburg, *Men, Years – Life: Vol. V, The War 1941–45* (London, 1964), p. 107.

81 Werth, *Russia at War*, p. 414.

82 Chuikov, *Beginning of the Road*, p. 78.

4 The Means to Victory
BOMBERS AND BOMBING

1   A. Werth, *Russia at War, 1941–1945* (London, 1964), p. 485.

2   USSR, Ministry of Foreign Affairs, *Correspondence between the Chairman of the Council of Ministers of the USSR and the Presidents of the USA and the Prime Ministers of Great Britain during the Great Patriotic War of 1941 to 1945: Correspondence with Winston S. Churchill and Clement R. Attlee* (Moscow, 1957), doc. 56: for Churchill's reply, W.F. Kimball (ed.), *Churchill and Roosevelt: The complete correspondence* (3 vols, Princeton, 1984), I, 529–32.

3   M. Matloff, E. M. Snell, *Strategic Planning for Coalition Warfare: 1941–1942* (Washington, 1953), pp. 177–87.

4   W. Averell Harriman with E. Abel, *Special Envoy to Churchill and Stalin 1941–1946* (London, 1976), p. 151; W. S. Churchill, *The Second World War* (6 vols, London, 1948–54), IV, pp. 428–9.

5   Harriman, *Special Envoy*, p. 152.

6   Ibid., p. 153.

7   Kimball (ed.), *Churchill and Roosevelt*, I, p. 561, letter from Churchill to Roosevelt, 13 August 1942.

8   Harriman, *Special Envoy*, p. 157.

9   For accounts of the final meeting: Churchill, *Second World War*, IV, pp. 448–9; A. H. Birse, *Memoirs of an Interpreter* (London, 1967), pp. 101–4. On the return journey, R. Beaumont, 'The Bomber Offensive as a Second Front', *Journal of Contemporary History* 22 (1987), p. 11.

10 J. Sweetman, 'Crucial Months for Survival: The Royal Air Force 1918–1919', *Journal of Contemporary History* 19 (1984), pp. 530–40.

11 Churchill, *Second World War*, II, p. 567.

12 Beaumont, 'Bomber Offensive as a Second Front', p. 12.

13 'Report by Mr Justice Singleton on the Bombing of Germany, 20 May 1942', appendix 17 in C. Webster and N. Frankland, *The Strategic Air Offensive Against Germany* (4 vols, London, 1961), IV, pp. 231–8.

14 A. van Ishoven, *The Fall of an Eagle: The life of fighter ace Ernst Udet* (London, 1977), pp. 159–71.

15 R.J. Overy, 'From "Uralbomber" to "Amerikabomber": The Luftwaffe and strategic bombing', *Journal of Strategic Studies* 1 (1978), pp. 156–7, 167–70.

16 H.G. Wells, *The War in the Air* (London, 1908), pp. 352–4.

17 *Liberty*, 5 December 1931, p. 52.

18 R.J. Overy, 'Air Power and the Origins of Deterrence Theory before 1939', *Journal of Strategic Studies* 14 (1992).

19 The use of biological metaphor was widespread. It can even be found in

the official manual of the RAF, *War Manual: Part I, Operations*, May 1935, p. 57, 'nerve centres, main arteries, heart and brain . . .'

20 Public Record Office (PRO), Kew, London, AIR 9/102, Air Targets Intelligence, Germany File, 13 October 1938.

21 N. Gibbs, *Grand Strategy: Vol. I, Rearmament Policy* (London, 1976), p. 598; J. Slessor, *The Central Blue* (London, 1956), pp. 203–5.

22 D. Richards, *Royal Air Force 1939–1945: Vol. 1, The Fight at Odds* (London, 1974), p. 124.

23 Kimball (ed.), *Churchill and Roosevelt*, I, p. 57, letter of 31 July 1940.

24 K. Maier *et al.*, *Germany and the Second World War: Vol. II, Germany's Initial Conquests in Europe* (Oxford, 1991), p. 386.

25 Details in F.J. Assersohn, 'Propaganda and Policy: The presentation of the strategic air offensive in the British mass media 1939–1945', Leeds University M.A. thesis, 1989.

26 M. Sherry, *The Rise of American Air Power: The creation of Armageddon* (New Haven, 1987), pp. 95–8.

27 H.L. Ickes, *The Secret Diary of Harold L. Ickes* (2 vols, London, 1955), II, p. 37.

28 Library of Congress, Arnold Papers, Box 246, G2 report, 22 January 1941, 'British Estimates of German Military Power'; G2 Report 16 January 1941: 'Germany was *capable* of *much greater* aircraft production' (italic in original).

29 Kimball (ed.), *Churchill and Roosevelt*, I, p. 224, letter from Churchill to Roosevelt, 25 July 1941.

30 'Report by Mr Butt to Bomber Command on his Examination of Night Photographs, 18 August 1941', appendix 13 in Webster and Frankland, *Strategic Air Offensive*, IV, pp. 205–13.

31 Churchill, *Second World War*, III, p. 451.

32 Air Ministry, *The Rise and Fall of the German Air Force 1933–1945* (London, 1947, reprinted 1983), pp. 185–8.

33 Richards, *Royal Air Force*, p. 381.

34 A. Harris, *Bomber Offensive* (London, 1947), chs 1–2. According to this account, Harris 'could see only one possible way of bringing serious pressure to bear on the Boche, and certainly only one way of defeating him; that was by air bombardment' (p. 15).

35 'Memorandum by Lord Trenchard on the Present War Situation, 19 May 1941', appendix 10 in Webster and Frankland, *Strategic Air Offensive*, IV, pp. 194–7.

36 'Air Ministry *directif*, 14 February 1942', appendix 8, xxii, in Webster and Frankland, *Strategic Air Offensive*, IV, pp. 143–5.

37 Harris, *Bomber Offensive*, p. 80.

38 On the build-up of the 8th air force see W. F. Craven, J.L. Cate, *The Army Air Forces in World War II* (6 vols, Washington, 1948–55, reissued 1983), I, pp. 612–54. On the B–17 bomber, see USAF Historical Study no. 6, *The Development of the Heavy Bomber 1918–1944* (Maxwell, Alabama, August 1951) and R.W. Krauskopf, 'The Army and the Strategic Bomber, 1930–1939', *Military Affairs* 22 (1958–9).

39 Craven, Cate, *Army Air Forces*, I, p. 664.

40 Ibid., pp. 242–73.

41 R.E. Sherwood, *The White House Papers of Harry L. Hopkins* (2 vols, London, 1949), II, pp 665–70.

42 Details of the directive in 'The Combined Bomber Offensive from the United Kingdom (Pointblank) as approved by the Combined Chiefs of Staff, 14 May 1943, in Webster and Frankland, *Strategic Air Offensive*, IV, pp. 273–83.

43 Air Ministry, *Rise and Fall*, pp. 191–2.

44 F. Gilbert (ed.), *Hitler Directs His War: The secret record of his daily military conferences* (New York, 1950), meeting of 25 July 1943, pp. 41–2.

45 R. V. Jones, *Most Secret War: British scientific intelligence 1939–1945* (London, 1978), pp. 297–9.

46 H.E. Nossack *The End: Hamburg 1943* (Chicago, 2004), p. 12.

47 A. Speer, *Inside the Third Reich* (London, 1971), p. 284. On the effects on Hamburg see U. Büttner, 'Hamburg im Luftkrieg: Die politischen und wirtschaftlichen Folgen des "Unternehmens Gomorrha"', in A. Hiller *et al* (eds), *Städte im 2. Weltkrieg* (Essen, 1991), pp. 272–94.

48 C. Bekker, *The Luftwaffe War Diaries* (London, 1967), pp. 400–1.

49 Webster and Frankland, *Strategic Air Offensive*, III, pp. 162–3.

50 Ibid., pp. 193, 208–9.

51 A. Galland, *The First and the Last* (London, 1955), p. 185; Craven, Cate, *Army Air Forces*, II, pp. 681–3; F. Golücke, *Schweinfurt und der strategische Luftkrieg 1943* (Paderborn, 1980).

52 Speer, *Inside the Third Reich*, p. 286; Craven, Cate, *Army Air Forces*, II, pp. 665–9.

53 Craven, Cate, *Army Air Forces*, III, p. 8.

54 Ibid., pp. 217–20 on the development of the P–51. See too S.L. McFarland, 'The Evolution of the American Strategic Fighter in Europe, 1942–44', *Journal of Strategic Studies* 10 (1987).

55 W. Murray, *Luftwaffe: Strategy for Defeat* (London, 1985), pp. 211–15.

56 Galland, *First and the Last*, p. 201.

57 H.H. Arnold, *Second Report of the Commanding General of the United States Army Air Forces* (London, 1945), p. 36.

58 H. Knoke, *I Flew for the Führer* (London, 1953), p. 139.

59 On oil, see R.C. Cooke, R.C. Nesbit, *Target: Hitler's Oil: Allied attacks on German oil supplies 1939–1945* (London, 1985); on the railway network, A.C. Mierzejewski, *The Collapse of the German War Economy 1944–1945: Allied air power and the German national railway* (Chapel Hill, North Carolina, 1988), esp. pp. 194–8.

60 Speer, *Inside the Third Reich*, p. 424. See too the evaluation in H.S. Hansell, *The Strategic Air War against Germany and Japan* (Office of Air Force History, Washington, 1986), pp. 119–30.

61 United States Strategic Bombing Survey (USSBS), Pacific Theatre, 'Summary Reports' (Washington, 1 July 1946), pp. 16–18, 20.

62 Background in Sherry, *Rise of American Air Power*, pp. 316–30; R. Rhodes, *The Making of the Atomic Bomb* (New York, 1986), pp. 679–91.

63 G.H. Roeder, *The Censored War: American visual experience during World War Two* (New Haven, 1993), p. 86.

64 USSBS, Pacific Theatre, 'The Effects of Strategic Bombing on Japanese Morale' (Washington, June 1947), pp. 35–6.

65 USSBS, 'Summary Report', pp. 22–4; USSBS, 'Effects of Strategic Bombing on Japanese Morale', pp. 91–3.

66 On the work of the Survey team see J.K. Galbraith, *A Life In Our Times: Memoirs* (London, 1981), pp. 209–40. His very negative conclusions on the effects of bombing influenced the tone of the Bombing Survey report, *The Effects of Strategic Bombing on the German War Economy*, USSBS Report 3, 31 October 1945.

67 On Italy see S.J. Harvey, 'The Italian War Effort and the Strategic Bombing of Italy', *History* 70 (1985), pp. 40–41.

68 PRO, Kew, AIR 10/3866, British Bombing Survey Unit, "The Strategic Air War Against Germany 1939–1945', pp. 38–9. This figure covered the whole war period; for the last two years of war the bombing offensive took 12 per cent of the war effort (measured in terms of production and combat man-hours).

69 S. Harvey, 'The Italian War Effort and the Strategic Bombing of Italy', *History* 70 (1985), pp. 32–45.

70 On the diversion of resources, Beaumont, 'Bomber Offensive as a Second Front', pp. 14–15; Air Ministry, *Rise and Fall*, pp. 283–6. On German aircraft production, Webster and Frankland, *Strategic Air Offensive*, IV, appendix 49, xxii, pp. 494–5.

71 Imperial War Museum, Speer Collection, Box 368, Report 67, p. 14, interrogation and notes of Karl-Otto Saur. See too Report 68, p. 18, interrogation of Ernst Blaicher, who confirmed that Germany could have produced 30,000 tanks in 1944, rather than the 19,000 actually produced.

72 Air Ministry, *Rise and Fall*, p. 298.

73 PRO, AIR 10/3873, British Bombing Survey Unit, 'German Experience in the Underground Transfer of War Industries', pp. 6–12.

74 See the general treatment of this question in E. Beck, *Under The Bombs: The German Home Front 1942–1945* (Lexington, Kentucky, 1986). For good case studies of German cities: D. Busch, *Der Luftkrieg im Raum Mainz während des Zweiten Weltkrieges 1939–1945* (Mainz, 1988); M. Hiller (ed.), *Stuttgart im Zweiten Weltkrieg* (Gerlingen, 1989).

75 G. Kirwin, 'Allied Bombing and Nazi Domestic Propaganda', *European History Quarterly* 15 (1985), pp. 341–62.

76 J. Stern, *The Hidden Damage* (New York, 1947, reissued London, 1990), p. 230.

77 USSBS, 'Effects of Strategic Bombing on Japanese Morale', pp. 20–1.

78 USSBS, European Theatre, Report 64B, 'The Effects of Strategic Bombing on German Morale', vol. I (Washington, May 1947), pp. 13–14.

79 USSBS, Report 18, 'Bayerische Motorenwerke AG (Munich)' (Washington, 22 October 1945), p. 5; USSBS, 'Effects of Strategic Bombing on German Morale', p. 20.

80 Quoted in M. Hastings, *Bomber Command* (London, 1979), p. 241.

5 Along a Good Road
THE INVASION OF FRANCE

1   S. Roskill, *The War at Sea 1939–1945* (3 vols, London, 1961), II, pp. 185–90; A. J. Marder, M. Jacobsen, J. Horsfield, *Old Friends, New Enemies: The Royal Navy and the Imperial Japanese Navy 1942–1945* (Oxford, 1990), pp. 155–9.

2   W.S. Churchill, *The Second World War* (6 vols, London, 1948–54), IV, p. 212.

3   J. A. Brown, *Eagles Strike: South African Forces in World War II, Vol. IV* (Cape Town, 1974), pp. 388–400; Marder *et al, Old Friends*, p. 158.

4   Churchill, *Second World War*, IV, pp. 197–8.

5   T. Parrish, *Roosevelt and Marshall: Partners in politics and war* (New York, 1989), p. 255; M. Matloff, E. Snell, *Strategic Planning for Coalition Warfare* (2 vols, Washington, 1953–9), I, pp. 161–77.

6   R.F. Weigley, *The American Way of War: A history of United States military strategy and policy* (London, 1973), pp. 317–21.

7   F. Morgan, *Overture to Overlord* (London, 1950), pp. 134–6.

8   R.M. Leighton, R.W. Coakley, *Global Logistics and Strategy: 1943–45* (Washington, 1968), pp. 835–6, appendices d-1, d-3.

9   Ibid., pp. 16–25, appendix B-1, pp. 826–8.

10 Ibid., pp. 10–25, 205–13, appendix B-2, p. 829.

11 Morgan, *Overture*, pp. 142–4; J. Ehrman, *Grand Strategy, Vol. V: August 1943 to September 1944* (London, 1946), pp. 54–6. The size of the assault force in the original plan was conditioned by the narrow beach exits, which it was estimated could handle only 12,100 vehicles in the first 24 hours, enough for only three divisions.

12 *The Secret History of World War II: The Ultra-secret wartime letters and cables of Roosevelt, Stalin and Churchill* (New York, 1986), Stalin to Roosevelt, 11 June 1943, pp. 106–7.

13 E. Morgan, *FDR: A biography* (London, 1985), p. 625.

14 M. Matloff, *Strategic Planning for Coalition Warfare 1943–44* (Washington, 1959), pp. 10–11. The remark was made by General Thomas T. Hardy. On British preferences, see M. Howard, *The Mediterranean Strategy in the Second World War* (London, 1968), chs 1–2. For a wartime analysis, B.L. Liddell Hart, *The British Way in Warfare* (London, 1942).

15 The contrast is discussed in Weigley, *American Way of War*, pp. 327–34. See too K. Greenfield, *American Strategy in World War II: A reconsideration* (Baltimore, 1963), esp. pp. 24–48.

16 Matloff, *Strategic Planning*, pp. 213–14.

17 M. A. Stoler, *The Politics of the Second Front: American military planning in coalition warfare, 1941–1943* (Westport, Conn., 1977), pp. 11–15.

18 Ibid., pp. 130–1; A. Bryant, *Triumph in the West: The war diaries of Field Marshal Viscount Alanbrooke* (London, 1959), p. 58; D. Dilks (ed.), *The Diaries of Sir Alexander Cadogan 1938–1945* (London, 1971), p. 570, entry for 26 October 1943.

19 S. Beria, *Beria, My Father: Inside Stalin's Kremlin* (London, 2001), pp. 192–3.

20 Details on Teheran from R.E. Sherwood (ed.), *The White House Papers of Harry L. Hopkins* (2 vols, London, 1949), II, pp. 771–90; Morgan, *FDR*, pp. 693–4,697–704; Dilks (ed.) *Diaries of Cadogan*, pp. 578–82; Stoler, *Politics of the Second Front*, pp. 146–9; J. Ehrman, *Grand Strategy Vol. V: August 1943 to September 1944* (London, 1956), pp. 174–81.

21 A. Birse, *Memoirs of an Interpreter* (London, 1967), pp. 160–1.

22 P. Brendon, *Ike: His life and times* (New York, 1986), p. 45 for quotation, chs 1–4 for Eisenhower's career.

23 Bryant, *Triumph in the West*, p. 114, Alanbrooke's diary entry for 24 January 1944: 'Eisenhower has got absolutely no strategical outlook. He makes up, however, by the way he works for good co-operation between allies'. See the full comments in A. Danchev, D. Todman (eds), *War Diaries 1939–1945: Field Marshal Lord Alanbrooke* (London, 2001), p. 351: 'I am afraid that Eisenhower as a general is quite hopeless.'

24 Details from B.L. Montgomery, *The Memoirs of Field-Marshal the Viscount Montgomery* (London, 1958), chs 1–3.

25 Montgomery *Memoirs*, pp. 210–12; Churchill, *Second World War*, V, p. 393.

26 D.D. Eisenhower, *Report by the Supreme Commander to the Combined Chiefs of Staff on the Operations in Europe* (London, 1946), pp. 11, 16–17; G. A. Harrison, *Cross-Channel Attack* (Washington, 1951), pp. 164–74.

27 Eisenhower, *Report by the Supreme Commander*, p. 7; S.E. Morison, *The Invasion of France and Germany 1944–1945, History of US Naval Operations*, vol. IX (London 1957), pp. 28–33.

28 Leighton, Coakley, *Global Logistics*, appendix D-5, p. 838, appendix D-3, p. 836; the figure of 350,000 from Eisenhower, *Report by the Supreme Commander*, p. 16.

29 Quotation from Morison, *Invasion of France and Germany*, p. 32; shipping details from L. Ellis, *Victory in the West: Vol. I, The Battle of Normandy* (London, 1962), pp. 34, 507; Royal Navy Historical Branch, Battle Summary No. 39, *Operation 'Neptune'* (London, 1994), p. 38 gives the following figures for the overall shipping allocated:

naval units	1,206
landing craft	4,127
ancillary craft	423
merchant ships	1,260
total	7,016

30 Roskill, *War at Sea*, III, pp. 25–8.

31 Eisenhower, *Report by the Supreme Commander*, pp. 13–14; W.F. Craven, J.L. Cate, *The Army Air Forces in World War II* (6 vols, Washington, 1948–55, reissued 1983), III, pp. 73–4.

32 R.G. Davis, *Carl A. Spaatz and the Air War in Europe* (Washington, 1992), p. 366; A. Harris, *Bomber Offensive* (London, 1947), pp. 197–201.

33 S.E. Ambrose, *Eisenhower: Soldier and President* (New York, 1990), p. 126.

34 Davis, *Carl A. Spaatz*, pp. 336–8.

35 Craven, Cate, *Army Air Forces*, III, pp. 75–81; Davis, *Carl A. Spaatz*, pp. 347–9.

36 Craven, Cate, *Army Air Forces*, III, p. 160; Churchill, *Second World War*, V, pp. 465–7 for the Churchill-Eisenhower-Roosevelt correspondence; Davis, *Carl A. Spaatz*, pp. 400–1.

37 Craven, Cate, *Army Air Forces*, III, p. 158; Eisenhower, *Report by the Supreme Commander*, p. 20.

38 Craven, Cate, *Army Air Forces*, III, pp. 159–60.

39 For details see J. C. Masterman, *The Double-Cross System 1939–1945* (London, 1972), esp. chs 10, 11; M. Howard, *Strategic Deception in the Second World War* (London, 1990), chs 5, 6; on air reconnaissance D.

Kahn, *Hitler's Spies: German military intelligence in World War II* (London, 1978), pp. 499–500. An average of one successful photo-reconnaissance flight every other day was the most the German air force could supply, far too little to provide a clear picture of Allied forces or movements.

40 Howard, *Strategic Deception*, 115–17; Kahn, *Hitler's Spies*, pp. 488–9; C. Cruickshank, *Deception in World War II* (London, 1979), pp. 99–113, 177–86.

41 Cruickshank, *Deception*, pp. 177–81; W.B. Breuer, *Hoodwinking Hitler: The Normandy deception* (Westport, Conn., 1993), pp. 110–17.

42 Cruickshank, *Deception*, pp. 176–7; Howard, *Strategic Deception*, pp. 106, 120.

43 Breuer, *Hoodwinking Hitler*, pp. 110–12; Kahn, *Hitler's Spies*, pp. 489–91.

44 Howard, *Strategic Deception*, p. 131; Kahn, *Hitler's Spies*, pp. 492–6.

45 Eisenhower, *Report by the Supreme Commander*, pp. 20, 24; on German reactions see Kahn, *Hitler's Spies*, pp. 488–9; Howard, *Strategic Deception*, pp. 130–1; Harrison, *Cross-Channel Attack*, pp. 257–60.

46 Cited in Kahn, *Hitler's Spies*, p. 497.

47 H. Trevor-Roper (ed.), *Hitler's War Directives 1939–1945* (London, 1964), pp. 218–20; F. Gilbert (ed.), *Hitler Directs His War: The secret records of his daily military conferences* (New York, 1950), pp. 76–7; B. Liddell Hart, *The Rommel Papers* (London, 1953), pp. 465–6.

48 A. Speer, *Inside the Third Reich* (London, 1971), pp. 352–4; Harrison, *Cross-Channel Attack*, pp. 142–7; M. Cooper, *The German Army, 1933–1945* (London, 1978), pp. 485–91; F. Ruge, 'The Invasion of Normandy', in H.-A. Jacobsen, J. Rohwer, *Decisive Battles of World War II: The German view* (London, 1965), pp. 322–5.

49 Details in D. Fraser, *Knight's Cross: A life of Field-Marshal Rommel* (London, 1993), pp. 452–60; Harrison, *Cross-Channel Attack*, pp. 241–64; Liddell Hart, *Rommel Papers*, pp. 454–60.

50 Morison, *Invasion of France and Germany*, pp. 43–6; Harrison, *Cross-Channel Attack*, pp 236–40.

51 Harrison, *Cross-Channel Attack*, pp. 154–5, 249–52; B. Liddell Hart, *The Other Side of the Hill* (London, 1948), pp. 387–9 for the views of Rundstedt and von Schweppenburg; E.F. Ziemke, 'Rundstedt', in C. Barnett (ed.) *Hitler's Generals* (London, 1989), pp. 198–9; F. Ruge, *Rommel und die Invasion: Erinnerungen von Friedrich Ruge* (Stuttgart, 1959), pp. 174–5.

52 Fraser, *Knight's Cross*, pp. 460–4.

53 Liddell Hart, *Other Side*, pp. 398–403; Cooper, *German Army*, pp. 500–1.

54 Fraser, *Knight's Cross*, p. 461.

55 R. Miller, *Nothing Less than Victory: The oral history of D-Day* (London,

1993), pp. 79–88, 90–8 on the mood of German soldiers; Kahn, *Hitler's Spies*, pp. 507–9 on the May scare.

56   Kahn, *Hitler's Spies*, pp. 505–11.

57   Roskill, *War at Sea*, III, p. 18; Morison, *Invasion of France and Germany*, pp. 69–70; on the meeting, see C. D'Este, *Decision in Normandy* (New York, 1983), pp. 82–90.

58   D.D. Eisenhower, *Crusade in Europe* (London, 1948), p. 269; on Churchill's remarks to Harriman, see W. Averell Harriman with E. Abel, *Special Envoy to Churchill and Stalin 1941–1946* (London, 1976), p. 311: '"He told me", Harriman recalled, "that if Overlord failed, the United States would have lost a battle, but for the British it would mean the end of their military capability."'

59   Morison, *Invasion of France and Germany*, p. 68; Roskill, *War at Sea*, III, p. 12.

60   Eisenhower, *Report by the Supreme Commander*, pp. 18–19; Royal Navy Historical Branch, *Operation 'Neptune'*, pp. 70–1.

61   Bryant, *Triumph in the West*, p. 157, Alanbrooke's diary entry for 27 May 1944; M. Blumenson (ed.), *The Patton Papers 1940–1945* (Boston, 1974), p. 454.

62   Ambrose, *Eisenhower*, p. 135.

63   Morison, *Invasion of France and Germany*, pp. 79–80; Royal Navy Historical Branch, *Operation 'Neptune'*, pp. 72–4. The Admiralty predicted a force eight gale in the Irish Sea, moving eastwards. See too Eisenhower, *Crusade*, pp. 273–4.

64   Ambrose, *Eisenhower*, pp. 138–40; Eisenhower, *Crusade*, pp. 274–5; Brendon, *Ike*, p. 145 has the final words as '"O.K. We'll go."' Perhaps he spoke less distinctly than those present recalled.

65   Morison, *Invasion of France and Germany*, pp. 84–7; Fraser, *Knight's Cross*, p. 485; Harrison, *Cross-Channel Attack*, pp. 275–6; Ruge, *Rommel und die Invasion*, p. 166.

66   Miller, *Nothing Less than Victory*, p. 221.

67   On the bombardments see Roskill, *War at Sea*, III, pp. 42–7; Morison, *Invasion of France and Germany*, pp. 11–12, 93–4; Craven, Cate, *Army Air Forces*, III, pp. 190–3.

68   Eisenhower, *Report by the Supreme Commander*, p. 32.

69   Ruge 'Invasion of Normandy', pp. 336, 342–3; Roskill, *War at Sea*, III, pp. 53–5; Royal Navy Historical Branch, *Operation 'Neptune'*, p. 132 for losses throughout the campaign. Weather claimed 153 craft of all kinds (mostly small vessels); enemy action (submarines, torpedo boats, aircraft, shore fire) claimed 108, including 25 warships and 18 cargo vessels.

70 British Air Ministry, *The Rise and Fall of the German Air Force 1919–1945* (London, 1947, reissued 1983), pp. 323–5, 327–32. By the end of June the Allies enjoyed an 11:1 superiority in fighter aircraft in France. On Rommel's views on the decisive effect of air attack, see Ruge, *Rommel und die Invasion*, pp. 169–70; L.F. Ellis, *Victory in the West: Vol. I, The Battle of Normandy* (London, 1962), p. 567, has figures of 466 serviceable combat aircraft for the whole 3rd air fleet territory.

71 Speer, *Inside the Third Reich*, pp. 354–5; Goebbeis diaries, entries for 5th, 7th and 8 June, *Daily Mail* 11 July 1992, pp. 6–7; H. Eberle, M. Uhl (eds) *The Hitler Book: The Secret Dossier Prepared for Stalin* (London, 2005) pp. 148–9.

72 *Fuehrer Conferences on Naval Affairs 1939–1945* (London, 1948, reissued 1990), pp. 395–6, conference of 12 June 1944; pp. 403–4, conference at the Berghof, 12 July 1944; Kahn, *Hitler's Spies*, pp. 514–19.

73 Bryant, *Triumph in the West*, p. 167, letter from Montgomery to Brooke, 13 June 1944; E. Bauer, *Der Panzerkrieg: Band II, Der Zusammenbruch des Dritten Reiches* (Bonn, 1966?), pp. 100–1. The balance of forces developed as follows:

	Allied	German
7.6.44	9	6
9.6.44	12	9
11.6.44	14	11
13.6.44	15	15

74 Roskill, *War at Sea*, III, pp. 63–6; C. Wilmot, *The Struggle for Europe* (London, 1952), pp. 321–2; Eisenhower, *Report by the Supreme Commander*, pp. 68–9.

75 Harrison, *Cross-Channel Attack*, pp. 422–49; Eisenhower, *Report by the Supreme Commander*, pp. 39–40; Morison, *Invasion of France and Germany*, pp. 176–9.

76 H.C. Butcher, *Three Years with Eisenhower* (London, 1946), p. 498, entry for 13 June 1944, and p. 520, entry for 1 July 1944.

77 Ambrose, *Eisenhower*, pp. 144–7; N. Hamilton, *Monty: Vol. II, Master of the Battlefield* (London, 1983), pp. 642–4, 657–61; Wilmot, *Struggle for Europe*, pp. 338–41.

78 Figures from D. Belchem, *Victory in Normandy* (London, 1981), p. 178; Bauer, *Panzerkrieg*, pp. 124–5. On the issue of Caen see D'Este, *Decision in Normandy*, pp. 249–50, who cites Montgomery's view that 'Ground did not matter so long as German divisions stayed on this flank'. See too O.N. Bradley, *A Soldier's Story of the Allied Campaigns from Tunis to the Elbe* (London, 1951), p. 317: 'while Montgomery held the pivot at Caen, the

whole Allied line was to wheel eastward . . .' For a more critical assessment of the failure to take Caen, see A. Horne with D. Montgomery, *The Lonely Leader: Monty 1944–1945* (London, 1994), pp. 123–6.

79   *Kriegstagebuch des Oberkommandos der Wehrmacht* (4 vols, Frankfurt, 1961), IV, pp. 316–17; Ruge 'Invasion of Normandy', pp. 339–40.

80   *Fuehrer Conferences on Naval Affairs*, p. 398, conference of 29 June to 1 July, Berghof; *Kriegstagebuch*, IV, pp. 323–4.

81   Wilmot, *Struggle for Europe*, pp. 356–7; H. von Luck, *Panzer Commander* (New York, 1989), pp. 150–1.

82   Cited in Horne, *Lonely Leader*, p. 207, from a report by General Dawnay to Brooke, delivered on 14 July; see too Wilmot, *Struggle for Europe*, pp. 353–5, who cites Montgomery's instructions to General O'Connor on 15 July that the object of Goodwood was 'generally to destroy German equipment and personnel'. Also see Ellis, *Victory in the West*, I, pp. 329–30. Eisenhower endorsed this reading of the strategy in *Report by the Supreme Commander*, p. 39: 'Our strategy . . . was to hit hard in the east in order to contain the enemy main strength there while consolidating our position in the west . . .'

83   *Kriegstagebuch*, IV, p. 326; Bauer, *Panzerkrieg*, pp. 104–5, 125–6; Wilmot, *Struggle for Europe*, p. 388; on Goodwood, Belchem, *Victory in Normandy*, pp. 153–6; J. A. English, *The Canadian Army and the Normandy Campaign* (New York, 1991), pp. 227–31; Cooper, *German Army*, pp. 506–7.

84   Butcher, *Three Years with Eisenhower*, pp. 528–37; Ambrose, *Eisenhower*, pp. 147–9; Montgomery, *Memoirs*, pp. 261–3; M. Blumenson, *Breakout and Pursuit* (US Army in World War II, European Theatre, Washington, 1961), pp. 194–6.

85   Wilmot, *Struggle for Europe*, pp. 390–2; G. Picot, *Accidental Warrior: In the front-line from Normandy till victory* (London, 1993), p. 109; Belchem, *Victory in Normandy*, p. 178.

86   On air tactics, R.P. Hallion, *Strike from the Sky: The history of battlefield air attack 1911–1945* (Washington, 1989), pp. 206–14; Blumenson, *Breakout and Pursuit*, pp. 238–9.

87   Blumenson (ed.), *Patton Papers*, p. 456.

88   Cooper, *German Army*, pp. 507–8; Bauer, *Panzerkrieg*, pp. 132–4; *Kriegstagebuch*, IV, pp. 328–9.

89   W. Warlimont, *Inside Hitler's Headquarters* (London, 1964), pp. 447–9; Liddell Hart, *Other Side*, pp. 418–22; Wilmot, *Struggle for Europe*, pp. 400–2.

90   Blumenson, *Breakout and Pursuit*, pp. 457–65; on Ultra, R. Bennett, *Ultra in the West: The Normandy campaign of 1944–1945* (London, 1979), pp. 112–16; Craven, Cate, *Army Air Forces*, III, pp. 248–52.

91 Cooper, *German Army*, pp. 510–11; Wilmot, *Struggle for Europe*, pp. 420–1.

92 M. Shulman, *Defeat in the West* (London, 1947), p. 175; von Luck, *Panzer Commander*, pp. 162–5.

93 W. Thornton, *The Liberation of Paris* (London, 1963), pp. 205–9.

94 Speer, *Inside the Third Reich*, pp. 488–9.

95 *Hitlers politisches Testament: Die Bormann Diktate vom Februar und April 1945* (Hamburg, 1981), pp. 121–5, 2 April 1945.

96 Eisenhower, *Report by the Supreme Commander*, pp. 143–5 on the surrender; on victory celebrations in Britain, R. Cross, *VE Day: Victory in Europe*, (London, 1985), pp. 92–3; on the Soviet Union, R. Parker, *Moscow Correspondent* (London, 1949), pp. 11–13; a rather different account in C. Porter, M. Jones, *Moscow in World War II* (London, 1987), p. 214.

97 Stoler, *Politics of the Second Front*, p. 158; Harriman, *Special Envoy*, p. 315.

98 Ambrose, *Eisenhower*, p. 134; Bryant, *Triumph in the West*, p. 165.

99 Butcher, *Three Years with Eisenhower*, p. 510.

100 Eisenhower, *Report by the Supreme Commander*, p. 141.

6 A Genius for Mass-Production
ECONOMIES AT WAR

1 A. Yakovlev, *Notes of an Aircraft Designer* (Moscow, 1961), pp. 144–9.

2 A. Werth, *Russia at War 1941–1945* (London, 1964), p. 216; *Great Patriotic War of the Soviet Union 1941–1945* (Moscow, 1970), pp. 77–80; M. Harrison, *Soviet Planning in Peace and War 1938–1945* (Cambridge, 1985), pp. 72, 78.

3 Harrison, *Soviet Planning*, pp. 77–9; L.V. Pozdeeva, 'The Soviet Union: Phoenix', in W. Kimball, D. Reynolds, A. O. Chubarian (eds), *Allies at War: The Soviet, American and British Experience 1939–1945* (New York, 1994), pp. 150–2.

4 Details in Harrison, *Soviet Planning*, pp. 81–5. The eastern zones possessed only 18.5 per cent of arms enterprises in 1941, but held 76 per cent by 1942. The east supplied only 39 per cent of the steel in 1941, but 83 per cent in 1942.

5 Harrison, *Soviet Planning*, pp. 250, 253; R. Wagenführ, *Die deutsche Industrie im Kriege* (Berlin, 1963), p. 182.

6 W. Moskoff, *The Bread of Affliction: The food supply in the USSR during World War II* (Cambridge, 1990), pp. 71–2.

7 A. Nove, *An Economic History of the USSR* (London, 1989), p. 262.

8 M. Harrison, 'Resource Mobilization for World War II: The USA, UK, USSR and Germany, 1938–1945', *Economic History Review* 41 (1988).

9   D. Dallin, B. I. Nicolaevsky, *Forced Labour in Soviet Russia* (London, 1947), pp. 262–75. There is valuable material on the work ethic in the camps in D. Panin, *The Notebooks of Sologdin* (New York, 1976), esp. pp. 92–6.

10 On Soviet planning see Harrison, *Soviet Planning*, pp. 14–19; E. Zaleski, *Stalinist Planning for Economic Growth 1933–1952* (Chapel Hill, 1980); P. Sutela, *Socialism, Planning and Optimality; A Study of Soviet Economic Thought* (Helsinki, 1984), pp. 13–15.

11 Harrison, *Soviet Planning*, pp. 174–5.

12 *Great Patriotic War*, pp. 70–81, 140–3.

13 Harrison, *Soviet Planning*, pp. 190–1,193–4.

14 Werth, *Russia at War*, p. 223; S. J. Zaloga, J. Grandsen, *Soviet Tanks and Combat Vehicles of World War II* (London, 1984), pp. 125–37; Von Hardesty, *Red Phoenix: The rise of Soviet air power 1941–1945* (London, 1982), pp. 250–1, appendix 8. The aircraft were the Il–2 fighter-bomber (36,000 produced), the Pe–2 bomber (11,500 produced), the Lagg–3 (6,500), the Yak–9 (16,700) and La–5 (10,000). These provided over two-thirds of all Soviet aircraft built.

15 M. Hindus, *Russia Fights on* (London, 1942), pp. 65–6; M. Dobb, *Soviet Economy and the War* (London, 1941), pp. 50–1.

16 W.L. White, *Report on the Russians* (New York, 1945), pp. 209–10, 212–13.

17 J. Barber, M. Harrison, *The Soviet Home Front 1941–1945* (London, 1991), p. 220.

18 Nove, *Economic History*, p. 272. On the restructuring of Soviet industry, see Dobb, *Soviet Economy*, pp. 48–56; Werth, *Russia at War*, pp. 622–3.

19 Barber, Harrison, *Soviet Home Front*, pp. 215–19.

20 Ibid., pp. 163–9; Stalin's appeal in J. Stalin, *The Great Patriotic War of the Soviet Union* (New York, 1945), p. 34, speech in Moscow, 6 November 1941.

21 White, *Report on the Russians*, pp. 211–12.

22 Moskoff, *Bread of Affliction*, pp. 138–9; Barber, Harrison, *Soviet Home Front*, p. 214.

23 J. Erickson, 'Soviet Women at War', in J. and C. Garrard (eds), *World War 2 and the Soviet People* (London, 1993), pp. 55–6; Nove, *Economic History*, pp. 269–70.

24 White, *Report on the Russians*, pp. 41–3; on the cult of over-achievement, see L.H. Siegelbaum, *Stakhanovism and the Politics of Productivity in the USSR, 1935–1941* (Cambridge, 1988). By 1940 there were almost three million so-called 'shock workers' in Soviet industry, who were supposed to achieve much more than their stated quotas.

25 *Great Patriotic War*, pp. 140–1.

26 Barber, Harrison, *Soviet Home Front*, pp. 83–4, 171–3; White, *Report on the Russians*, pp. 51–4.

27 White, *Report on the Russians*, p. 104.

28 I. Ehrenburg, *Men, Years – Life: The war years 1941–1945* (London, 1964), p. 123.

29 Speech by Sir William Layton, 17 October 1940, to the Associated Industries of Massachusetts, in *The American Speeches of Lord Lothian* (Oxford, 1941), p. 128.

30 W.F. Craven, J.L. Cate, *The Army Air Forces in World War II* (6 vols, Washington, 1948–1955, reissued 1983), I, p. 104.

31 R.E. Smith, *The Army and Economic Mobilization* (Washington, 1958), p. 4. On the Victory Programme, R.M. Leighton, R.W. Coakley, *Global Logistics and Strategy* (2 vols, Washington, 1955–68), I, pp. 126–36.

32 On American production, G. Simonson, 'The Demand for Aircraft and the Aircraft Industry', *Journal of Economic History* 20 (1960); Leighton, Coakley, *Global Logistics*, II, pp. 832–3.

33 Calculated from Leighton, Coakley, *Global Logistics*, II, appendix B2, p. 829; *Statistical Abstract of United States Historical Statistics* (Washington, 1947); United States Strategic Bombing Survey (USSBS), Pacific Theatre, Report 46, 'Japanese Naval Shipbuilding' (Washington, 15 November 1946), p. 2.

34 A. Clive, *State of War: Michigan in World War II* (Chicago, 1979), p. 25.

35 T. A. Wilson, 'The United States: Leviathan', in Kimball *et al*, *Allies at War*, pp. 176–8.

36 A.P. Sloan, *My Years with General Motors* (London, 1986), pp. 379, 381.

37 F.C. Lane, *Ships for Victory: A history of shipbuilding under the US Maritime Commission in World War II* (Baltimore, 1951), p. 67; F. Walton, *Miracle of World War II* (New York, 1956), pp. 75–7.

38 Lane, *Ships for Victory*, p. 224ff.

39 Walton, *Miracle of World War II*, p. 79; P. Fearon, *War, Prosperity and Depression: The US economy 1917–1945* (London, 1986), p. 274; Lane, *Ships for Victory*, pp. 53–4.

40 Walton, *Miracle of World War II*, p. 559; Clive, *State of War*, p. 22.

41 A. Nevins, F.E. Hill, *Ford: Decline and rebirth 1933–1961* (New York, 1962), p. 226. Ford produced 277,896 jeeps, 93,217 trucks, 8,685 bombers, 57,851 aero-engines, 2,718 tanks, 12,500 armoured cars, 26,954 tank engines, and so on. Italy produced approximately 7,000 combat aircraft, 2,800 tanks and self-propelled guns, 46,600 light vehicles and approximately 90,000 heavy vehicles – details in Istituto centrale di statistica, *Sommario di Statistiche Storiche Italiane 1861–1955* (Rome, 1958).

42 J. Rae, *Climb to Greatness: The story of the American aircraft industry* (Cambridge, Mass., 1968), pp. 143–4, 157–61; Sloan, *My Years with General Motors*, pp. 377–81; Clive, *State of War*, pp. 27–8.

43 Nevins, *Ford*, pp. 115, 186–7.

44 Walton, *Miracle of World War II*, pp. 306–9; Nevins, *Ford*, pp. 187–9; Craven, Cate, *Army Air Forces*, VI, pp. 329–30.

45 Clive, *State of War*, p. 30.

46 On migration patterns, H.S. Shrycock, 'Internal Migration in Peace and War', *American Sociological Review* 12 (1947); wages and strikes in H. Vatter, *The US Economy in World War II* (New York, 1985), pp. 20–1; R. Polenberg, *War and Society: The United States 1941–1945* (Philadelphia, 1972), pp. 159–72.

47 Walton, *Miracle of World War II*, pp. 555–6.

48 Polenberg, *War and Society*, p. 13.

49 E. Ludendorff, *Der totale Krieg* (Munich, 1935). The expression first appeared in his war memoirs. See the general discussion of the concept in I. Beckett 'Total War', in C. McInnes, G. Sheffield, *Warfare in the Twentieth Century* (London, 1988), pp. 1–21.

50 Details in R.J. Overy, 'Hitler's War Plans and the Economy', in R. Boyce (ed.), *Paths to War: New Essays on the Origins of the Second World War* (London, 1990), pp. 96–127.

51 L. Stoddard, *Into the Darkness: Nazi Germany today* (London, 1941), pp. 90–1. On the wartime rationing and shortages see R.J. Overy, *War and Economy in the Third Reich* (Oxford, 1994), pp. 274–86.

52 Imperial War Museum (IWM), London, FD 3056/49, 'Statistical Material on the German Manpower Position during the War Period 1939–1944', 31 July 1945, table 7. The figures were based on the official statistics collected by German industry during the war.

53 In general W.A. Boelcke, *Die Kosten von Hitlers Krieg* (Paderborn, 1988); W. Abelshauser, 'Germany' in M. Harrison (ed.), *The Economics of World War II: Six great powers in international comparison* (Cambridge, 1998).

54 R.J. Overy, *Goering: The 'Iron Man'* (London, 1984), p. 193.

55 H. Trevor-Roper (ed.), *Hitler's Table Talk 1941–1944* (London, 1973), p. 633.

56 Office of Air Force History, Washington, microfilm collection, roll R5003, 'German Air Force Policy during the 2nd World War', a memorandum by engineers at the Rechlin Aircraft Experimental Station, 15 August 1944, p. 1.

57 P. Kluke, 'Hitler und das Volkswagenprojekt', *Vierteljahrshefte für Zeitgeschichte* 8 (1960), pp. 349ff; K. Hopfinger, *Beyond Expectation: The Volkswagen story* (London, 1954), p. 70.

58 USSBS, European Theatre, Report 88, 'Volkswagen-Werke, Fallersleben' (Washington, September 1945), pp. 3–8, and exhibit J; W. Nelson, *Small Wonder: The amazing story of the Volkswagen* (London, 1967), pp. 73–6; K.-J. Siegfried, *Rüstungsproduktion und Zwangsarbeit im Volkswagenwerk 1939–1945* (Frankfurt, 1986), pp. 37, 43.

59 Overy, *Goering*, pp. 99, 161, on Opel. On the industry as a whole, see USSBS, European Theatre, Report 77, 'German Motor Vehicles Industry Report' (Washington, 1947), pp. 5–11; British Intelligence Objectives Sub-Committee, Report 21, 'The Motor Car Industry in Germany during the period 1939–1945' (London, 1949), pp. 7–12.

60 IWM, Cabinet Office collection of German documents, EDS Mi 14/433 file 2, Führer decree, 3 December 1941.

61 IWM EDS Mi 14/133, Army High Command, 'Studie Über Rüstung 1944', 15 January 1944. On rationalisation in general, see Overy, *War and Economy*, pp. 356–66.

62 A. Speer, *Inside the Third Reich* (London, 1970), p. 213; IWM Speer Collection, Box 368, Report 83, 'Relationship between the Army Ordnance Board and the Speer Ministry', p. 2.

63 IWM Speer Collection, Box 368, Report 52, interrogation of G. von Heydekampf, 6 October 1945, p. 11.

64 On forced labour, U. Herbert, *Fremdarbeiter: Politik und Praxis des 'Ausländer-Einsatzes' in der Kriegswirtschaft des Dritten Reiches* (Bonn, 1985), pp. 270–3; on social life under bombing, E. Beck, *Under the Bombs: The German home front 1942–1945* (Lexington, Kentucky, 1986); on the SS, A. Speer, *The Slave State: Heinrich Himmler's master plan for SS supremacy* (London, 1981), esp. chs 13–17.

65 H. Gatzke, *Germany and the United States* (Cambridge, Mass., 1980), p. 113; F. Taylor (ed.), *The Goebbels Diaries 1939–1941* (London, 1982), p. 414, entry for 14 June 1941.

66 J. Colville, *The Fringes of Power: 10 Downing Street diaries 1939–1955* (London, 1985), p. 346, entry for 26 January 1941; *Foreign Relations of the United States, 1941, Vol. I* (Washington, 1959), p. 840; E. Roosevelt (ed.), *The Roosevelt Letters: Vol. III, 1928–1945* (London, 1952), p. 385, letter from Roosevelt to Stimson, 30 August 1941.

67 White, *Report on the Russians*, preface.

68 Speer, *Inside the Third Reich*, p. 213.

7 A War of Engines

TECHNOLOGY AND MILITARY POWER

1 H. Ritgen, *Die Geschichte der Panzer-Lehr Division im Westen 1944–1945* (Stuttgart, 1979), pp. 164–6; W.F. Craven, J.L. Cate, *The Army Air Forces in World War II* (6 vols, Washington, 1948–55, reissued 1983), III, pp. 231–4.

2 R. Hallion, *Strike from the Sky: The history of battlefield air attack 1911–1945* (Washington, 1989), pp. 212–13; M. Blumenson, *Breakout and Pursuit* (Washington, 1961), pp. 240–6, 251.

3 Ritgen, *Panzer-Lehr*, p. 187, appendix 5: 'Personnel, strength and losses'. From June to August 1944 the division suffered 8,525 casualties, but received only 3,357 replacements. On German losses of motorised equipment, see J. Piekalkiewicz, *Tank War 1939–1945* (New York, 1986), pp. 239, 287–91.

4 R. Steiger, *Armour Tactics in the Second World War: Panzer army campaigns of 1939–41 in German war diaries* (Oxford, 1991), p. 10.

5 J.Stalin, *The Great Patriotic War of the Soviet Union* (New York, 1945), speech on the 24th anniversary of the October revolution, 6 November 1941, pp. 25–6.

6 M. and S. Harries, *Soldiers of the Sun: The rise and fall of the Imperial Japanese Army 1868–1945* (London, 1991), p. 296.

7 General H. von Manteuffel, 'Fast, Mobile and Armoured Troops', in D. Detweiler (ed.), *World War II German Military Studies* (24 vols, New York, 1979), VI, MS B-036, pp. 5–7; on radios, see M. Messerschmidt, 'The Political and Strategic Significance of Advances in Armament Technology: Developments in Germany and the "Strategy of Blitzkrieg"', in R. Ahmann, A.M. Birke, M. Howard (eds), *The Quest for Stability: Problems of West European security 1918–1957* (Oxford, 1993), pp. 258–9.

8 S.J. Zaloga, J. Grandsen, *Soviet Tanks and Combat Vehicles of World War II* (London, 1984), pp. 126–7; Von Hardesty, *Red Phoenix: The rise of Soviet air power 1941–1945* (London, 1982), pp. 13–15.

9 Zaloga, Grandsen, *Soviet Tanks*, pp. 146–9, 160–2; R.M. Ogorkiewicz, *Armoured Forces: A history of armoured forces and their vehicles* (London, 1970), pp. 123–4.

10 *Main Front: Soviet leaders look back on World War II* (London, 1987), pp. 109, 121.

11 Von Hardesty, 'Roles and Missions: Soviet tactical air power in the second period of the Great Patriotic War', in C. Reddel (ed.), *Transformations in Russian and Soviet Military History* (Office of Air Force

History, Washington, 1990), pp. 154–5; A. Boyd, *The Soviet Air Force since 1918* (London, 1977), pp. 109–11.

12 Von Hardesty, *Red Phoenix*, pp. 83–8; Von Hardesty, 'Roles and Missions', pp. 155–7.

13 R.J. Overy, *The Air War 1939–1945* (London, 1980), pp. 52–6.

14 Von Hardesty, 'Roles and Missions', pp. 161–7; K. Uebe, *Russian Reactions to German Air Power in World War II* (New York, 1964), pp. 29–42; *The Soviet Air Force in World War II* (Soviet official history, ed. R. Wagner, London, 1974), pp. 54–5, 127–8, 145–6.

15 Zaloga, Grandsen, *Soviet Tanks*, p. 206. Britain and America supplied an additional 20,900 armoured vehicles, including 12,365 tanks.

16 H.P. van Tuyll, *Feeding the Bear; American aid to the Soviet Union 1941–1945* (New York, 1989), pp. 156–7; J. Beaumont, *Comrades in Arms: British aid to Russia 1941–1945* (London, 1980), pp. 210–12. British aid included 247,000 telephones and one million miles of field telephone wire.

17 British Air Ministry, *The Rise and Fall of the German Air Force* (reissued London, 1983), pp. 357–9; Zaloga, Grandsen, *Soviet Tanks*, p. 223.

18 M. Cooper, *The German Army 1933–1945* (London, 1978), pp. 485–90; Ogorkiewicz, *Armoured Forces*, pp. 74–9.

19 M. van Creveld, *Supplying War: Logistics from Wallenstein to Patton* (Cambridge, 1977), pp. 150–3; R.L. Di Nardo, *Mechanized Juggernaut or Military Anachronism: Horses and the German Army in World War II* (London, 1991), pp. 37–40; von Manteuffel, 'Fast, Mobile and Armoured troops', p. 5: 'When in 1939 the war began, the motorization of the great mass of the army *not* belonging to the fast mobile formations had *consciously been neglected.*' (Italics in original.)

20 R. Stolfi, *Hitler's Panzers East: World War II reinterpreted* (Norman, Oklahoma, 1991), pp. 157–9.

21 Di Nardo, *Mechanized Juggernaut*, pp. 49–50; Steiger, *Armour Tactics*, pp. 95–105.

22 Di Nardo, *Mechanized Juggernaut*, pp. 50–6; United States Strategic Bombing Survey (USSBS), European Theatre, Report 77, 'German Motor Vehicle Industry Report' (Washington, 1947), p. 9.

23 Steiger, *Armour Tactics*, p. 127; van Creveld, *Supplying War*, pp. 150–1.

24 *World War II German Military Studies*, XXIII, 'German Tank Maintenance in World War II', pp. 2–4, 23–4; A. Speer, *Inside the Third Reich* (London, 1971), p. 234.

25 Speer, *Inside the Third Reich*, pp. 233–4, 241.

26 *World War II German Military Studies*, XXIII, p. 25.

27 Ibid., p. 26.

28 Di Nardo, *Mechanized Juggernaut*, pp. 92–7; Cooper, *German Army*, pp.

487–8; Piekalkiewicz, *Tank War*, p. 246. By the end of 1943 the German tank arm had been reduced temporarily from 3,000 tanks in the east to 300. On the process of 'de-modernisation', see O. Bartov, *Hitler's Army: Soldiers, Nazis and war in the Third Reich* (Oxford, 1991), ch. 2.

29　A. van Ishoven, *The Fall of an Eagle: The life of fighter ace Ernst Udet* (London, 1977), pp. 143–58.

30　R.J. Overy, 'From "Uralbomber" to "Amerikabomber": The Luftwaffe and strategic bombing', *Journal of Strategic Studies* 1 (1978), pp. 167–71.

31　British Air Ministry, *Rise and Fall of the German Air Force*, pp. 212–18.

32　W. Murray, *Luftwaffe* (London, 1985), p. 148; British Air Ministry, *Rise and Fall of the German Air Force*, pp. 222–3. On the Udet/Milch crisis, D. Irving, *The Rise and Fall of the Luftwaffe: The life of Erhard Milch* (London, 1973), pp. 124–49.

33　J.J. Sweet, *Iron Arm: The mechanization of Mussolini's army 1920–1940* (Westport, Conn., 1980), pp. 3–15, 175–86.

34　USSBS, Pacific Theatre, Report 47, 'Japanese Motor Vehicle Industry' (Washington, November 1946), p. 2; J. Cohen, *Japan's Economy in War and Reconstruction* (Minneapolis, 1949), p. 237; Harries, *Soldiers of the Sun*, pp. 300–1.

35　USSBS, Report 47, 'Japanese Motor Vehicle Industry', pp. 1–5.

36　Harries, *Soldiers of the Sun*, pp. 297–302; J. Scofield, 'The Japanese Soldier's Arms and Weapons', in *How the Jap Army Fights* (London, 1943), pp. 24–31; P. Warner, *Firepower: From slings to Star Wars* (London, 1988), pp. 145–6.

37　H. Doud, 'Peace-time Preparation: Six months with the Japanese infantry', in *How the Jap Army Fights*, pp. 34–46.

38　USSBS, Pacific Theatre, Report 46, 'Japanese Naval Shipbuilding' (Washington, November 1946), pp. 2–3; USSBS, Pacific Theatre, Report 72, 'Interrogation of Japanese Officials', vol. II (Washington, 1946), p. 86.

39　Ogorkiewicz, *Armoured Forces*, pp. 87–92; K. Greenfield, R.R. Palmer, B. Wiley, *The Organization of Ground Combat Troops* (Washington, 1947), pp. 64–72.

40　Society of Motor Manufacturers and Traders, *The Motor Industry of Great Britain, 1938* (London, 1939); League of Nations, *World Production and Prices 1925–1935* (Geneva, 1936), p. 90.

41　USSBS, Report 77, 'German Vehicle Industry Report', pp. 8–10, 13.

42　H.H. Arnold, *Global Mission* (New York, 1949), p. 199.

43　On intelligence on Germany, see Library of Congress, Arnold Papers, Box 246 for the G2 report, 'Germany, Domestic Production, Capacity and Sources of Aviation Equipment' (16 January 1941), and the 'Memorandum for Chief of Intelligence, Estimates of German Air

Strength' (no date). It was calculated that German front-line air forces would total 36,000 by June 1941, when the true figure was 3,451.

44 D. Syrett, 'The Tunisian Campaign, 1942–1943', in B.F. Cooling (ed.), *Case Studies in the Development of Close Air Support* (Office of Air Force History, Washington, 1990), pp. 158–70; Hallion, *Strike from the Sky*, pp. 149–58; A. Tedder, *Air Power in War* (London, 1948), pp. 38–40.

45 Hallion, *Strike from the Sky*, p. 161.

46 Ibid., pp. 181–2.

47 Di Nardo, *Mechanized Juggernaut*, pp. 91–5; Piekalkiewicz, *Tank War*, p. 310. At the same time on the eastern front there were only 318 serviceable tanks, and 616 serviceable self-propelled guns.

48 D.D. Eisenhower, *Report by the Supreme Commander to the Combined Chiefs of Staff* (London, 1946), p. 17.

49 E. Bauer, *Der Panzerkrieg: Band II, der Zusammenbruch des Dritten Reiches* (Bonn, 1966), p. 247; Piekalkiewicz, *Tank War*, p. 310; Craven, Cate, *Army Air Forces*, VI, p. 423; R. Kilmarx, *A History of Soviet Air Power* (London, 1962), pp. 188–92; A. Seaton, *The Russo-German War 1941–1945* (London, 1971), pp. 436, 527.

50 C.F. Jones, 'Industrial Capacity and Supplies of Raw Materials', in S. van Valkenburg (ed.), *America at War: A geographical analysis* (New York, 1943), p. 162.

51 M. A. Barnhart, *Japan Prepares for Total War: The search for economic security 1919–1941* (Ithaca, New York, 1987), p. 146; M.A. Barnhart, 'Japan's Economic Security and the Origins of the Pacific War', *Journal of Strategic Studies* 4 (1981), pp. 114–19; D. Yergin, *The Prize: The epic quest for oil, money and power* (New York, 1991), pp. 355–66; 'Oil in the Far East', *Petroleum Press Service* 9 (1942), pp. 1–7.

52 USSBS, Pacific Theatre, Report 48, 'Japanese Merchant Shipbuilding' (Washington, January 1947), pp. 1, 18; J. Cohen, *Japan's Economy in War and Reconstruction* (Minneapolis, 1949), pp. 133–46.

53 USSBS, Report 72, 'Interrogation of Japanese Officials', no. 53, pp. 214–16.

54 USSBS, Pacific Theatre, Report 46, 'Japanese Naval Shipbuilding', (Washington, November 1946), p. 9 (there were 6,197 suicide boats, and 419 human torpedoes or *kaiten*); T.A. Bisson, *Japan's War Economy* (New York, 1945), pp. 164–6.

55 P. Hayes, *Industry and Ideology: IG Farben in the Nazi era* (Cambridge, 1987), pp. 36–42, 66–7, 133–5; W. Birkenfeld, *Der synthetische Treibstoff 1933–1945* (Göttingen, 1963).

56 USSBS, Report 109, 'Oil Division Final Report' (Washington, August

1945), pp. 14–18; M. Pearton, *Oil and the Romanian State* (Oxford, 1971), pp. 249, 259–61.

57 O. Nissen, *Germany: Land of substitutes* (London, 1944), pp. 77–8, 82; details on the conversion of German vehicles in *Petroleum Press Service* 9 (1942), p. 55. By the summer of 1941 some 150,000 commercial vehicles had been converted to gas generators, 60,000 to liquid gas.

58 Van Creveld, *Supplying War*, pp. 196–201.

59 H. Trevor-Roper (ed.), *Hitler's War Directives* (London, 1964), p. 179, also pp. 151–2.

60 'The Threat to Russia's Oil Supplies', *Petroleum Press Service* 9 (1942), pp. 105–7; 'German Drive for Maikop Oil', *Petroleum Times* 47 (1943), p. 56.

61 Overy, *Goering*, pp. 216–17; A. Fredborg, *Behind the Steel Wall: Berlin 1941–43* (London, 1944), p. 126; Yergin, *The Prize*, pp. 336–7.

62 C.R. Richardson, 'French Plans for Allied Attacks on the Caucasus Oilfields Jan-Apr 1940', *French Historical Studies* 8 (1973).

63 R. Cooke, R. Nesbit, *Target: Hitler's Oil: Allied attacks on German oil supplies 1939–1945* (London, 1985), pp. 102–7.

64 C. Webster, N. Frankland, *The Strategic Air Offensive against Germany* (4 vols, London, 1961), IV, pp. 508–16; USSBS, Report 109, 'Oil Division Final Report', pp. 18–26; Piekalkiewicz, *Tank War*, p. 291. The total needed per month was calculated to be 327,000 tons.

65 G.D. Nash, *United States Oil Policy 1890–1964* (Pittsburgh, 1968), pp. 157–65; Yergin, *The Prize*, pp. 371–2. On the cooperation of the big oil businesses, see H. Larson, E. Knowlton, C. Popple, *History of Standard Oil Company: Vol. Ill, New horizons 1927–50* (New York, 1971), pp. 522–36.

66 *Petroleum Press Service* 10 (1943), pp. 1, 31–2; *Petroleum Press Service* 11 (1944), pp. 45–6; A.M. Johnson, *Petroleum Pipelines and Public Policy 1905–1959* (Cambridge, Mass., 1967), pp. 320–5.

67 Yergin, *The Prize*, pp. 383–4.

68 Ibid., p. 382.

69 Harries, *Soldiers of the Sun*, p. 304.

70 R. Rhodes, *The Making of the Atomic Bomb* (New York, 1986), pp. 580–2, 612; Harries, *Soldiers of the Sun*, p. 303.

71 G. Brooks, *Hitler's Nuclear Weapons* (London, 1992), p. 46.

72 M. Walker, *German National Socialism and the Quest for Nuclear Power 1939–1949* (Cambridge, 1989), pp. 17–21; Brooks, *Hitler's Nuclear Weapons*, pp. 20–3.

73 Brooks, *Hitler's Nuclear Weapons*, pp. 43–6; Rhodes, *Making of the Atomic Bomb*, pp. 455–7, 513–17.

74 Brooks, *Hitler's Nuclear Weapons*, pp. 102–5; Walker, *German National Socialism and the Quest for Nuclear Power*, pp. 81–7.

75 P. Joubert de la Ferté, *Rocket* (London, 1957), pp. 17–18; M.J. Neufeld, 'The Guided Missile and the Third Reich: Peenemünde and the forging of a technological revolution', in M. Walker, M. Renneberg (eds), *Science, Technology and National Socialism* (Cambridge, 1993), pp. 52–8.

76 De la Ferté, *Rocket*, pp. 21–2; A. Speer, *The Slave State: Heinrich Himmler's masterplan for SS supremacy* (London, 1981), pp. 203–7.

77 Speer, *Slave State*, pp. 210–11; on problems in development, Neufeld, 'Guided Missile', pp. 62–5.

78 B. Collier, *The Defence of the United Kingdom* (London, 1957), pp. 523, 527; T.H. O'Brien, *Civil Defence* (London, 1955), pp. 652–68, 682.

79 Brooks, *Hitler's Nuclear Weapons*, pp. 120–2; USSBS, Report 60, 'V Weapons (Crossbow) Campaign' (Washington, 24 September 1945), pp. 35–6; R. Karlsch *Hitlers Bombe: Die geheime Geschichte der deutschen Kernwaffenversuche* (Munich, 2005).

80 R.W. Clark, *The Birth of the Bomb* (London, 1961), pp. 135–9, 155–6, 168–70; Rhodes, *Hitler's Atomic Bomb*, pp. 368–9.

81 F.M. Szasz, *British Scientists and the Manhattan Project* (London, 1992), appendix iv, p. 152; L.R. Groves, *Now It Can Be Told* (New York, 1962), pp. 40, 295–8.

82 Groves, *Now It Can Be Told*, p. 298.

83 Brooks, *Hitler's Nuclear Weapons*, p. 127.

84 J. Erickson, 'New Thinking about the Eastern front in World War II', *Journal of Military History* 56 (1992), p. 284.

8 Impossible Unity

ALLIES AND LEADERS IN WAR

1 C.E. Bohlen, *Witness to History 1929–1969* (London, 1973), p. 131.

2 A.H. Birse, *Memoirs of an Interpreter* (London, 1967), p. 154; Bohlen, *Witness to History*, p. 142.

3 S. Butler (ed.), *My Dear Mr. Stalin: The Complete Correspondence of Franklin D. Roosevelt and Joseph V. Stalin* (New Haven, 2005), pp. 129, 172.

4 H.H. Arnold, *Global Mission* (New York, 1949), p. 465; Birse, *Memoirs of an Interpreter*, pp. 154–5.

5 Bohlen, *Witness to History*, p. 142.

6 A. Perlmutter, *FDR and Stalin: A not so Grand Alliance, 1943–1945* (Columbia, Missouri, 1993), p. 69 (the reference is to the diary of Joseph E. Davies, who acted as go-between for Roosevelt).

7 Arnold, *Global Mission*, p. 468; Perlmutter, *FDR and Stalin*, p. 159.

8 F. Perkins, *The Roosevelt I Knew* (London, 1947), pp. 70–1; K. Eubank, *Summit at Teheran* (New York, 1985), pp. 350–1.

9   W. Averell Harriman with E. Abel, *Special Envoy to Churchill and Stalin 1941–1946* (London, 1976), pp. 273–4.

10 K. Sainsbury, *The Turning Point* (London, 1986), p. 266; A. Bryant, *Triumph in the West: The war diaries of Field Marshal Viscount Alanbrooke* (London, 1959), p. 83.

11 Eubank, *Summit at Teheran*, p. 351; Harriman, *Special Envoy*, p. 278; Birse, *Memoirs of an Interpreter*, pp. 160–1.

12 Sainsbury, *Turning Point*, p. 322, appendix B, 'The Teheran Communiqué'.

13 J. Colville, *The Fringes of Power: 10 Downing Street diaries 1939–1955* (London, 1985), pp. 331–2, entry for 11 January 1941; *Foreign Relations of the United States, 1941, Vol. 3* (Washington, 1959), p. 38, memorandum by Dean Acheson, 3 October 1941.

14 J.C. Schneider, *Should America Go to War? The debate over foreign policy in Chicago 1939–1941* (Chapel Hill, North Carolina, 1989), p. 61; on the general background, see W. Heinrichs, *Threshold of War: Franklin D. Roosevelt and American entry into World War II* (Oxford, 1988); W.F. Kimball, *The Most Unsordid Act: Lend-Lease 1939–1941* (Baltimore, 1969).

15 W.S. Churchill, *The Second World War* (6 vols, London, 1948–54), III, p. 539.

16 A.H. Vandenberg (ed.), *Private Papers of Senator Vandenberg* (London, 1953), p. 28.

17 R. A. Divine, *Roosevelt and World War II* (London, 1969), p. 31. The quotations are from a speech by Roosevelt on 10 June 1940 at the University of Virginia, Charlottesville.

18 D. Dimbleby, D. Reynolds, *An Ocean Apart: The relationship between Britain and America in the twentieth century* (London, 1988), p. 139.

19 Perkins, *Roosevelt I Knew*, p. 307.

20 A. Danchev (ed.), *Establishing the Anglo-American Alliance: The Second World War diaries of Brigadier Vivian Dykes* (London, 1990), p. 80, entry for 28 December 1941; A. Danchev, *Very Special Relationship: Field Marshal Sir John Dill and the Anglo-American Alliance 1941–1944* (London, 1986), pp. 13–24.

21 H. G. Nicholas (ed.), *Washington Dispatches 1941–1945: Weekly political reports from the British Embassy* (Chicago, 1981), report of 9 October 1943, pp. 257–8.

22 A.C. Wedemeyer, *Wedemeyer Reports!* (New York, 1958), pp. 211–17.

23 Memorandum on 'Soviet-American Relations' by Andrei Gromyko, 14 July 1944, reproduced in Perlmutter, *FDR and Stalin*, appendix 4, p. 259.

24 I. Maisky, *Memoirs of a Soviet Ambassador: The war 1939–1943* (London, 1967), pp. 159–60; Colville, *Fringes of Power*, pp. 404–5, diary entries for

21 and 22 June 1941; M. Kitchen, *British Policy Towards the Soviet Union During the Second World War* (London, 1986), pp. 57–8.

25 Perlmutter, *FDR and Stalin*, p. 72; W. Kimball, *The Juggler: Franklin Roosevelt as wartime statesman* (Princeton, 1991), pp. 21–4; R.H. Dawson, *The Decision to Aid Russia, 1941* (Chapel Hill, North Carolina, 1959), pp. 116–22.

26 Kitchen, *British Policy*, p. 64; on the State Department, see S.E. Ambrose, *Rise to Globalism: American foreign policy since 1938* (London, 1988), p. 34, who quotes George Kennan from the Russian desk: 'We should do nothing at home to make it appear that we are following the course Churchill seems to have entered upon . . .'

27 Kitchen, *British Policy*, p. 65.

28 Colville, *Fringes of Power*, p. 344. Churchill's comment was not meant to be taken seriously.

29 E. Roosevelt (ed.), *The Roosevelt Letters 1928–1945* (3 vols, London, 1952), III, p. 385, Roosevelt to Stimson, 30 August 1941.

30 Harriman, *Special Envoy*, p. 75; Perlmutter, *FDR and Stalin*, p. 73; R. Sherwood (ed.), *The White House Papers of Harry L. Hopkins* (2 vols, London, 1948), I, pp. 326–45.

31 *Foreign Relations of the United States, 1941, Vol. I*, pp. 815–16, Sumner Welles to Ambassador Umansky, 2 August 1941, and memorandum, Division of European Affairs, 19 August 1941.

32 Dawson, *Decision to Aid Russia*, pp. 282–9.

33 W.P. and Z. Coates, *A History of Anglo-Soviet Relations* (London, 1944), p. 696.

34 R.B. Levering, *American Opinion and the Russian Alliance 1939–1945* (Chapel Hill, North Carolina, 1976), p. 61.

35 S.M. Miner, *Between Churchill and Stalin: The Soviet Union, Great Britain, and the origins of the Grand Alliance* (Chapel Hill, North Carolina, 1988), p. 155.

36 Miner, *Between Churchill and Stalin*, p. 158; D. Reynolds, W. Kimball, A. Chubarian (eds), *Allies at War: The Soviet, American and British experience 1939–1945* (New York, 1994), p. 210.

37 Kimball, *Juggler*, p. 195; Perlmutter, *FDR and Stalin*, p. 82; on the cynical motives behind Britain's agreements with the Soviet Union, see G. Ross, 'Foreign Office attitudes to the Soviet Union 1941–1945', in W. Laqueur (ed.), *The Second World War: Essays in military and political history* (London, 1982), pp. 256–8.

38 P. Winterton, *Report on Russia* (London, 1945), p. 2.

39 J.R. Deane, *The Strange Alliance: The story of American efforts at wartime co-operation with Russia* (London, 1947), pp. 95–7

40 Perlmutter, *FDR and Stalin*, pp. 231–2; Kitchen, *British Policy*, p. 80.

41 Deane, *Strange Alliance*, pp. 93–4; A. Werth, *Russia at War* (London, 1964), p. 574.

42 B. Sokolov, 'Lend Lease in Soviet Military Efforts 1941–1945', *Journal of Slavic Military Studies*, 7 (1994), p. 567–8; J.L. Schecter, V.V. Luchkov (eds), *Khrushchev Remembers: The Glasnost Tapes* (Boston, 1990), p. 84.

43 Sherwood (ed.), *White House Papers of Harry L. Hopkins*, II, p. 777.

44 M. Kitchen, 'Winston Churchill and the Soviet Union during the Second World War', *Historical Journal* 30 (1987), p. 421.

45 Butler (ed.), *My Dear Mr. Stalin*, p. 63.

46 S. Bialer (ed.), *Stalin and His Generals: Soviet military memoirs of World War II* (New York, 1969), pp. 347–8.

47 M. Carver, 'Churchill and the Defence Chiefs', in R. Blake, W.R. Louis (eds), *Churchill* (Oxford, 1993), pp. 356–7.

48 I. Deutscher, *Stalin* (Oxford, 1966), pp. 22–3.

49 Miner, *Between Churchill and Stalin*, p. 140; A. Bullock, *Hitler and Stalin: Parallel lives* (London, 1991), pp. 804–5.

50 For a recent Russian account V.P. Yampolsky (ed.), *Organy Gosudarstvennoi Bezopasnosti SSSR v Velikoi Otechestvennoi voine* (Moscow, 2000) vol ii, pp. 98–104; see too E. Mawdsley, *Thunder in the East; the Nazi–Soviet War* (London, 2006), pp. 63–4.

51 Admiral Kuznetsov, 'Command in Transition', in Bialer, *Stalin and His Generals*, p. 348; Schecter, Luchov (eds), *Khrushchev Remembers*, pp. 63–4.

52 In Bialer, *Stalin and His Generals:* N.N. Voronov, 'The Vexations of Centralization', p. 368, S.M. Shtemenko, 'Stalin, the Taskmaster', pp. 352–4, and A.M. Vasilevski, 'Chief of the General Staff', pp. 350–1.

53 Bryant, *Triumph in the West*, p. 77.

54 Arnold, *Global Mission*, p. 468.

55 R.T. Goldberg, *The Making of Franklin D. Roosevelt: Triumph over disability* (Cambridge, Mass., 1981), p. 169.

56 Perkins, *Roosevelt I Knew*, p. 29.

57 Lord Halifax, *Fulness of Days* (London, 1957), p. 241.

58 A.H. Vandenberg (ed.), *The Private Papers of Senator Vandenberg* (London, 1953), p. 10, diary entry 8 March 1941.

59 Bohlen, *Witness to History*, p. 136; Danchev, *Very Special Relationship*, p. 57. This was the view of Field Marshal Dill, who also wrote, less charitably, that the more he got to know Roosevelt, 'the more superficial and selfish I think him'.

60 E. Morgan, *FDR: A biography* (London, 1985), p. 632.

61 R.H. Ferrel (ed.), *The Eisenhower Diaries* (New York, 1981), p. 40, entry for 4 January 1942.

62 R. Weigley, *History of the US Army* (London, 1968), pp. 453–5; R.A. Divine, *Roosevelt and World War II* (London, 1969), pp. 3–5; Butler (ed.), *My Dear Mr. Stalin*, pp. 6-7 on Leahy, Hopkins, and the Map Room.

63 Morgan, *FDR*, p. 550.

64 Bryant, *Triumph in the West*, p. 77; Lord Ismay, *The Memoirs of Lord Ismay* (London, 1960), p. 338.

65 Harriman, *Special Envoy*, pp. 440–1; Halifax, *Fulness of Days*, p. 241.

66 Details in Blake, Louis (eds), *Churchill*, and J. Charmley, *Churchill: The end of glory* (London, 1993).

67 G.T. Eggleston, *Roosevelt, Churchill and the World War II Opposition* (Old Greenwich, Conn., 1979), p. 130.

68 Halifax, *Fulness of Days*, pp. 218–20, recalled the situation more accurately than Churchill, in his *Second World War*, I, pp. 572–4. See too Charmley, *Churchill*, pp. 393–5; Blake, Louis (eds), *Churchill*, pp. 265–70; D. Reynolds, *In Command of History: Churchill Fighting and Writing the Second World War* (London, 2004), pp. 127-8.

69 Churchill, *Second World War*, I, pp. 524–5.

70 I. Berlin, *Mr. Churchill in 1940* (London, 1950), p. 12.

71 Halifax, *Fulness of Days*, pp. 221–2; Harriman, *Special Envoy*, p. 205.

72 Lord Moran, *Winston Churchill: The struggle for survival 1940–1965* (London, 1966), p. 269.

73 Ibid., p. 102.

74 Halifax, *Fulness of Days*, p. 222.

75 J. Wheeler-Bennett (ed.), *Action This Day: Working with Churchill* (London, 1988), pp. 17–28, 245.

76 Bryant, *Triumph in the West*, p. 77; Wedemeyer, *Wedemeyer Reports!*, p. 79.

77 Wheeler-Bennett, *Action This Day*, pp. 27, 245; K. Robbins, *Churchill* (London, 1992), pp. 127–31.

78 M. Howard, 'Churchill and the First World War', in Blake, Louis, *Churchill*, p. 142.

79 Public Record Office (PRO), Kew, London, AIR 8/463, letter from Chief of Air Staff to Churchill, 20 March 1941; PRO, AIR 9/102, Air Ministry, 'Appreciation of the attack on German war industry', 18 February 1941.

80 Moran, *Winston Churchill*, p. 134.

81 Ibid., p. 102.

82 Churchill, *Second World War*, I, p. 526; R. Lewin, *Churchill as Warlord* (London, 1973), p. 264.

83 A. Bryant, *The Turn of the Tide 1939–1943* (London, 1957), p. 35.

84 D. Fraser, *Alanbrooke* (London, 1987), p. 202.

85 Churchill, *Second World War*, II, p. 234; Ismay, *Memoirs*, p. 318.

86 J. Erickson, *The Road to Berlin* (London, 1983), pp. 53–5.

87 S. Shtemenko, 'Profile of a Staff Officer', in Bialer, *Stalin and His Generals*, pp. 355–9.

88 H. Salisbury (ed.), *Marshal Zhukov's Greatest Battles* (London, 1969), p. 214.

89 Shtemenko, 'Profile of a Staff Officer', p. 359; Zhukov, *Greatest Battles*, pp. 221, 262.

90 Details in R. Payne, *General Marshall* (London, 1952), and M. Stoler, *George C. Marshall: Soldier-statesman of the American century* (Boston, 1989).

91 Stoler, *George C. Marshall*, p. 110.

92 L. Bland (ed.), *The Papers of George Catlett Marshall: Vol. 2, 'We Cannot Delay', July 1 1939 to December 6 1941* (Baltimore, 1986), p. 633, letter from Marshall to Harry Woodring, 8 October 1941.

93 Bland (ed.), *The Papers of George Catlett Marshall*, p. 249, 'Speech to the Veterans of Foreign Wars', 19 June 1940.

94 Weigley, *History of the US Army*, pp. 422–32; Stoler, *George C. Marshall*, pp. 93–5.

95 Stoler, *George C. Marshall*, pp. 101, 109.

96 W. Warlimont, *Inside Hitler's Headquarters* (London, 1964), pp. 243–4.

97 P.E. Schramm, *Hitler the Man and the Military Leader* (London, 1972), p. 11.

98 Ibid., pp. 100–11, 149.

99 'Memorandum dictated by General Alfred Jodl on Hitler's Leadership, 1946', in Schramm, *Hitler the Man*, p. 198, appendix II.

100 Schramm, *Hitler the Man*, p. 137; Speer, *Inside the Third Reich*, p. 241; W. Warlimont, 'The German High Command during World War II', in D. Detweiler (ed.), *World War II German Military Studies* (24 vols, New York, 1979), VI, MS T–101, pp. 6–8. Halder quotation in Nuremberg Trials, background documents, case xi, Körner Defence Doc. Book, 1B, p. 81.

101 Speer, *Inside the Third Reich*, p. 166; Warlimont, 'German High Command', pp. 27–9, 57–9.

102 R.J. Overy, *Interrogations: The Nazi Elite in Allied Hands, 1945* (London, 2001), p. 371, interrogation of Keitel, 27 June 1945.

103 Ibid., pp 276–9, interrogation of Alfred Jodl, 29 June 1945. See too K. Zeitzler 'The German Army High Command', December 1949 in *World War II Military Studies*, vol VI, pp. 100-1.

104 Warlimont, *Inside Hitler's Headquarters*, pp. 249–51.

105 Ibid., p. 260.

106 Zeitzler, 'German Army High Command', p. 100.

107 B. Liddell Hart, *The Other Side of the Hill* (London, 1948), p. 468; see too

J. Strawson, *Hitler as Military Commander* (London, 1971), pp. 228–32.

108 Schramm, *Hitler the Man*, appendix II, pp. 194, 205.

109 J. Steinhoff, *The Last Chance: The pilots' plot against Göring 1944–45* (London, 1977).

110 Speer, *Inside the Third Reich*, pp. 471–3; *Fuehrer Conferences on Naval Affairs 1939–1945* (London, 1990), p. 487.

111 Speer, *Inside the Third Reich*, p. 463.

112 R. Reuth, *Goebbels* (London, 1993), pp. 352–3; text of appeal in Strawson, *Hitler as Military Commander*, pp. 223–4.

113 H. A. DeWeend (ed.), *Selected Speeches and Statements of General of the Army George C. Marshall* (New York, 1973), p. 249, speech at Yale University, 16 February 1944.

114 F. Genoud (ed.), *The Testament of Adolf Hitler: The Hitler-Bormann documents February–April 1945* (London, 1961), pp. 51–2, 76–7, 104–5.

115 Heinz Guderian, 'Unification and Co-ordination – The Armed Forces Problem', MS T–113, October 1948, *German Military Studies* VI, pp. 6, 18.

9 Evil Things, Excellent Things
THE MORAL CONTEST

1 M. Spinka, *The Church in Soviet Russia* (Oxford, 1956), pp. 82–6.

2 A. Werth, *Russia at War 1941–1945* (London, 1964), pp. 429–38.

3 Spinka, *Church in Soviet Russia*, p. 85; Werth, *Russia at War*, p. 435.

4 W.P and Z. Coates, *A History of Anglo-Soviet Relations* (London, 1944), pp. 696–7; G.K.A. Bell, *The Church and Humanity 1939–1946* (London, 1946), p. 217, sermon, 7 September 1941.

5 F. Perkins, *The Roosevelt I Knew* (London, 1947), pp. 28–9.

6 R.H. Dawson, *The Decision to Aid Russia 1941* (Chapel Hill, North Carolina, 1959), pp. 188–9.

7 Dawson, *Decision to Aid*, pp. 87–8.

8 H. Trevor-Roper (ed.), *Hitler's Table Talk 1941–1944* (London, 1953), pp. 144–5, entry for 13 December 1941; p. 722, entry for 29 November 1944.

9 S. Roberts, *The House that Hitler Built* (London, 1937), pp. 275–6; H. Höhne, *The Order of the Death's Head* (London, 1969), pp. 163–72.

10 Dawson, *Decision to Aid*, pp. 266–9.

11 W.S. Churchill, *The Second World War* (6 vols, London, 1948–54), III, p. 331. See Roosevelt's views in S. Butler (ed.), *My Dear Mr. Stalin: The Complete Correspondence of Franklin D. Roosevelt and Joseph V. Stalin* (New Haven, 2005), p. 9: 'The principles and doctrines of Communist dictatorship are as intolerable and alien to American beliefs as are the principles and doctrines of Nazi dictatorship.'

12 Churchill, *Second World War*, III, p. 332.

13 J. Colville, *The Fringes of Power: 10 Downing Street Diaries 1939–1955* (London, 1985), p. 404, entry for 21 June 1941; Lord Moran, *Winston Churchill: The struggle for survival 1940–1965* (London, 1966), p. 79, entry for 15 August 1942, '. . . the P.M.'s only interest was to defeat Hitler; that was an obsession with him.'

14 A. Perlmutter, *FDR and Stalin: A not so Grand Alliance, 1943–1945* (Columbia, Missouri, 1993), appendix 2, pp. 232–40.

15 G. Orwell, *The Collected Essays, Journalism and Letters of George Orwell: Volume 3, As I Please 1943–1945* (London, 1968), pp. 178–9.

16 Dawson, *Decision to Aid*, p. 99; H. Cantril (ed.), *Public Opinion 1935–1946* (Princeton, 1951), p. 1187.

17 Colville, *Fringes of Power*, pp. 363, 404.

18 B. von Everen, 'Franklin D. Roosevelt and the Problem of Nazi Germany', in C.L. Egan, A. W. Knott (eds), *Essays in Twentieth Century American International History* (Lanham, Maryland, 1982), pp. 138–9; E. Morgan, *FDR: A biography* (New York, 1985), pp. 44–5.

19 Von Everen, 'Franklin D. Roosevelt and the Problem of Nazi Germany', pp. 145–8; H. Ickes, *The Secret Diary of Harold L. Ickes* (3 vols, London, 1955), III, pp. 468–9, entry for 18 September 1938; Morgan, *FDR*, pp. 734–5; W. Kimball, *The Juggler: Franklin Roosevelt as wartime statesman* (Princeton, 1991), p. 199, who records Roosevelt's remark after the Teheran conference that Abyssinian doctors should be let loose on Germany to practise 'their particular brand of surgery'.

20 H. Butcher, *Three Years with Eisenhower* (London, 1946), p. 524, entry for 10 July 1944. On the general question of Allied treatment of German POWs, see J. Bacque, *Other Losses: An investigation into the mass deaths of German prisoners of war after World War II* (London, 1989), esp. pp. 23, 31–2 for Eisenhower's attitude.

21 For example Werth, *Russia at War*, p. 176, who saw posters on the walls of Moscow in 1941 portraying Hitler as a giant crab, crushed by a Soviet tank, or as a giant rat under the slogan 'Crush the Fascist vermin!'

22 Werth, *Russia at War*, p. 417; Colville, *Fringes of Power*, p. 344.

23 V. Gollancz, *Shall Our Children Live or Die?: A reply to Lord Vansittart on the German problem* (London, 1942), pp. 5, 7.

24 *The Background and Issues of the War* (Oxford, 1940), pp. 92–3.

25 Gollancz, *Shall Our Children Live or Die?* pp. 62–3.

26 Von Everen, 'Franklin D. Roosevelt and the Problem of Nazi Germany', p. 141.

27 E. Tennant, *True Account* (London, 1957), p. 183; Bell, *Church and Humanity*, pp. 80–5, talk on 'The Threat to Civilization', Canterbury, 15

October 1942.

28 H.G. Wells, *The War in the Air* (London, 1908), pp. 352–4.

29 W. Lippmann, *U.S. War Aims* (Boston, 1944), p. 47.

30 R.J. Overy, 'Air Power and the Origins of Deterrence Theory before 1939', *Journal of Strategic Studies* 14 (1992), p. 82.

31 M. Rader, *No Compromise: The conflict between two worlds* (London, 1939), p. v.

32 Bell, *Church and Humanity*, p. 51.

33 J. Stalin, *The Great Patriotic War of the Soviet Union* (New York, 1945), p. 43, Order of the Day, 23 February 1942.

34 J. and C. Garrard, *World War 2 and the Soviet People* (London, 1993), pp. 16–19. Order 227 was more an enabling decree than an order to kill regardless. Out of the 34 million Soviet citizens mobilised, 442,000 were sent to penal units, and 436,000 imprisoned. See J. Erickson, 'Soviet War Losses', in J. Erickson, D. Dilks, *Barbarossa: The Axis and the Allies* (Edinburgh, 1994), p. 262.

35 J. Barber, 'The Image of Stalin in Soviet Propaganda and Public Opinion during World War 2', in Garrard, *Soviet People*, p. 46.

36 R. Parker, *Moscow correspondent* (London, 1949), pp. 21–2.

37 I. Ehrenburg, *Men, Years – Life: The war years 1941–1945* (London, 1964), p. 123.

38 P. Grigorenko, *Memoirs* (London, 1983), p. 139.

39 Werth, *Russia at War*, p. 416.

40 On attitudes to death, A. Sella, *The Value of Human Life in Soviet Warfare* (London, 1992), esp. pp. 151–71; the 'public language' in Stalin, *Great Patriotic War*, pp. 33–4, speech in Moscow, 6 November 1941, and pp. 45–6, Order of the Day, 23 February 1942.

41 G.H. Roeder, *The Censored War: American visual experience during World War Two* (New Haven, 1993), pp. 20–1.

42 Roeder, *Censored War*, p. 25.

43 O. Bartov, *Hitler's Army: Soldiers, Nazis and war in the Third Reich* (Oxford, 1991), p. 96.

44 M. Wells, 'Aviators and Air Combat: A Study of the US Eighth Air Force and RAF Bomber Command', University of London PhD thesis, 1992, p. 137.

45 A. Calder, *The Myth of the Blitz* (London, 1991), esp. chs 5 and 7.

46 H.A. DeWeend (ed.), *Selected Speeches and Statements of General of the Army George C. Marshall* (New York, 1973), p. 251, address to the American legion, 18 September 1944.

47 J.W. Dower, *War Without Mercy: Race and power in the Pacific war* (New York, 1986), pp. 83–4.

48 A. Wedemeyer, *Wedemeyer Reports!* (New York, 1958), p. 107.

49 Dower, *War Without Mercy*, pp. 16–17. The image of the good war inspired Studs Terkel's book of the same title: *The 'Good War': An oral history of World War Two* (New York, 1984).

50 See the discussion in S. Garrett, *Ethics and Airpower in World War Two* (New York, 1993), and C.C. Crane, *Bombs, Cities and Civilians: American airpower strategy in World War II* (Kansas University Press, 1993).

51 Dower, *War Without Mercy*, p. 16; P. Winterton, *Report on Russia* (London, 1945), pp. 6–15; Perlmutter, *FDR and Stalin*, pp. 69–75.

52 W.J. West (ed.), *Orwell: The war commentaries* (London, 1985), pp. 20–1.

53 West (ed.), *Orwell*, p. 22; Orwell, *Collected Essays*, pp. 168–70: 'But Stalin seems to be becoming a figure rather similar to what Franco used to be, a Christian gent whom it is not done to criticize' (p. 170).

54 Directorate of Army Education, *The British Way and Purpose* (London, 1944), appendix B, pp. 570–1.

55 *Letters of Thomas Mann 1889–1955*, trans. R. and C. Winston (2 vols, London, 1970), I, p. 377, letter to Agnes Meyer, 7 October 1941.

56 Colville, *Fringes of Power*, p. 355, entry for 10 February 1941.

57 Stalin, *Great Patriotic War*, p. 30; Wedemeyer, *Wedemeyer Reports!*, p. 106.

58 W. Shirer, *Berlin Diary* (London, 1941), pp. 161–2, entry for 3 September 1939.

59 M. Muggeridge (ed.), *Ciano's Diary 1930–1943* (London, 1947), pp. 263–4, entry for 10 June 1940.

60 United States Strategic Bombing Survey (USSBS), Pacific Theatre, 'The Effects of Strategic Bombing on Japanese Morale' (Washington, June 1947), p. 15.

61 USSBS, Pacific Theatre, Report 72, 'Interrogation of Japanese Officials' (2 vols, Washington, 1946). Midway was taken by most of those interrogated as the turning-point. For example vol. I, p. 6, 'After Midway I thought it was defensive holding from that time on'; p. 169, 'it was the opinion of most officers that the loss of the aircraft carriers during the summer of 1942 stopped the expansion . . .'; p. 266, 'I think the failure of the Midway Campaign was the beginning of total failure . . .'; etc.

62 T. Kase, *Eclipse of the Rising Sun* (London, 1951), p. 66.

63 USSBS, 'Effects of Strategic Bombing on Japanese Morale', pp. 16–17, 24

64 *Fading Victory: The diary of Admiral Matome Ugaki 1941–1945* (Pittsburgh, 1991), p. 33, entry for 2 December 1941.

65 USSBS, 'Effects of Strategic Bombing on Japanese morale', p. 25.

66 R. Nagatsaku, *I Was a Kamikaze* (London, 1973), p. 22ff; USSBS, Pacific Theatre, Report 1, 'Summary Report' (Washington, 1 July 1946), p. 12.

67 Kase, *Eclipse*, pp. 78–80; I. Nish, 'Japan', in J. Noakes (ed.), *The Civilian*

in War: The home front in Europe, Japan and the USA in World War II (Exeter, 1992), pp. 98–101.

68 USSBS, 'Summary Report', p. 23; Kase, Eclipse of the Rising Sun, pp. 67–8.

69 USSBS, 'Effects of Strategic Bombing on Japanese Morale', p. vi.

70 USSBS, 'Summary Report', pp. 17–21; Nagatsaku, I Was a Kamikaze, p. 43.

71 USSBS, 'Effects of Strategic Bombing on Japanese Morale', p. 21; USSBS, 'Summary Report', p. 21; USSBS, Pacific Theatre, Report 2, 'Japan's Struggle to End the War' (Washington, 1 July 1946), pp. 21–2, memorandum of meeting between Prince Konoye and Emperor Hirohito 14 February 1945; Konoye told the emperor, 'I think that there is no longer any doubt about our defeat . . . The greatest obstacle to ending the war is the existence of the military group which has been propelling the country into the present state ever since the Manchurian Incident . . .'

72 W.A. Boelcke (ed.), The Secret Conferences of Dr. Goebbels Oct. 1939 to March 1943 (London, 1967), pp. 175–6, conference of 22 June 1941; pp. 192–5, conferences of 7th, 11th and 19 December 1941; R.G. Reuth, Goebbels (London, 1993), pp. 291–3.

73 F. Taylor (ed.), The Goebbels Diaries 1939–1941 (London, 1982), p. 415, entry for 16 June 1941.

74 M. Jonas, The United States and Germany: A diplomatic history (New York, 1984), pp. 232–3.

75 N. Rich, Hitler's War Aims: The establishment of the New Order (London, 1974), pp. 351–2, 327; Bartov, Hitler's Army, pp. 66–72, 84–8.

76 J. Forster, 'The Relation between Operation Barbarossa as an Ideological War of Extermination and the Final Solution', in D. Cesarani (ed.), The Final Solution: Origins and implementation (London, 1994), p. 92; O. Bartov, 'Operation Barbarossa and the origins of the Final Solution', in Cesarani, The Final Solution, pp. 127–9.

77 C. Browning, Ordinary Men: Reserve Police Battalion 101 and the Final Solution in Poland (New York, 1993).

78 Taylor (ed.), Goebbels Diaries, p. 426.

79 M. Messerschmidt 'Deserteure in Zweiten Weltkrieg' in W. Wette (ed.), Deserteure dei Wehrmacht: Feiglinge – Opfer – Hoffnungsträger (Essen, 1995), pp. 61–2; O. Hennicke, F. Wüllner 'Über die barbarischen Vollstreckungsmethoden von Wehrmacht und Justiz im Zweiten Weltkrieg' in Wette (ed.), Deserteure, pp. 80–1.

80 Bartov, Hitler's Army, pp. 96–9. The Soviet figures are calculated from J. Erickson, D. Dilks (eds), Barbarossa: the Axis and the Allies (Edinburgh, 1994), p. 262.

81 Taylor (ed.), *Goebbels Diaries*, p. 425, entry for 23 June 1941; Spanish ambassador in W. Kimball (ed.), *Churchill and Roosevelt: The complete correspondence* (3 vols, London, 1984), II, Churchill to Roosevelt, 1 January 1943, enclosing Special Intelligence Serial No. 75.

82 U. von Hassell, *The von Hassell Diaries, 1938–1944* (London, 1948), p. 182, entry for 13 July 1941.

83 See in particular K. von Klemperer, *German Resistance against Hitler* (Oxford, 1992); P. Hoffmann, *The German Resistance to Hitler* (Cambridge, Mass., 1988).

84 Morale figures from USSBS, European Theatre, Report 64B, 'The Effects of Strategic Bombing on German Morale', vol. I (Washington, May 1947), p. 16 (including 16 per cent who considered the war lost from the outset); Goebbels remark in Z. Zeman, *Nazi Propaganda* (Oxford, 1973), p. 164.

85 A. Fredborg, *Behind the Steel Wall: Berlin 1941–43* (London, 1944), p. 164.

86 Ibid., pp. 218–19.

87 Boelcke, *Secret Conferences*, pp. 314–26, conferences of 6th, 21st and 25–7 January 1943.

88 USSBS, 'Effects of Strategic Bombing on German Morale', pp. 14, 16; von Hassell, *Diaries*, p. 301, entry for 2 January 1944, p. 303, entry for 23 February 1944.

89 F. von Schlabrendorff, *The Secret War Against Hitler* (London, 1966), p. 223.

90 Details from von Schlabrendorff, *Secret War Against Hitler*, pp. 229–40.

91 M. Baigent, R. Leigh, *Secret Germany* (London, 1994).

92 Details from von Schlabrendorff, *Secret War Against Hitler*, pp. 276–92; Baigent, Leigh, *Secret Germany*, pp. 46–58; H. B. Gisevius, *To the Bitter End* (London, 1948), pp. 526–56.

93 USSBS, 'Effects of Strategic Bombing on German Morale', pp. 14–16. Of those willing to fight to the end 24 per cent claimed never to have wavered, 5 per cent to have wavered and then recovered confidence.

94 USSBS, 'Effects of Strategic Bombing on Japanese Morale', p. 23.

95 Ugaki, *Fading Victory*, pp. 665–6; *kessen* in Nish, 'Japan', pp. 99–100.

96 G. Tempel, *Speaking Frankly About the Germans* (London, 1963), pp. 43–4.

97 P. Padfield, *Himmler: Reichsführer SS* (London, 1990), pp. 606–11.

98 T. Taylor, *The Anatomy of the Nuremberg Trials* (London, 1993), pp. 168–9.

99 Taylor, *Anatomy of the Nuremberg Trials*, p. 211.

100 G.M. Gilbert, *Nuremberg Diary* (London, 1948), pp. 29–31.

10 Why the Allies Won

1 F. Genoud (ed.), *The Testament of Adolf Hitler: The Hitler-Bormann documents February–April 1945* (London, 1961), p. 70, 17 February 1945.

2 Ibid., pp. 63, 79, 98–9, 15th, 20th and 26 February 1945.

3 Ibid., pp. 101–2, 26 February 1945.

4 Ibid., pp. 52, 82–3, 87–8, 13th, 21st and 24 February 1945.

5 W. Averell Harriman with E. Abel, *Special Envoy to Churchill and Stalin 1941–1946* (London, 1976), pp. 90–1.

6 R.H. Ferrell (ed.), *The Eisenhower Diaries* (New York, 1981), p. 53, entry for 5 March 1942.

7 United States Strategic Bombing Survey (USSBS), Pacific Theatre, Report 72, 'Interrogation of Japanese Officials', vol. I (Washington, 1946), p. 285.

8 F. Gilbert (ed.), *Hitler Directs His War: The secret records of his daily military conferences* (New York, 1950), p. 24, conference of 5 March 1943.

9 H.A. DeWeend (ed.), *Selected Speeches and Statements of General of the Army George C. Marshall* (New York, 1973), p. 249, remarks at the award of the Howland Memorial Prize, Yale University, 16 February 1944.

10 D. Glantz (ed.), *Soviet Documents on the Use of War Experience* (3 vols, London, 1993), III, pp. vii–ix.

11 R.J. Overy, *The Air War 1939–1945* (London, 1980), pp. 134–7, 145–8; A. Nielsen, *The German Air Force General Staff* (New York, 1959), pp. 46–8; A. Coox, 'The Effectiveness of the Japanese Military Establishment in the Second World War', in A. Millett, W. Murray (eds), *Military Effectiveness in World War II* (London, 1988), pp. 27, 34–5.

12 USSBS, Pacific Theatre, Report 1, 'Summary Report' (Washington, 1 July 1946), pp. 10–11.

13 M. Stoler, *George C. Marshall: Soldier-statesman of the American century* (Boston, 1989), pp. 93–5.

14 D.D. Eisenhower, *Report by the Supreme Commander to the Combined Chiefs of Staff on the Operations in Europe* (London, 1946), p. 17.

15 A. V. Khrulev, 'Quartermaster at Work', in S. Bialer (ed.), *Stalin and His Generals: Soviet military memoirs of World War II* (New York, 1969), pp. 370–5.

16 Some measure of this can be found in R.G. Ruppenthal, *Logistical Support of the Armies* (2 vols, Washington, 1953). On the role of the military in Germany see, for example, Nielsen, *German Air Force General Staff*, or H. Boog, *Die deutsche Luftwaffenführung 1935–1945* (Stuttgart, 1982). On Japan, R. Smethurst, *A Social Basis for Pre-war Japanese Militarism*

(Berkeley, 1974), or R. Benedict, *The Chrysanthemum and the Sword* (London, 1967).

17 Lord Moran, *Winston Churchill: The struggle for survival 1940–1965* (London, 1966), p. 273.

18 S.E. Morison, *American Contributions to the Strategy of World War II* (London, 1958), p. 47.

19 British Air Ministry, *The Rise and Fall of the German Air Force* (reissued London, 1983), pp. 273–5, 307–9.

20 Ibid., p. 407.

21 Lord Tedder, *Air Power in War* (London, 1948), p. 82.

22 M. Cooper, *The German Army 1933–1945: Its political and military failure* (London, 1978), p. 506.

23 J.C. Gaston, *Planning the American Air War: Four men and nine days in 1941* (National Defense University, Washington, 1982), pp. 13–15, 22, 67–8, 98–102. The plan called for a total air force of 21,813 combat aircraft. By December 1943 American combat air strength was 20,185 overseas, and by the end of 1944 32,957. German air strength was 5,536 and 6,297 on the same dates.

24 M. Rader, *No Compromise: The conflict between two worlds* (London, 1939), p. viii.

25 USSBS, Report 72, 'Interrogation of Japanese Officials', vol. I, no. 43, interrogation of Captain Ohmae, South-east Area Fleet Staff.

Epilogue

1 D.W. White, 'The Nature of World Power in American History: An evaluation at the end of World War II', *Diplomatic History* 11 (1987), pp. 182–4.

2 R.J. Bosworth, *Explaining Auschwitz and Hiroshima: History writing and the Second World War* (London, 1993), p. 159.

3 L. Lazarev, 'Russian Literature on the War and Historical Truth', in J. and C. Garrard (eds), *World War 2 and the Soviet People* (London, 1993), pp. 33–4. See too A. Weiner, *Making Sense of War: The Second Word War and the Fate of the Bolshevik Revolution* (Princeton, 2001), esp. pp. 18–21.

4 Air Policy Commission, *Survival in the Air Age* (Washington, 1 January 1948), pp. 18–21.

5 W. Horsley, R. Buckley, *Nippon: Japan since 1945* (London, 1990), p. 39.

SELECTED READING

The following booklist includes all those books and articles referred to directly in writing *Why the Allies Won*. It is not intended to be a general bibliography of the Second World War. Where possible I have included English translations of foreign-language works.

Air Ministry, *The Rise and Fall of the German Air Force 1933–1945* (London 1947, reprinted 1983)

Ambrose, S.E., *D-Day, June 6th 1944: The climactic battle of World War II* (New York, 1994)

Ambrose, S.E., *Eisenhower: Soldier and President* (New York, 1990)

Ambrose, S.E., *Rise to Globalism: American foreign policy since 1938* (London, 1988)

Anders, W., *Hitler's Defeat in Russia* (Chicago, 1951)

Andrew, C, Gordievsky, O., *KGB: The inside story* (London, 1990)

Armstrong, R.A., 'Stalingrad: Ordeal or turning point', *Military Review* 72 (1992)

Arnold, H.H., *Global Mission* (New York, 1949)

Arnold, H.H., *Second Report of the Commanding General of the USAAF* (London, 1945)

Assersohn, F. J., 'Propaganda and Policy: The presentation of the strategic air offensive in the British mass media 1939–1945', Leeds University M.A. thesis, 1989

Axell, A. *Stalin's War through the Eyes of his Commanders* (London, 1997)

The Background and Issues of the War (Oxford, 1940)

Bacque, J., *Other Losses: The investigation into the mass deaths of German prisoners of war after World War II* (London, 1989)

Baigent, M., Leigh, R., *Secret Germany* (London, 1994)

Barber, J., Harrison, M., *The Soviet Home Front 1941–1945* (London, 1991)

Barnett, C, *Engage the Enemy More Closely: The Royal Navy in the Second World War* (London, 1991)

Barnett, C. (ed.), *Hitler's Generals* (London, 1989)

Barnhart, M. A., *Japan Prepares for Total War: The search for economic security 1919–1941* (Cornell University Press, 1987)

Barnhart, M.A., 'Japan's Economic Security and the Origins of the Pacific War', *Journal of Strategic Studies* 4 (1981)

Bartov, O., *Hitler's Army: Soldiers, Nazis and war in the Third Reich* (Oxford, 1991)

Bauer, E., *Der Panzerkrieg* (2 vols, Bonn, 1966)

Beachley, D., 'Soviet Radio-Electronic Combat in World War II', *Military Review* 61 (1981)

Beaumont, A., 'The Bomber Offensive as a Second Front', *Journal of Contemporary History* 22 (1987)

Beaumont, J., *Comrades in Arms: British aid to Russia 1941–1945* (London, 1980)

Beck, E., *Under the Bombs: The German Home Front 1942–1945* (Lexington, Kentucky, 1986)

Beesly, P., *Very Special Intelligence: The story of the Admiralty Operational Intelligence Centre 1939–1945* (London, 1977)

Behrens, B.A., *Merchant Shipping and the Demands of War* (London, 1955)

Bekker, C, *The Luftwaffe War Diaries* (London, 1967)

Belchem, D., *Victory in Normandy* (London, 1981)

Bell, G.K., *The Church and Humanity 1939–1946* (London, 1946)

Benedict, R., *The Chrysanthemum and the Sword* (London, 1967)

Bennett, R., *Behind the Battle: Intelligence in the war with Germany 1939–1945* (London, 1994)

Bennett, R., *Ultra in the West: The Normandy campaign of 1944–1945* (London, 1979)

Beria, S., *Beria, My Father: Inside Stalin's Kremlin* (London, 2001)

Berle, B.B., Jacobs, T.B. (eds), *Navigating the Rapids 1918–1971: From the papers of Adolph A. Berle* (New York, 1973)

Best, G., *Churchill at War* (London, 2005)

Beyerchen, A., *Scientists Under Hitler* (New Haven, 1977)

Bialer, S. (ed.) *Stalin and His Generals: Soviet military memoirs of World War II* (New York, 1969)

Bialer, U., *The Shadow of the Bomber: The fear of air attack and British politics 1932–1939* (London, Royal Historical Society, 1980)

Birkenfeld, W., *Der synthetische Treibstoff 1933–1945* (Göttingen, 1963)

Birse, A., *Memoirs of an Interpreter* (London, 1967)

Bisson, T.A., *Japan's War Economy* (New York, 1945)

Blake, R., Louis, W.R., (eds), *Churchill* (Oxford, 1993)

Bland, L. (ed.), *The Papers of George Catlett Marshall: Vol. 2, 'We cannot delay', July 1 1939 to December 6 1941* (Baltimore, 1986)

Blumenson, M., *Breakout and Pursuit* (Washington, 1961)

Blumenson, M. (ed.), *The Patton Papers 1940–1945* (Boston, 1974)

Boberach, H. (ed.), *Meldungen aus dem Reich: Auswahl aus den geheimen Lageberichten der Sicherheitsdienst der SS, 1939–1945* (Berlin, 1965)

Boelcke, W.A. (ed.), *The Secret Conferences of Dr. Goebbels, Oct. 1939 to March 1943* (London, 1967)

Bohlen, C.E., *Witness to History 1929–1969* (London, 1973)

Bond, B. (ed.), *Chief of Staff: The diaries of Lieutenant-General Sir Henry Pownall: Vol. 1, 1933–1940* (London, 1972)

Boog, H. *et al.*, *Germany and the Second World War; Volume IV: The Attack on the Soviet Union* (Oxford, 1998)

Boog, H. (ed.), *The Conduct of Air War in the Second World War: An international comparison* (Oxford, 1992)

Boog, H., *Die deutsche Luftwaffenführung 1935–1945* (Stuttgart, 1982)

Bosworth, R. J., *Explaining Auschwitz and the Holocaust: History writing on the Second World War* (London, 1973)

Boyd, A., *The Soviet Air Force since 1918* (London, 1977)

Boyd, C., *The Extraordinary Envoy: General Hiroshi Oshima and diplomacy in the Third Reich 1934–1939* (Washington, 1980)

Bradley, O., *A Soldier's Story of the Allied Campaign from Tunis to the Elbe* (London, 1951)

Bragadin, M.A., *The Italian Navy in World War II* (US Naval Institute, Annapolis, 1957)

Braithwaite, R., *Moscow 1941: A City and its People at War* (London, 2006)

Brendon, P., *Ike: His life and times* (New York, 1986)

Breuer, W., *Hoodwinking Hitler: The Normandy deception* (Westport, Conn., 1993)

Brown, J. A., *Eagles Strike: South African forces in World War II, Vol. IV* (Cape Town, 1974)

Browning, C., *Ordinary Men: Reserve Police Battalion 101 and the Final Solution in Poland* (New York, 1993)

Bryant, A., *Triumph in the West: The war diaries of Field Marshal Viscount Alanbrooke* (London, 1959)

Buckley, J., 'Air Power in the Battle of the Atlantic', *Journal of Contemporary History* 28 (1993)

Bullock, A., *Hitler and Stalin: Parallel lives* (London, 1991)

Burdick, C., Jacobsen H.-A., (eds), *Halder Diary 1939–1942* (London, 1988)

Busch, D., *Der Luftkrieg im Raum Mainz während des Zweiten Weltkriegs 1939–1945* (Mainz, 1988)

Butler, S. (ed.), *My Dear Mr. Stalin: The Complete Correspondence of Franklin D. Roosevelt and Joseph V. Stalin* (New Haven, 2005)

Calder, A., *The Myth of the Blitz* (London, 1991)

Calder, A., *The People's War: Britain 1939–1945* (London, 1969)

Cantril, H. (ed.), *Public Opinion 1935–1946* (Princeton, 1951)

Carell, P., *Hitler's War on Russia* (2 vols, London, 1970)

Cassidy, H.C., *Moscow Dateline 1941–1943* (London, 1944)

Cesarani, D. (ed.), *The Final Solution: Origins and implementation* (London, 1994)

Chalmers, W.S., *Max Horton and the Western Approaches* (London, 1954)

Chaney, O.P., *Zhukov* (Norman, Oklahoma, 1972)

Charmley, J., *Churchill: The end of glory* (London, 1993)

Chickering, R., Förster, S., Greiner, B. (eds), *A World at Total War: Global Conflict and the Politics of Destruction, 1937–1945* (Cambridge, 2005)

Chuikov, V.I., *The Beginning of the Road: The story of the battle for Stalingrad* (London, 1963)

Churchill, W.S., *The Second World War* (6 vols, London, 1948–55)

Clark, R.W., *The Birth of the Bomb* (London, 1961)

Clive, A., *State of War: Michigan in World War II* (Chicago, 1979)

Coates, W.P. and Z., *A History of Anglo-Soviet Relations* (London, 1944)

Cohen, J., *Japan's Economy in War and Reconstruction* (Minneapolis, 1949)

Collier, B., *The Defence of the United Kingdom* (London, 1957)

Colville, J., *The Fringes of Power: 10 Downing Street diaries 1939–1955* (London, 1985)

Constantini, A., *L'Union soviétique en guerre 1941–1945* (Paris, 1968)

Conway, J., *The Nazi Persecution of the Churches* (London, 1968)

Cooke, R., Nesbit, R., *Target Hitler's Oil: Allied attacks on German oil supplies 1939–1945* (London, 1985)

Cooling, B.F. (ed.), *Case Studies in the Development of Close Air Support* (Office of Air Force History, Washington, 1990)

Cooper, M., *The German Army 1933–1945* (London, 1978)

Crane, C.C., *Bombs, Cities and Civilians: American Airpower Strategy in World War II* (Kansas University Press, 1993)

Craven, W.F., Cate, J.L., *The Army Air Forces in World War II* (6 vols, Washington, 1948–55, reissued 1983)

Creveld, M. van, *Supplying War: Logistics from Wallenstein to Patton* (Cambridge, 1977)

Creveld, M. van, *Fighting Power: German and US army performance 1939–1945* (London, 1983)

Cross, R., *VE Day: Victory in Europe* (London, 1985)

Cruikshank, C., *Deception in World War II* (London, 1979)

Cunningham, Viscount A., *A Sailor's Odyssey* (London, 1951)

Dallek, R., *Franklin D. Roosevelt and American Foreign Policy 1932–1945* (Oxford, 1979)

Dallin, A., *German Rule in Russia* (2nd ed., London, 1981)

Dallin, D., Nicolaevsky, O., *Forced Labour in Soviet Russia* (London, 1947)

Danchev, A. (ed.), *Establishing the Anglo-American Alliance: The Second World War diaries of Brigadier Vivian Dykes* (London, 1990)

Danchev, A., *Very Special Relationship: Field Marshal Sir John Dill and the Anglo-American Alliance 1941–1944* (London, 1986)

Danchev, A., Todman, D. (eds), *War Diaries 1939–1945: Field Marshal Lord Alanbrooke* (London, 2001)

Davis, R.G., *Carl A. Spaatz and the Air War in Europe* (Washington, 1992)

Dawson, R.H., *The Decision to Aid Russia, 1941* (Chapel Hill, North Carolina, 1959)

Deane, J.R., *The Strange Alliance: The story of American efforts at wartime co-operation with Russia* (London, 1947)

D'Este, C., *Decision in Normandy* (New York, 1983)

Detweiler, D. (ed.), *World War II German Military Studies* (24 vols, New York, 1979)

DiNardo, R.L., Bay, A., 'Horse-Drawn Transport in the German Army', *Journal of Contemporary History* 23 (1988)

DiNardo, R.L., *Mechanized Juggernaut or Military Anachronism: Horses and the German Army in World War II* (London, 1991)

Directorate of Army Education, *The British Way and Purpose* (London, 1944)

Divine, R.A., *Roosevelt and World War II* (London, 1969)

Dobb, M., *Soviet Economy and the War* (London, 1941)

Dönitz, K., *Memoirs: Ten Years and Twenty Days* (London, 1959)

Doughty, M., *Merchant Shipping at War* (London, 1982)

Douglas, R., *The World War 1939–1945: The cartoonists' vision* (London, 1990)

Dower, J., *War without Mercy: Race and power in the Pacific War* (New York, 1986)

Dull, P.S., *A Battle History of the Imperial Japanese Navy 1941–1945* (Cambridge, 1978)

Eggleston, G.T., *Roosevelt, Churchill and the World War II Opposition* (Old Greenwich, Conn., 1979)

Ehrenburg, I., *Men, Years – Life: The war 1941–1945* (London, 1964)

Ehrman, J., *Grand Strategy: Vol. V, August 1943 to September 1944* (London, 1946)

Ellis, J., *Brute Force: Allied strategy and tactics in the Second World War* (London, 1990)

Ellis, L., *Victory in the West: Vol. I, The battle for Normandy* (London, 1962)

Eisenhower, D.D., *Crusade in Europe* (London, 1948)

English, J.A., *The Canadian Army and the Normandy Campaign* (New York, 1991)

Erickson, J., Dilks, D. (eds), *Barbarossa: The Axis and the Allies* (Edinburgh, 1994)

Erickson, J., 'New Thinking about the Eastern Front in World War II', *Journal of Military History* 56 (1992)

Erickson, J., *The Road to Berlin: Stalin's war with Germany* (London, 1983)

Erickson, J., *The Road to Stalingrad* (London, 1975)

Eubank, K., *Summit at Teheran* (New York, 1985)

Everen, B. van, 'Franklin D. Roosevelt and the Problem of Nazi Germany', in Egan, C.C., Knott, A.W. (eds), *Essays in Twentieth Century American International History* (Lanham, Maryland, 1982)

Faber, H. (ed.), *Luftwaffe: An analysis by former Luftwaffe generals* (London, 1979)

Fearon, P., *War, Prosperity and Depression in the US Economy, 1917–1945* (London, 1986)

Ferrel, R.H. (ed.), *The Eisenhower Diaries* (New York, 1981)

Ferro, M., *The Great War 1914–1918* (London, 1973)

Fischer, J., 'Ober den Entschluss zur Luftversorgung Stalingrads. Ein Beitrag zur militärischen Führung im Dritten Reich', *Militärgeschichtliche Mitteilungen* 6 (1969)

François-Poncet, A., *The Fateful Years: Memoirs of a French ambassador in Berlin, 1931–1938* (London, 1949)

Fraser, D., *And We Shall Shock Them: The British army in the Second World War* (London, 1983)

Fraser, D., *Knight's Cross: A life of Field-Marshal Rommel* (London, 1993)

French, D., *The British Way in Warfare* (London, 1989)

Fredborg, A., *Behind the Steel Wall: Berlin 1941–43* (London, 1944)

Frye, A., *Nazi Germany and the American Hemisphere 1933–1941* (New Haven, 1987)

Fuchida, M., Okumiya, M., *Midway: The battle that doomed Japan* (Annapolis, 1955)

Fuehrer Conferences on Naval Affairs 1939–1945 (London, 1948)

Galbraith, J.K., *A Life in Our Times: Memoirs* (London, 1981)

Galland, A., *The First and the Last* (London, 1955)

Garrard, C., Garrard, J. (eds), *World War 2 and the Soviet People* (London, 1993)

Garrett, S., *Ethics and Airpower in World War II* (New York, 1993)

Gaston, J.C., *Planning the American Air War: Four men and nine days in 1941* (National Defense University, Washington, 1982)

Gatzke, H., *Germany and the United States* (Cambridge, Mass., 1980)

Gebhardt, J., 'World War II: The Soviet side', *Military Review* 72 (1992)

Gibbs, N., *Grand Strategy: Vol. I, Rearmament policy* (London, 1976)

Gilbert, F. (ed.), *Hitler Directs His War: The secret records of his daily military conferences* (New York, 1950)

Gilbert, G.M., *Nuremberg Diary* (London, 1948)

Gisevius, H.B., *To the Bitter End* (London, 1948)

Glantz, D., *From the Don to the Dnepr* (London, 1991)

Glantz, D., *The Military Strategy of the Soviet Union: A history* (London, 1992)

Glantz, D., *The Role of Intelligence in Soviet Military Strategy in World War II* (Novato, California, 1990)

Glantz, D. (ed.), *Soviet Documents on the Use of War Experience in World War II* (3 vols, London, 1993)

Goldberg, R.J., *The Making of Franklin D. Roosevelt: Triumph over disability* (Cambridge, Mass., 1981)

Gollancz, V., *Shall Our Children Live or Die? A Reply to Lord Vansittart on the German Problem* (London, 1942)

Golücke, F., *Schweinfurt und der strategische Luftkrieg 1943* (Paderborn, 1980)

Gorodetsky, G., *Grand Delusion: Stalin and the German Invasion of Russia* (New Haven, 1999)

Grayling, A.C., *Among the Dead Cities* (London, 2006)

Great Patriotic War of the Soviet Union 1941–1945: A general outline (Moscow, 1970)

Greenfield, K., *American Strategy in World War II: A reconsideration* (Baltimore, 1963)

Groehler, O., *Bombenkrieg gegen Deutschland* (Berlin, 1990)

Grigorenko, P., *Memoirs* (London, 1983)

Groves, L.R., *Now It Can Be Told* (New York, 1962)

Grunburg, J.A., *Divided and Conquered: The French High Command and the defeat in the west, 1940* (Westport, Conn., 1979)

Guderian, H., *Erinnerungen eines Soldaten* (Heidelberg, 1951)

Gueritz, E.F., 'Nelson's Blood: Attitudes and action of the Royal Navy 1939–1945', *Journal of Contemporary History* 16 (1981)

Hadley, M.L., *U-Boats against Canada* (London, 1990)

Halifax, Lord, *Fulness of Days* (London, 1957)

Hallion, R.P., *Strike from the Sky: The history of battlefield air attack, 1911–1945* (Washington, 1989)

Hallendorf, J., Jordan, R.S. (eds), *Maritime Strategy and the Balance of Power* (London, 1989)

Hamilton, N., *Monty: Vol. II, Master of the battlefield* (London, 1983)

Hansell, H.S., *The Strategic Air War against Germany and Japan* (Washington, 1986)

Hardesty, Von, *Red Phoenix: The rise of Soviet airpower* (London, 1982)

Harries, M., Harries, S., *Soldiers of the Sun: The rise and fall of the Imperial Japanese Army 1868–1945* (London, 1991)

Harriman, W. Averell, with Abel, E., *Special Envoy to Churchill and Stalin, 1941–1946* (London, 1976)

Harris, A., *Bomber Offensive* (London, 1947)

Harrison, G., *Cross-Channel Attack* (Washington, 1951)

Harrison, M. (ed), *The Economics of World War II: Six great powers in international comparison* (Cambridge, 1998)

Harrison, M., 'Resource Mobilisation for World War II: The USA, UK, USSR and Germany 1938–1945', *Economic History Review* 41 (1988)

Harrison, M., *Soviet Planning in Peace and War 1938–1945* (Cambridge, 1985)

Hartcup, G., *The Challenge of War: Scientific and engineering contributions to World War II* (Newton Abbot, 1970)

Harvey, A., 'The Italian War Effort and the Strategic Bombing of Italy', *History* 70 (1985)

Hassel, U. von, *The von Hassell Diaries 1938–1944* (London, 1948)

Hastings, M., *Bomber Command* (London, 1979)

Hayes, G.P., *The History of the Joint Chiefs of Staff in World War II* (London, 1982)

Hayes, P., *Industry and Ideology: IG Farben in the Nazi Era* (Cambridge, 1987)

Hayward, J., *Stopped at Stalingrad: The Luftwaffe and Hitler's Defeat in the East, 1942–1943* (Lawrence, Kans., 1998)

Heer, H., *Tote Zonen: Die deutsche Wehrmacht an der Ostfront* (Hamburg, 1999)

Heiber, H., Glantz, D.G., *Hitler and his Generals: Military Conferences 1942–1945* (London, 2002)

Heinrichs, W., *Threshold of War: Franklin D. Roosevelt and American entry into World War II* (Oxford, 1988)

Herbert, U., *Fremdarbeiter: Politik und Praxis des 'Ausländer-Einsatzes' in der Kriegswirtschaft des Dritten Reiches* (Bonn, 1985)

Herwig, H., 'The Failure of German Sea Power 1914–1945: Mahan, Tirpitz and Raeder reconsidered', *International History Review* 10 (1988)

Higham, R., 'The Ploesti Ploy: British considerations on the idea of bombing the Romanian oilfields 1940–41', *War and Society* 5 (1987)

Higham, R., Kagan, K.W. (eds), *The Military History of the Soviet Union* (New York, 2002)

Hiller, M. (ed.), *Städte im Zweiten Weltkrieg* (Essen, 1991)

Hiller, M. (ed.), *Stuttgart im Zweiten Weltkrieg* (Gerlingen, 1989)

Hindus, M., *Russia Fights On* (London, 1942)

Hinsley, F.H. *et al.*, *British Intelligence in the Second World War* (4 vols, London, 1979–90)

Hoffmann, P., *The German Resistance to Hitler* (Cambridge, Mass., 1988)

Höhne, H., *The Order of the Death's Head* (London, 1969)

Holt, T., *The Deceivers: Allied Military Deception in the Second World War* (London, 2004)

Hopfinger, K., *Beyond Expectation: The Volkswagen story* (London, 1954)

Horne, A., Montgomery, D., *The Lonely Leader: Monty 1944–1945* (London, 1994)

Horsley, W., Buckley, R., *Nippon: Japan since 1945* (London, 1990)

Howard, M., *The Mediterranean Strategy in the Second World War* (London, 1968)

Howard, M., *Strategic Deception in the Second World War* (London, 1990)

Howse, D., *Radar at Sea: The Royal Navy in World War II* (London, 1993)

How the Jap Army Fights (London, 1943)

Hoyt, E.P., *Japan's War: The great Pacific conflict* (London, 1986)

Hull, C, *Memoirs* (2 vols, London, 1948)

Ickes, H., *The Secret Diary of Harold L. Ickes* (3 vols, London, 1955)

Irving, D. (ed.), *Adolf Hitler: The medical diaries* (London, 1983)

Irving, D., *The Rise and Fall of the Luftwaffe: The life of Field Marshal Erhard Milch* (London, 1973)

Ishoven, A. van, *The Fall of an Eagle: The life of fighter ace Ernst Udet* (London, 1977)

Ismay, Lord H., *Memoirs* (London, 1960)

Jackson, A., *The British Empire and the Second World War* (London, 2006)

Jacobs, W.A., 'Strategic Bombing and American National Strategy 1941–1943', *Military Affairs* 50 (1986)

Jacobsen, H.-A., Rohwer, J. (eds), *Decisive Battles of World War II: The German view* (London, 1965)

Jonas, M., *The United States and Germany: A diplomatic history* (New York, 1984)

Jones, C.F., 'Industrial Capacity and Supplies of Raw Materials', in Valkenburg, S. van (ed.), *America and War: A geographical analysis* (New York, 1943)

Jones, F.C., *Japan's New Order in East Asia* (Oxford, 1954)

Jones, R.V., *Most Secret War: British scientific intelligence 1939–1945* (London, 1978)

Johnson, B., *The Secret War* (London, 1978)

Joubert de la Ferté, P., *Rocket* (London, 1957)

Junge, T., *Until the Final Hour: Hitler's Last Secretary* (London, 2003)

Kahn, D., 'Codebreakers in World Wars I and II: The major successes and failures, their causes and their effects', *Historical Journal* 23 (1980)

Kahn, D., *Hitler's Spies: German military intelligence in World War II* (London, 1978)

Kahn, D., *Seizing the Enigma: The race to break the German U-boat codes* (Boston, 1991)

Karlsch, R., *Hitlers Bombe: Die geheime Geschichte der deutschen Kernwaffenversuche* (Munich, 2005)

Kase, T., *Eclipse of the Rising Sun* (London, 1951)

Kater, M., 'Hitler in a Social Context', *Central European History* 14 (1981)

Keitel, W., *The Memoirs of Field Marshal Keitel* (London, 1965)

Kennedy, P., *The Rise and Fall of British Naval Mastery* (London, 1976)

Kennedy, P., *The Rise and Fall of the Great Powers* (London, 1988)

Kilmarx, R., *A History of Soviet Air Power* (London, 1962)

Kimball, W. (ed.), *Churchill and Roosevelt: The complete correspondence* (3 vols, Princeton, 1984)

Kimball, W., *The Juggler: Franklin Roosevelt as wartime statesman* (Princeton, 1991)

Kimball, W., *The Most Unsordid Act: Lend-Lease 1939–1941* (Baltimore, 1969)

King, E.J., Whitehill, W.M., *Fleet Admiral King: A naval record* (New York, 1952)

Kirwin, G., 'Allied Bombing and Nazi Domestic Propaganda', *European History Quarterly* 15 (1985)

Kitchen, M., *British Policy Towards the Soviet Union During the Second World War* (London, 1986)

Kluke, P., 'Hitler und das Volkswagenprojekt, *Vierteljahrshefte für Zeitgeschichte* 8 (1960)

Knoke, H., *I Flew for the Führer* (London, 1953)

Koch, H., 'Operation Barbarossa: The current state of the debate', *Historical Journal* 31 (1988)

Koch, H., 'The Strategic Air Offensive against Germany: The early phase May–September 1940', *Historical Journal* 34 (1991)

Krauskopf, R. W., 'The Army and the Strategic Bomber 1930–1939', *Military Affairs* 22 (1958–9)

Kriegstagebuch des Oberkommandos der Wehrmacht (4 vols, Frankfurt, 1961)

Krivosheev, G. F., *Soviet Losses and Combat Casualties in the Twentieth Century* (London, 1997)

Lane, F.C., *Ships for Victory: A history of shipbuilding under the US Maritime Commission in World War II* (Baltimore, 1951)

Laqueur, W. (ed.), *The Second World War: Essays in military and political history* (London, 1982)

Larson, H., Knowlton, E., Popple, C., *History of Standard Oil Company: Vol. III, New Horizons 1927–50* (New York, 1971)

Leahy, W.D., *I Was There: The personal story of the Chief-of-Staff to Presidents Roosevelt and Truman* (London, 1950)

Leighton, R.M., Coakley, R.W., *Global Logistics and Strategy* (2 vols, Washington, 1955–68)

Levering, R.B., *American Opinion and the Russian Alliance 1939–1945* (Chapel Hill, 1976)

Levine, A.J., *The Strategic Bombing of Germany 1940–1945* (Westport, Conn., 1992)

Levine, A.J., 'Was World War II a Near Run Thing', *Journal of Strategic Studies* 8 (1985)

Lewin, R., *The American Magic: Codes, ciphers and the defeat of Japan* (New York, 1982)

Lewin, R., *Churchill as Warlord* (London, 1973)

Lewin, R., 'A Signal-Intelligence War?', *Journal of Contemporary History* 16 (1981)

Lewin, R., *Ultra Goes to War: The secret story* (London, 1978)

Liddell Hart, B., *The Other Side of the Hill* (London, 1948)

Liddell Hart, B., *The Rommel Papers* (London, 1953)

Lippmann, W., *U.S. War Aims* (Boston, 1944)

Loza, D. (ed), *Fighting for the Soviet Motherland: Recollections from the Eastern Front* (Lincoln, Nebr., 1998)

Luck, H. von, *Panzer Commander* (New York, 1989)

Ludendorff, E., *Her totale Krieg* (Munich, 1935)

MacDonald, C.B., *The Mighty Endeavour: American armed forces in the European theater in World War II* (New York, 1969)

Macfarland, S.L., 'The Evolution of the American Strategic Fighter in Europe 1942–1944', *Journal of Strategic Studies* 10 (1987)

Maier, K., et al., *Germany in the Second World War: Vol. II, Germany's initial conquests in Europe* (Oxford, 1991)

Main Front: Soviet leaders look back on World War II (London, 1987)

Maisky, I., *Memoirs of a Soviet Ambassador: The war 1939–1943* (London, 1967)

Mann, T., *The Letters of Thomas Mann 1943–1955* (2 vols, London, 1970)

Manstein, E. von, *Verlorene Siege* (Bonn, 1955)

Marder, A.J., Jacobsen, M., Horsfield, J., *Old Friends, New Enemies: The Royal Navy and the Imperial Japanese Navy 1942–1945* (Oxford, 1990)

Maser, W. (ed.), *Hitler's Letters and Notes* (New York, 1974)

Masterman, J.J., *The Double-Cross System 1939–1945* (London, 1972)

Matloff, M., Snell, E., *Strategic Planning for Coalition Warfare* (2 vols, Washington, 1953–9)

Mawdsley, E., *Thunder in the East; the Nazi-Soviet War* (London, 2006)

McInnes, C., Sheffield, G. (eds), *Warfare in the Twentieth Century* (London, 1988)

McLaine, I., *Ministry of Morale* (London, 1979)

McKale, D.M., *Hitler: The survival myth* (New York, 1981)

Mellenthin, F. von, *Panzer-Schlachten* (Neckargemünd, 1963)

Messerschmidt, M., 'The Political and Strategic Significance of Advances in Armament Technology. Developments in Germany and the Strategy of "Blitzkrieg"', in Ahmann, R., Birke, A. M., Howard, M. (eds), *The Quest for Stability: Problems of West European security 1918–1957* (Oxford, 1993)

Michaelis, M., 'World Power Status or World Dominion?', *Historical Journal* 15 (1972)

Mierzejewski, A.C., *The Collapse of the German War Economy 1944–1945: Allied air power and the German national railway* (Chapel Hill, North Carolina, 1988)

Miller, R., *Nothing Less Than Victory: An oral history of D-Day* (London, 1993)

Millett, A., Murray, W. (eds), *Military Effectiveness in World War II* (London, 1988)

Milner, M., 'The Battle of the Atlantic', *Journal of Strategic Studies* 13 (1990)

Milward, A.S., *War, Economy and Society 1939–1945* (London, 1987)

Miner, S.M., *Between Churchill and Stalin: The Soviet Union, Great Britain and the origins of the Grand Alliance* (Chapel Hill, North Carolina, 1988)

Montgomery, B.L., *The Memoirs of Field-Marshal the Viscount Montgomery* (London, 1958)

Moran, Lord, *Winston Churchill: The struggle for survival 1940–1965* (London, 1966)

Morgan, E., *FDR: A biography* (London, 1985)

Morgan, F., *Overture to Overlord* (London, 1950)

Morison, S.E., *American Contributions to the Strategy of World War II* (London, 1958)

Morison, S.E., *History of US Naval Operations in World War II* (15 vols, Boston, 1947–62)

Morton, H.V., *Atlantic Meeting* (London, 1943)

Moskoff, W., *The Bread of Affliction: The food supply in the USSR during World War II* (Cambridge, 1990)

Müller, R.-D., Volkmann, H.-E. (eds), *Die Wehrmacht: Mythos und Realität* (Munich, 1999)

Mulligan, T.P., 'Spies, ciphers and "Zitadelle": Intelligence and the Battle of Kursk, 1943', *Journal of Contemporary History* 22 (1987)

Muggeridge, M. (ed.), *Ciano's Diary 1939–1943* (London, 1947)

Murray, W., *Luftwaffe* (London, 1985)

Nagatsuka, R., *I Was a Kamikaze* (London, 1973)

Nash, G.D., *United States Oil Policy 1890–1964* (Pittsburgh, 1968)

Nekrich, A., *Pariahs, Partners, Predators: German-Soviet Relations 1922–1941* (New York, 1997)

Nelson, W., *Small Wonder: The amazing story of the Volkswagen* (London, 1967)

Nevins, A., Hill, F.E., *Ford: Decline and rebirth 1933–1961* (New York, 1962)

Nevins, A., *This Is England Today* (New York, 1941)

Neufeld, M., *The Rocket in the Reich: Peenemünde and the coming of the ballistic missile era* (Washington, 1994)

Nicholas, H.G. (ed.), *Washington Dispatches 1941–1945: Weekly political reports from the British embassy* (Chicago, 1981)

Nielsen, A., *The German Air Force General Staff* (New York, 1959)

Nissen, O., *Germany: Land of substitutes* (London, 1944)

Noakes, J. (ed.), *The Civilian in War: The home front in Europe, Japan and the USA in World War II* (Exeter, 1992)

Nossack, H.E., *The End: Hamburg 1943* (Chicago, 2004)

Nove, A., *An Economic History of the USSR* (London, 1989)

O'Ballance, E., *The Red Army* (London, 1964)

O'Brien, T.H., *Civil Defence* (London, 1955)

Olson, M., *The Economics of Wartime Shortage* (Durham, North Carolina, 1963)

Ogorkiewicz, R.M., *Armoured Forces: A history of armoured forces and their vehicles* (London, 1970)

Orwell, G., *The Collected Essays, Journalism and Letters of George Orwell: Vol. 3, As I please, 1943–1945* (London, 1968)

Overy, R.J., 'Air Power and the Origins of Deterrence Theory before 1939', *Journal of Strategic Studies* 14 (1992)

Overy, R.J., *The Air War 1939–1945* (London, 1980)

Overy, R.J., 'From "Uralbomber" to "Amerikabomber": The Luftwaffe and strategic bombing', *Journal of Strategic Studies* 1 (1978)

Overy, R.J., *Goering: The 'iron man'* (London, 1984)

Overy, R.J., *War and Economy in the Third Reich* (Oxford, 1994)

Overy, R.J., *Interrogations: The Nazi Elite in Allied Hands, 1945* (London, 2001)

Overy, R.J., *The Dictators: Hitler's Germany and Stalin's Russia* (London, 2004)

Padfield, P., *Dönitz: The last Führer* (London, 1984)

Padfield, P., *Himmler: Reichsführer SS* (London, 1990)

Panin, D., *The Notebooks of Sologdin* (New York, 1976)

Papen, F. von, *Memoirs* (London, 1952)

Parker, R., *Moscow Correspondent* (London, 1949)

Parker, R.A.C., *Struggle for Survival: The history of the Second World War* (Oxford, 1989)

Parrish, T., *Roosevelt and Marshall: Partners in politics and war* (New York, 1989)

Payne, R., *General Marshall* (London, 1952)

Payton-Smith, H., *Oil: A study in wartime policy and administration* (London, 1971)

Pearton, M., *Oil and the Romanian State* (Oxford, 1971)

Perkins, F., *The Roosevelt I Knew* (London, 1947)

Perlmutter, A., *FDR and Stalin: Not so Grand Alliance 1943–1945* (Columbia, Missouri, 1993)

Petracarro, D., 'The Italian Army in Africa 1940–1943', *War and Society* 9 (1991)

Picot, G., *Accidental Warrior: In the front-line from Normandy till victory* (London, 1993)

Polenberg, *War and Society: The United States 1941–1945* (Philadelphia, 1972)

Polmar, N., *Aircraft Carriers* (London, 1969)

Pons, S., *Stalin and the Inevitable War 1936–1941* (London, 2002)

Poolman, K., *Focke-Wulf Condor: Scourge of the Atlantic* (London, 1978)

Porten, E. van der, *The German Navy in World War II* (London, 1969)

Porter, C., Jones, M., *Moscow in World War II* (London, 1987)

Powers, T., *Heisenberg's War: The secret history of the German bomb* (London, 1993)

Rader, M., *No Compromise: The conflict between two worlds* (London, 1939)

Rae, J.J., *Climb to Greatness: The story of the American aircraft industry* (Cambridge, Mass., 1968)

Raeder, E., *Mein Leben* (2 vols, Tübingen, 1957)

Reddel, C. (ed.), *Transformations in Russian and Soviet Military History* (Office of Air Force History, Washington, 1990)

Reuth, R., *Goebbels* (London, 1993)

Reynolds, D., Kimball, W., Chubarian, A. (eds), *Allies at War: The Soviet, American and British experience 1939–1945* (New York, 1994)

Reynolds, D., *From Munich to Pearl Harbor: Roosevelt's America and the Origins of the Second World War* (Chicago, 2001)

Reynolds, D., *In Command of History: Churchill Fighting and Writing the Second World War* (London, 2004)

Rhodes, R., *The Making of the Atomic Bomb* (New York, 1986)

Rich, N., *Hitler's War Aims: The establishment of the New Order* (London, 1974)

Richards, D., *Royal Air Force 1939–1945: Vol. I, The fight at odds* (London, 1974)

Richardson, C.R., 'French Plans for Allied Attacks on the Caucasus Oilfields Jan-Apr 1940', *French Historical Studies* 8 (1973)

Richardson, W., Frieden, S. (eds), *The Fatal Decisions* (London, 1956)

Ritgen, H., *Die Geschichte der Panzer-Lehr Division im Westen 1944–1945* (Stuttgart, 1979)

Robbins, K., *Churchill* (London, 1972)

Roberts, S., *The House that Hitler Built* (London, 1937)

Roeder, G.H., *The Censored War: American visual experience during World War Two* (New Haven, 1933)

Rohwer, J., *The Critical Convoy Battles of March 1943* (London, 1977)

Rokossovsky, K., *A Soldier's Duty* (Moscow, 1970)

Roosevelt, E. (ed.), *The Roosevelt Letters: Vol. Ill, 1928–1945* (London, 1952)

Roskill, S.W., *The Navy at War 1939–1945* (London, 1960)

Roskill, S.W., *The War at Sea 1939–1945* (4 vols, London, 1954–61)

Rotundo, L. (ed.), *Battle of Stalingrad: The 1943 Soviet General Staff study* (Washington, 1989)

Rotundo, L., 'The Creation of Soviet Reserves and the 1941 Campaign', *Military Affairs* 49 (1985)

Ruge, F., *Rommel und die Invasion: Erinnerungen von Friedrich Ruge* (Stuttgart, 1959)

Ruge, F., *Der Seekrieg: The German Navy's story 1939–1945* (Annapolis, 1957)

Rupperthal, R.G., *Logistical Support of the Armies* (2 vols, Washington, 1953)

Quester, G., *Deterrence Before Hiroshima* (New York, 1966)

Sadkovich, J.J., 'Of Myths and Men: Rommel and the Italians in North Africa 1940–1942', *International History Review* 13 (1991)

Sainsbury, K., *The Turning Point* (London, 1986)

Salisbury, H. (ed.), *Marshal Zhukov's Greatest Battles* (London, 1969)

Salter, A., *Slave of the Lamp: A public servant's notebook* (London, 1967)

Schechter, J.L. (ed.), *Khrushchev Remembers: The Glasnost tapes* (Boston, 1990)

Schellenberg, W., *The Schellenberg Memoirs* (London, 1956)

Schlabrendorff, F. von, *The Secret War against Germany* (London, 1966)

Schneider, J.C., *Should America Go to War? The Debate over Foreign Policy in Chicago 1939–1941* (Chapel Hill, North Carolina, 1989)

Schofield, B.B., 'The Defeat of the U-Boats during World War II', *Journal of Contemporary History* 16 (1981)

Schramm, P.E., *Hitler: The man and the military leader* (London, 1972)

Seaton, A., *The Russo-German War 1941–1945* (London, 1971)

Seaton, A., *Stalin as Warlord* (London, 1976)

The Secret History of World War II: The ultra-secret wartime letters and cables of Roosevelt, Stalin and Churchill (New York, 1986)

Sella, A., *The Value of Human Life in Soviet Warfare* (London, 1992)

Sherry, M., *The Rise of American Air Power: The creation of Armageddon* (New Haven, 1987)

Sherwood, R.E., *The White House Papers of Harry L. Hopkins* (2 vols, London, 1949)

Shirer, W., *Berlin Diary* (London, 1941)

Shrycock, H.S., 'Industrial Migration in Peace and War', *American Sociological Review* 12 (1947)

Shtemenko, S.M., *The Soviet General Staff at War* (Moscow, 1970)

Shulman, M., *Defeat in the West* (London, 1947)

Siegelbaum, L.H., *Stakhanovism and the Politics of Productivity in the USSR 1935–1941* (Cambridge, 1988)

Siegfried, K.-J., *Rüstungsproduktion und Zwangsarbeit im Volkswagenwerk 1939–1945* (Frankfurt, 1988)

Simonson, G., 'The Demand for Aircraft and the Aircraft Industry', *Journal of Economic History* 20 (1960)

Slessor, J., *The Central Blue* (London, 1956)

Sloan, A.P., *My Years with General Motors* (London, 1986)

Smethurst, R., *A Social Basis for Pre-War Japanese Militarism* (Berkeley, 1974)

Smith, H.K., *Last Train from Berlin* (London, 1942)

Smith, R.E., *The Army and Economic Mobilization* (Washington, 1958)

Soviet Air Force in World War II (ed. R. Wagner, Soviet official history, London, 1974)

Speer, A., *Inside the Third Reich* (London, 1970)

Speer, A., *The Slave State: Heinrich Himmler's master plan for SS supremacy* (London, 1981)

Speer, A., *Spandau: The secret diaries* (London, 1976)

Spinka, M., *The Church in Soviet Russia* (Oxford, 1956)

Statistical Abstract of United States Historical Statistics (Washington, 1947)

Steiger, R., *Armour Tactics in the Second World War: Panzer army campaigns of 1939–1941 in German war diaries* (Oxford, 1991)

Steinhoff, J., *The Last Chance: The pilots' plot against Göring 1944–1945* (London, 1977)

Stern, J., *The Hidden Damage* (New York, 1947)

Stoakes, G., *Hitler and the Quest for World Dominion* (Leamington Spa, 1986)

Stoddard, L., *Into the Darkness: Nazi Germany today* (London, 1941)

Stoler, M., *George C. Marshall: Soldier-statesman of the American century* (Boston, 1989)

Stoler, M., *The Politics of the Second Front: American military planning and diplomacy in coalition warfare 1941–1943* (Westport, Conn., 1977)

Stolfi, R., *Hitler's Panzers East: World War II reinterpreted* (Norman, Oklahoma, 1991)

Strawson, J., *Hitler as Military Commander* (London, 1971)

Suchenwirth, R., *Command and Leadership in the German Air Force* (USAF Historical Study 179, New York, 1969)

Sullivan, B.R., 'The Rise and Fall of Italian Sea Power', *International History Review* 10 (1988)

Sweetman, J., 'Crucial Months for Survival: The Royal Air Force 1918–1919', *Journal of Contemporary History* 19 (1984)

Sydnor, C., *Soldiers of Destruction: The SS Death's Head Division 1933–1945* (Princeton, 1977)

Szasz, F.M., *British Scientists and the Manhattan Project* (London, 1992)

Tanin, O., Yohan, E., *When Japan Goes to War* (London, 1936)

Taylor, F. (ed.), *The Goebbels Diaries 1939–1941* (London, 1982)

Taylor, T., *The Anatomy of the Nuremberg Trials* (London, 1993)

Tedder, A., *Air Power in War* (London, 1948)

Tempel, G., *Speaking Frankly about the Germans* (London, 1963)

Tennant, E., *True Account* (London, 1957)

Terkel, S., *The 'Good War': An oral history of World War Two* (New York, 1984)

Terraine, J., *Business in Great Waters: The U-boat wars 1916–1945* (London, 1989)

The Testament of Adolf Hitler: The Hitler-Bormann Documents, February–April 1945 (London, 1961)

Thomas, C.S., *The German Navy in the Nazi Era* (London, 1990)

Thorne, C., *Allies of a Kind: The United States, Britain and the war against Japan 1941–1945* (Oxford, 1978)

Thorne, C., *The Issue of War: States, societies and the Far Eastern conflict of 1941–1945* (London, 1985)

Thornton, W., *The Liberation of Paris* (London, 1963)

Thurston, R., Bonwetsch, B. (eds), *The People's War; Responses to World War II in the Soviet Union* (Chicago, 2000)

Toland, J., *Adolf Hitler* (New York, 1976)

Toscano, M., *The Origins of the Pact of Steel* (Baltimore, 1967)

Tress, H.B., *British Strategic Bombing Policy through 1940* (Lampeter, 1988)

Treue, W. (ed.), 'Der Denkschrift Hitlers über die Aufgaben eines Vierjahresplan', *Vierteljahrshefte für Zeitgeschichte* 3 (1954)

Trevor-Roper, H. (ed.), *Hitler's Table Talk 1941–1944* (London, 1973)

Trevor-Roper, H. (ed.), *Hitler's War Directives 1939–1945* (London, 1964)

Trevor-Roper, H., *The Last Days of Hitler* (London, 1947)

Tuyll, H. van, *Feeding the Bear: American aid to the Soviet Union 1941–1945* (New York, 1989)

Uebe, K., *Russian Reactions to German Air Power in World War II* (New York, 1964)

Ueberschär, G.R., *Generaloberst Franz Halder* (Göttingen, 1991)

Ugaki, M., *Fading Victory: The diary of Admiral Matome Ugaki 1941–1945* (Pittsburgh, 1991)

United States Air Force, *The Development of the Heavy Bomber 1918–1945* (Maxwell, Alabama, 1961)

Vandenberg, A.H. (ed.), *The Private Papers of Senator Vandenberg* (London, 1953)

Vanse, J., *U-Boat Ace: The story of Wolfgang Lütz* (Annapolis, 1990)

Vat, D. van der, *The Pacific Campaign* (London, 1992)

Vatter, H., *The US Economy in World War II* (New York, 1985)

Walker, M., Renneberg, M. (eds), *Science, Technology and National Socialism* (Cambridge, 1993)

Walker, M., *German National Socialism and the Quest for Nuclear Power 1939–1949* (Cambridge, 1989)

Walton, F., *Miracle of World War II* (New York, 1956)

Warlimont, W., *Inside Hitler's Headquarters* (London, 1964)

Warner, P., *Firepower: From slings to star wars* (London, 1988)

Watt, D.C., *How War Came* (London, 1989)

Watt, D.C., *Succeeding John Bull: America in Britain's place* (Cambridge, 1984)

Wedemeyer, A.C., *Wedemeyer Reports!* (New York, 1958)

Wegner, B. (ed), *From Peace to War: Germany, the Soviet Union and the World, 1939–1941* (Providence, RI, 1997)

Weigley, R.F., *The American Way of War: A history of United States military strategy and policy* (London, 1973)

Weigley, R.F., *History of the US Army* (London, 1968)

Weinberg, G., *World in the Balance* (New England University Press, 1981)

Weiner, A., *Making Sense of War: The Second World War and the Fate of the Bolshevik Revolution* (Princeton, 2001)

Wells, H., *The War in the Air* (London, 1908)

Werth, A., *Russia at War 1941–1945* (London, 1964)

West, W.J. (ed.), *Orwell: The war commentaries* (London, 1985)

Wette, W. (ed.), *Deserteure der Wehrmacht; Feiglinge – Opfer – Hoffnungsträger* (Essen, 1995)

White, D.W., 'The Nature of World Power in American History: An evaluation of the end of World War II', *Diplomatic History* 11 (1987)

White, W.L., *Report on the Russians* (New York, 1945)

Wheeler-Bennett, J. (ed.), *Action This Day: Working with Churchill* (London, 1968)

Wieder, J., *Stalingrad und die Verantwortung des Soldaten* (Munich, 1962)

Willmott, H.P., *The Great Crusade* (London, 1989)

Wilmot, C., *Struggle for Europe* (London, 1952)

Wilson, T.A., *The First Summit: Roosevelt and Churchill at Placentia Bay* (London, 1969)

Winant, J.G., *A Letter from Grosvenor Square* (London, 1947)

Winterton, P., *Report on Russia* (London, 1945)

Winton, J., *Ultra at Sea* (London, 1988)

Wooldridge, E.T. (ed.), *Carrier Warfare in the Pacific: An oral history collection* (Washington, 1993)

Yakovlev, A., *Notes of an Aircraft Designer* (Moscow, 1961)

Yergin, D., *The Prize: The epic quest for oil, money and power* (New York, 1991)

Zaloga, S.J., Grandsen, J., *Soviet Tanks and Combat Vehicles of World War Two* (London, 1984)

Zeman, Z., *Nazi Propaganda* (Oxford, 1973)

Zhukov, G.K., *Reminiscences and Reflections* (Moscow, 1985)